Intellectual Property Issues in Life Sciences

Intellectual Property Issues in Life Sciences: Disputes and Controversies highlights emerging legal, social, and regulatory issues pertaining to various areas of life sciences. Patents occupy a prominent position in the innovation systems in the life sciences, but to what extent they support, or hinder innovation is widely disputed. Life science is a broad subject including agriculture, ecology, microbiology, plant and animal sciences, health and diseases, biotechnology, etc. However, despite the broad applications of biotechnology and molecular biology techniques, profits on investments are surprisingly low. Thus, it is vitally important for universities, public research organizations, and private enterprises to protect their innovations. There are vast differences of opinion on patentability of living organisms, which are largely barred from patent protection. However, mind-sets are rapidly shifting and IP issues in life sciences are receiving increasing attention. To compete with progressive bio-based economies the developing countries are amending their IP laws to encourage investment.

An effort has been made to avoid considering policy in isolation, but rather to emphasize the interplay between the policy mix, the wider institutional setting, market forces, and system organization solutions. Both empirical and conceptual chapters are included to bring them together and to yield facts and interpretations for the readers.

This book presents expert opinions by frontier academicians, researchers, and attorneys on the recent challenges in the rapidly evolving life science industry. The present book offers comprehensive knowledge on the contemporary issues in life sciences to a wide range of audiences including students, scholars, researchers, legal practitioners, policymakers, and others interested in emerging intellectual property issues.

Features

- The only compilation available on the contemporary intellectual property issues in life sciences in the post-COVID era.
- Focuses on the commercial, regulatory, bioethical, and socio-legal implications of patents in life sciences.
- Describes an integrated approach for sustained innovations in various areas of life sciences.
- Discusses the recent IP controversies in a pan-global context.
- Presents viewpoints to front-line practitioners, viz attorneys, researchers, etc.

Intellectual Property Issues in Life Sciences

Disputes and Controversies

Edited by
Chetan Keswani and Cristina Possas

CRC Press
Taylor & Francis Group
Boca Raton London New York

CRC Press is an imprint of the
Taylor & Francis Group, an **informa** business

Designed cover image: Shutterstock

First edition published 2025
by CRC Press
2385 NW Executive Center Drive, Suite 320, Boca Raton FL 33431

and by CRC Press
4 Park Square, Milton Park, Abingdon, Oxon, OX14 4RN

CRC Press is an imprint of Taylor & Francis Group, LLC

Library of Congress Cataloging-in-Publication Data
Names: Keswani, Chetan, editor. | Possas, Cristina, editor.
Title: Intellectual property issues in life sciences : disputes and controversies /
edited by Chetan Keswani and Cristina Possas
Description: First edition | Boca Raton, FL : CRC Press, 2025 |
Includes bibliographical references and index.
Identifiers: LCCN 2024026791 (print) | LCCN 2024026792 (ebook) |
ISBN 9781032522319 (hardback) | ISBN 9781032522333 (paperback) |
ISBN 9781003405665 (ebook)
Subjects: LCSH: Life sciences–Research–Law and legislation. |
Intellectual property (International law) | Patents (International law)
Classification: LCC K1401 .I559415 2024 (print) | LCC K1401 (ebook) |
DDC 346.04/8–dc23/eng/20240724
LC record available at https://lccn.loc.gov/2024026791
LC ebook record available at https://lccn.loc.gov/2024026792

ISBN: 9781032522319 (hbk)
ISBN: 9781032522333 (pbk)
ISBN: 9781003405665 (ebk)

DOI: 10.1201/9781003405665

Typeset in Times
by Newgen Publishing UK

Dedication

(1938–2020)

This volume is dedicated to the life and the memory of Professor Ananda Mohan Chakrabarty who in 1981 obtained the first ever U.S. patent on a living organism, shaping legal opinion around ownership of genetically engineered organisms and unknowingly spurring unprecedented levels of biotech innovation and commercialization for years to come. The U.S. biotech industry would not be where it is today without Prof. Chakrabarty's work, which was at the frontier of both life sciences and law.

Dr. Chetan Keswani owes deep gratitude to the late Prof. Chakrabarty for receiving guidance, encouragement, and support in all his previously edited volumes on intellectual property.

Contents

Preface

Intellectual property rights (IPRs) describe the set of rules and regulations imposed to protect the innovations of commercial value in a geo-temporal space. Patents occupy a prominent position in the innovation systems in the life sciences; but to what extent they support, or hinder innovation is widely disputed. Life Science is a broad subject including agriculture, ecology, microbiology, plant and animal sciences, health and diseases, biotechnology, etc. In the area of human health and medicine, diagnostic kits, monoclonal antibodies, vaccines, gene therapy and lifesaving drugs are some of the forefront achievements of pharmaceutical biotechnology. Microorganisms (including genetically modified microorganisms) have been used as biological factories for industrial scale production of antibiotics, enzymes, probiotics, biochemicals, etc. While contemporary agricultural research accentuates the development of novel disease resistant and climate resilient transgenic crop varieties for ensuring enhanced crop production and food security. However, despite broad applications of biotechnology and molecular biology techniques, profits on investments are surprisingly low. Thus, it is vitally important for the universities, public research organizations, and private enterprises to protect their innovations. There are vast differences of opinions on patentability of living organisms, which are largely barred from patent protection. However, mind-sets are rapidly shifting and IP issues in life sciences are receiving massive attention. To compete with progressive bio-based economies the developing countries are amending their IP laws to encourage investment. The present book highlights emerging legal, social, and regulatory issues pertaining to various areas of life sciences. This book presents expert opinions by frontier academicians, researchers, and attorneys on the recent challenges in the rapidly evolving life science industry. The present book offers comprehensive knowledge on the contemporary issues in life sciences to a wide range of audiences including students, scholars, researchers, legal practitioners, policymakers, and others interested in IP issues.

An effort has been made to avoid considering policy in isolation, but rather to emphasize the interplay between the policy mix, the wider institutional setting, market forces, and system organization solutions. Both empirical and conceptual chapters are included with an attempt to bring them together and to yield facts and interpretations for the readers.

Chetan Keswani
Cristina Possas
Editors

Acknowledgments

The editors express sincere gratitude to all the contributors and the reviewers for their valuable contribution to this volume.

The editors express enormous gratitude to Dr. Renu Upadhyay, Senior Commissioning Editor, Taylor & Francis Group and Ms. Jyotsna Jangra, also from Taylor & Francis Group for their sustained support and efficient management of this project.

Lastly, Chetan Keswani gratefully acknowledges the financial support from the project of the Ministry of Science and Higher Education of Russian Federation on the development of Young Scientist Laboratory within the framework of the Interregional Scientific and Educational Center of the South of Russia (no. FENW-2024-0001) and to the Strategic Academic Leadership Program of the Southern Federal University ("Priority 2030").

About the Editors

Chetan Keswani is the Head of Agro-Biotechnology Laboratory in the Academy of Biology and Biotechnology, Southern Federal University, Rostov-on-Don, Russia. He has a keen interest in regulatory, commercialization, and intellectual property aspects of agriculturally important microorganisms. He is an elected Fellow of the Linnaean Society of London, UK and received Best Ph.D. Thesis Award from the Uttar Pradesh Academy of Agricultural Sciences, India in 2015. He is an editorial board member of several reputed agricultural microbiology journals and has edited ten books on the subject. For his editorial endeavors he was awarded the Publons Top Peer Review Award in Agricultural Sciences (2018), Outstanding Editor Award (2019) by Archives of Phytopathology and Plant Protection, Taylor and Francis, and Springer-Society Award (2020) for Excellent contribution to the journal *Environmental Sustainability*, Springer–Nature. In addition, he has been listed among the World's Top 2% Scientists for the single year 2022 (sub-fields: Plant Biology & Botany and Biotechnology).

Cristina Possas is a Full Professor and Researcher at the Oswaldo Cruz Foundation-Fiocruz, Rio de Janeiro, Brazil. Scientific Advisor for Fiocruz´ President she directed the Advisory Unit for Special Studies for the Presidency. She has been for 10 years Director of the Research and Technological Development Unit of the Brazilian National AIDS Program in the Ministry of Health and for two years the Executive Secretary of the National Biosafety Commission – CTNBio, responsible for evaluation of genetically modified organisms (GMOs) projects nominated by the Brazilian Minister of Science and Technology. She has a post-doctorate in International Health from Harvard University, School of Public Health, where she stayed for 10 years as a Takemi Fellow, Fulbright Fellow and Visiting Scientist at the Harvard New Diseases Group. She coordinated FIOCRUZ Graduate Programs in Public Health and in Clinical Research on Infectious Diseases (Master's, Doctoral, and Postdoctoral). She developed at Fiocruz her academic activities and research projects at the National School of Public Health and at the Evandro Chagas National Institute of Infectious Diseases-INI, in the Laboratory of AIDS Clinical Research-LAPCLIN-AIDS, and at Bio-Manguinhos, where she is Senior Scientific Advisor and Professor at the Multi-institutional Doctoral Program in Nanobiosystems.

Contributors

Elisabeth Abergel, Department of Sociology and Environmental Science Institute, Université du Québec à Montréal (UQAM), Montréal, Québec, Canada.

Adam Azlan, Department of Biomedical Sciences, Advanced Medical and Dental Institute, Universiti Sains Malaysia,13200 Kepala Batas, Pulau Pinang, Malaysia and School of General and Foundation Studies, AIMST University, 08100, Bedong, Kedah, Malaysia.

Alejandro Barragán-Ocaña, National Polytechnic Institute (Instituto Politécnico Nacional – IPN). Center for Economic, Administrative and Social Research (Centro de Investigaciones Económicas, Administrativas y Sociales – CIECAS). Lauro Aguirre 120. Col. Agricultura, Miguel Hidalgo, C.P. 11360. Ciudad de México.

Alane Beatriz Vermelho, Bioinovar Laboratory, General Microbiology Department, Institute of Microbiology Paulo de Goes, Federal University of Rio de Janeiro, Rio de Janeiro, Brazil.

Michael Blakeney, University of Western Australia, Perth, Australia.

Verônica S. Cardoso, Bioinovar Laboratory, General Microbiology Department, Institute of Microbiology Paulo de Goes, Federal University of Rio de Janeiro, Rio de Janeiro, Brazil.

Neil Davey, Harvard Law School, Cambridge, MA, United States.

Adelaide Antunes, School of Chemistry, Federal University of Rio de Janeiro, Brazil and National Institute of Industrial Property, Rio de Janeiro, Brazil.

Alessandra Moreira de Oliveira, School of Chemistry, Federal University of Rio de Janeiro, Brazil.

Flavia Lima do Carmo, Molecular Microbial Ecology Laboratory, General Microbiology Department, Institute of Microbiology Paulo de Goes, Federal University of Rio de Janeiro, Rio de Janeiro, Brazil.

Maria Do Socorro Santos Rosa, Bioinovar Laboratory, General Microbiology Department, Institute of Microbiology Paulo de Goes, Federal University of Rio de Janeiro, Rio de Janeiro, Brazil.

Reda El Boukhari, Chemical Science and Engineering Research Team, Department of Chemistry, Polydisciplinary Faculty of Beni Mellal, Sultan Moulay Slimane University, Beni Mellal, Morocco.

Ahmed Fatimi, Chemical Science and Engineering Research Team, Department of Chemistry, Polydisciplinary Faculty of Beni Mellal, Sultan Moulay Slimane University, Beni Mellal, Morocco.

Homa Fatma, Molecular Cancer Genetics & Translational Research Lab, Section of Genetics, Department of Zoology, Aligarh Muslim University, Aligarh, India.

Akira Homma, Bio-Manguinhos, Oswaldo Cruz Foundation, Rio de Janeiro, Brazil.

Sadaf R. Khan, Molecular Cancer Genetics & Translational Research Lab, Section of Genetics, Department of Zoology, Aligarh Muslim University, Aligarh, India.

Misael Linares-Bravo, National Polytechnic Institute (Instituto Politécnico Nacional – IPN). Center for Economic, Administrative and Social Research (Centro de Investigaciones Económicas, Administrativas y Sociales – CIECAS). Lauro Aguirre 120. Col. Agricultura, Miguel Hidalgo, C.P. 11360. Ciudad de México.

Flavia Maria Lins Mendes, Fiocruz Mato Grosso do Sul, Oswaldo Cruz Foundation, Campo Grande, Brazil.

Felipe R. P. Mansoldo, Bioinovar Laboratory, General Microbiology Department, Institute of Microbiology Paulo de Goes, Federal University of Rio de Janeiro, Rio de Janeiro, Brazil.

Ernesto TA Marques, Aggeu Magalhães Institute, Oswaldo Cruz Foundation, Recife, Brazil and Department of Infectious Diseases and Microbiology, University of Pittsburgh, Pittsburgh, USA.

Jean Vinícius Moreira, Bioinovar Laboratory, General Microbiology Department, Institute of Microbiology Paulo de Goes, Federal University of Rio de Janeiro, Rio de Janeiro, Brazil.

Rabiatul Basria S. M. N. Mydin, Department of Biomedical Sciences, Advanced Medical and Dental Institute, Universiti Sains Malaysia, Kepala Batas, Pulau Pinang, Malaysia.

Hanane Nahid, Private Law and Economic Law Laboratory, Art Department, National Higher School of Art and Design, Hassan II University, Casablanca, Morocco.

Cristina Possas, Bio-Manguinhos, Oswaldo Cruz Foundation, Rio de Janeiro, Brazil.

Suzanne Schumacher, School of Chemistry, Federal University of Rio de Janeiro, Brazil.

Marilda M Siqueira, Oswaldo Cruz Institute, Oswaldo Cruz Foundation, Rio de Janeiro, Brazil.

Hifzur R. Siddique, Molecular Cancer Genetics & Translational Research Lab, Section of Genetics, Department of Zoology, Aligarh Muslim University, Aligarh, India.

Tamara Arizbe Virgilio-Virgilio, National Polytechnic Institute (Instituto Politécnico Nacional – IPN). Center for Economic, Administrative and Social Research (Centro de Investigaciones Económicas, Administrativas y Sociales -CIECAS). Lauro Aguirre 120. Col. Agricultura, Miguel Hidalgo Miguel Hidalgo, C. P. 11360. Ciudad de México.

1 The Need for Global Regulation of Gene Editing in the Wake of CRISPR–Cas9

Neil Davey

1.1 INTRODUCTION

1.1.1 Breakthrough Technology

In early 2013, the Zhang Lab at the Broad Institute of MIT and Harvard documented for the first time that CRISPR–Cas9 gene-editing technology could be employed in human cells.[1] While the Doudna Lab at the University of California (UC) Berkeley first published in 2012 the ability of CRISPR–Cas9 to identify and cut at specific sites in the genome, never before had the technology been shown to function successfully in human genes.[2] Prior gene-editing techniques—including zinc finger proteins (ZFNs) and transcription activator-like effector nucleases (TALENs)—lacked incision accuracy and were both time-intensive and costly.[3] In contrast, the Cas9 enzyme showed unprecedented levels of specificity, and CRISPR guide RNAs could be manufactured rapidly and inexpensively.[4] The clinical possibilities were endless; for the first time in history, deleterious genes could be precisely removed from human beings.[5] However, while scientists have shown great enthusiasm for this breakthrough discovery, numerous questions have been raised in the realms of bioethics, law, and policy. On the one hand, CRISPR has the incredible potential to delete harmful mutations from the human genome to save patients' lives, but on the other hand, eugenic use of the technology to improve certain traits in select populations could be incredibly dangerous.[6]

As scientific research in this area is progressing at a groundbreaking rate, with the number of gene-editing publications quadrupling since 2013, national lawmaking bodies and international organizations are attempting to create legislation and set standards to prevent unethical use of revolutionary technology without dampening necessary biomedical innovation.[7] Given the immense consequences for human health, governments ought to quickly address the numerous policy challenges resulting from the rise of gene-editing technologies. While federal governments are competing to incentivize innovation in this area through specific R&D programs and unique intellectual property (IP) regimes, international regulatory agencies are pushing back and

DOI: 10.1201/9781003405665-1

hope to establish critical global restraints. This tension is particularly problematic, as it prevents the international community from having any control regarding gene editing and reduces competition and consequently access, as various countries have differing policies on the issue.

1.1.2 Bioethical and Human Rights Concerns About Gene Editing

The potential for employing a highly specific technique such as CRISPR–Cas9 is limitless; it could be used to remove allergens from nutritious foods, make livestock and crops resistant to infections to increase yields, and even eradicate malaria by preventing mosquitos from breeding.[8] In the medical context, somatic gene editing could remove fatal genes in diseases such as Huntington's disease to improve a patient's health. However, as Doudna and her colleagues note in their seminal 2015 *Science* paper, there are serious dangers of CRISPR if the technology is used at the germline level, implying that changes to the genome would be hereditary.[9] Not only would changes be passed on inorganically, meaning that offspring have little control over their own genomic material, gene editing in this form could be used for enhancement as opposed to treatment.[10] This would mean that wealthier or more powerful individuals with greater access to the technology could use it to make their children stronger and smarter, further exacerbating existing inequalities in society.[11]

Given these concerns, leading experts in the genome editing field recommended at the 2015 Innovative Genomics Institute's Bioethics Forum Addressing Concerns about CRISPR Research to "strongly discourage, even in those countries with lax jurisdictions where it might be permitted, any attempts at germline genome modification for clinical applications in humans."[12] While this forum was composed exclusively of Americans, the 2015 International Summit on Human Gene Editing hosted by the US National Academies of Sciences and Medicine reiterated a similar message: progress should be made on somatic gene editing as research can be conducted within existing policy frameworks, but germline editing should be halted until further safety and efficacy issues can be resolved by forums of experts in the scientific and bioethics communities.[13] Both the UK Royal Society and the Chinese Academy of Sciences cohosted the summit and agreed on this consensus opinion. The committee also included in their statement uncertainty about changes that could result from interactions between the environment and widespread genetic alterations in the population, including the drastic impact these interactions could have on human evolution.[14]

1.1.3 The Importance of Continued Research

At the same time, however, experts at both the forum and summit agreed that state governments ought to continue funding gene-editing research.[15] In particular, although CRISPR–Cas9 is quite powerful, technical improvements are still needed before gene therapy can be approved for safe injection into humans. Not only are there concerns about the gene-editing platform mistakenly cleaving off-target portions of the genome, there have been issues in delivering the technology into specific cells and tissues, as well as limitations in reaching hidden genomic sites.[16] These barriers must be overcome even for somatic gene therapies to enter the market.[17] Additionally, the question of germline gene editing is tied to that of human embryonic stem cell (hESC) research. Namely, to properly understand the repercussions of editing at the germline level, scientists must be able to conduct experimentation on hESCs. In 2018, Japan began to allow gene editing in human embryos, joining only the US and UK in this regard.[18] Importantly, the United States currently allows the patenting of research on embryos up to 14 days postfertilization, while the European Union rejects such intellectual property outright, effectively killing lucrative research in this area.[19]

When stem cell technologies accelerated to the frontier 20 years ago, regulation had to play significant catch-up. As a result, countries currently have very different policies regarding embryonic stem cell research.[20] Not only does this prevent the international community from setting constraints on nations that may be acting irresponsibly, it limits innovation to a small number of countries, thereby decreasing competition and reducing access to technology around the world. While the global community seems willing to reel in the pace of questionable CRISPR–Cas9 inventions, individual countries still have unique innovation policies, shedding light on their own competitive interests. To prevent such heterogeneity from occurring again, it is important for the international community to come together and make collective decisions on gene editing.

1.2 NATIONAL INNOVATION FRAMEWORKS AND IP REGIMES

1.2.1 United States

The United States is undoubtedly the scientific leader in cutting-edge gene-editing technologies, with both Doudna and Zhang's pivotal CRISPR discoveries to have occurred in the country. Furthermore, three of the five largest public gene-editing companies (Editas Medicine, Intellia Therapeutics, and Sangamo Therapeutics) are located within the US, and America publishes more CRISPR-related articles than any other nation.[21] Much of this leadership can be attributed to generous innovation policy by the federal government, including both strong R&D policy and a flexible intellectual property regime. More specifically, the United States has been historically supportive of the biopharmaceutical industry, as witnessed by large tax credits for innovation in the sector, with more national R&D spending than any other country (USD 476 billion in 2017) and relatively lenient policies for biomedical experimentation.[22,23] For example, the United States is one of few countries allowing for hESC research, as previously noted.[24]

Furthermore, the US has expedited FDA approval processes for breakthrough therapies with profound lifesaving potential or treatments for rare diseases, accelerating innovation in the gene-editing space.[25] The White House Office of Science and Technology Policy is also in the process of developing a regulatory oversight committee for gene editing, and while the National Institutes of Health is not accepting proposals for germline alterations, it is actively encouraging somatic gene-editing applications.[26] In addition to strong government-industry links, durable academia-industry collaborations made possible through a flexible intellectual property regime have pushed the frontier in genome editing. Such innovation policy, which is increasingly harmonized with the biotechnology industry, has been shown to be effective, as most of the essential improvements to CRISPR–Cas9 since 2013 have occurred in the US, and these are being harnessed by various American biotechnology companies. For one, researchers at the University of Pennsylvania were the first to edit CAR-T immune cells using CRISPR to create *ex vivo* treatments for a number of conditions, leading the way for a whole new class of therapies.[27] In addition, the Broad Institute and Editas Medicine have performed extensive research on a novel enzyme, Cpf1 (also known as Cas12a), that reduces the off-target effects of Cas9.[28] Finally, MIT recently discovered the enzyme scCas9, which is able to reach "harder-to-access" portions of the genome.[29] This research overcomes many prior CRISPR-related challenges and brings the globe closer to additional approved gene therapies.

Regarding intellectual property, the United States is quite liberal in granting patents to biotechnology inventions. In the landmark decision *Diamond v. Chakrabarty,* 447 U.S. 303 (1980), genetically modified organisms were found to be patent-eligible under 35 U.S.C. § 101, spurring innovation in the biotechnology industry by permitting IP protection for humanmade biological products.[30] While it was decided in the Supreme Court case *Association for Molecular Pathology v. Myriad Genetics, Inc.*, 569 U.S. 576 (2013) that naturally occurring DNA could not be patented, any artificially created piece of DNA (e.g., DNA that is altered using CRISPR–Cas9) is patent-eligible.[31,32] Importantly, the US Patent and Trademark Office (USPTO) granted patents for both Broad and UC, stating that the former's patent covers the use of CRISPR in multicellular organisms, while the latter can employ the technology for unicellular organisms, despite the UC filing first and believing that Broad infringed on its patent.[33] Of course, Broad's patent is more powerful given that it covers applications in humans, but the USPTO nonetheless decided that the intellectual property is sufficiently different. Granting additional patents based on narrow claims allows for additional filings and increases competition in the space by preventing monopolistic behavior. Notably, Broad's patent is the backbone of Editas Medicine, while UC's patent is the basis of Intellia Therapeutics.[34]

1.2.2 European Union

The European Union falls behind the US in biotechnology innovation in terms of gene-editing publications and spending.[35] Furthermore, according to the European Commission, "the average university in Europe generates far fewer inventions and patents" compared to America, and the US has more international biotechnology patents than the European Union.[36,37] Underlying the EU's policies regarding research in the biomedical sector is the precautionary principle, which states that research should not be done unless it is proven that there is close to no risk involved.[38] While such an approach may prevent any potentially unethical R&D from occurring, it also dampens innovation and stifles economic growth. For decades, the European Union has had a hardline policy on genetically modified crops, despite scientists' confirmation that they are perfectly safe.[39] Not only does this endorsement of a protectionist mentality restrict competition, it prevents the EU from integrating novel gene-editing products into its economy and thereby limits scientific progress. For example, with the UK no longer part of the EU, no EU country allows for gene-editing research to be conducted on human embryos on moral grounds.[40] Without the ability to apply CRISPR–Cas9 on hESCs, it will be impossible for Europe to compete with America on gene therapies hoping to address hereditary diseases or on genetically edited *ex vivo* stem cell therapies. As previously discussed, Europe also falls behind the US on innovation in the context of somatic gene editing given less generous policies to stimulate R&D. Interestingly, the highly innovative CRISPR Therapeutics, which is working on stem cell therapies employing gene editing, is a Swiss company that does not fall under such stringent European regulation.[41] Similarly, the pharmaceutical giant Novartis, which has created lipid nanoparticle delivery systems for CRISPR platforms, is based in Switzerland and falls outside of traditional EU jurisdiction as well.[42]

Ironically, as per the Biotechnology Directive that serves as the pillar of biotechnology-related policy in the European Union, the EU actually does allow for naturally isolated gene products to be patented, unlike the United States.[43] However, both Broad and UC are still having trouble with their CRISPR patents in the EU. More specifically, both companies' European patents have been confronted with a barrage of opposition proceedings due to trivial issues: Broad's European patent is being attacked as it lacks an inventor who was on the original application, and UC's patent was questioned extensively for containing overlapping matter with Broad's initial patent.[44] Such strict IP standards by the European Patent Office are not harmonized with the rest of the world and will likely cripple scientific discovery by reducing incentives to innovate in Europe.

In addition to the patent hurdles that are dampening biotechnology innovation, Europe's funding infrastructure for R&D innovation in this sphere is also lacking. Namely, the program at the cornerstone of EU innovation policy is Horizon 2020, which is the eighth iteration of the Framework Programs for Research and Technological Development and allocates a European R&D budget.[45] The European Research Council notes that Horizon 2020 grant applications prohibit: "research activities intended to modify the genetics of human beings that could make such changes heritable and research activities intended to create human embryos solely for the purposes of research or stem cell procurement."[46] Although much scientific work has been done since the Biotechnology Directive to demonstrate that hESC research is vital for the discovery of new cell therapies, today's funding body in Europe continues to limit such activity and observe a general principle of precaution. To understand how the EU has reached this position, it is critical to discuss historical legislative developments in this space.

1.2.2.1 The Oviedo Convention

In April 1997, the Convention on Human Rights and Biomedicine was signed in Oviedo, Spain.[47] Put into effect to protect human dignity in the context of biomedicine, the Convention was the result of efforts initiated by the Council of Europe in response to the rapid rate of biomedical innovation. While the Council of Europe is entirely distinct from the European Union, its 47 member states include all 28 members of the European Union.[48] Although the Council of Europe cannot create binding legislation (unlike the EU), it is able to enforce multilateral European agreements such as the Oviedo Convention, and therefore, the international treaty represents a significant measure to hold European nations to higher bioethical standards.

Of particular interest in the Oviedo Convention is Article 13: "Interventions on the Human Genome – An intervention seeking to modify the human genome may only be undertaken for preventive, diagnostic or therapeutic purposes and only if its aim is not to introduce any modification in the genome of descendants."[49] Article 13 effectively narrows genome editing to nonhereditary somatic cells and prohibits its use for enhancement. Furthermore, Article 18 of the Convention expressly prohibits the creation of human embryos for research purposes, a key provision that further limits the possibility of germline genome editing and eventually informs European regulation of human embryonic stem cells.[50] Embryonic stem cell research is highly intertwined with germline gene editing, as scientists need to experiment on embryos and their derivatives when developing heritable therapies for various diseases. Articles 13 and 18 represent a critical juncture from US policy.

Interestingly, only 29 member states of the Council of Europe have actually ratified the Oviedo Convention, with notable exceptions being the United Kingdom and Germany.[51] Many of the states that have failed to ratify the convention are rather technology-intensive, demonstrating that while the Oviedo Convention is a first step in attempting to limit any potential abuse of biotechnology, it is not sufficient, as the Council of Europe cannot create binding laws.[52]

1.2.2.2 EU Biotechnology Directive

In July 1998, the European Parliament and Council of the European Union passed Directive 98/44/EC on the legal protection of biotechnological innovations.[53] This landmark directive, initially proposed by the European Commission, aimed to balance appeasing ethical concerns with spurring innovation in the context of biotechnology. Importantly, the European Union acutely recognized that "in the field of genetic engineering, research and development require a considerable amount of high-risk investment and therefore only adequate legal protection can make them profitable."[54] While the European Union traditionally abides by the precautionary principle, often deciding to avoid supporting R&D that presents even a small health or safety risk, the Biotechnology Directive represents an acknowledgment that Europe must recognize uncertain investments in biotechnology through the patent system to encourage the development of lifesaving clinical therapies.[55] In addition to stating that biotechnology inventions are "of fundamental importance for the Community's industrial development," 98/44/EC specifically notes that member states have a duty to encourage "research into and production of biotechnological medicines which are needed to combat rare or orphan diseases" through a robust intellectual property regime.[56]

In the context of gene editing, the Biotechnology Directive lays out a number of essential provisions. First and foremost, the document states that gene patents—or applications on isolated DNA segments from the genome—are possible so long as the sequences demonstrate clear industrial application or clinical significance.[57] By extension, the Directive allows for intellectual property protection of DNA sequences that are modified by gene-editing technology, opening the door for innovation in the gene therapy space. However, while the Biotechnology Directive supports somatic gene editing, it reinforces the Oviedo Convention with regards to germline gene-editing and embryonic research. To prevent R&D in areas that the EU considers contrary to the *ordre public* or morality, Article 6 of 98/44/EC explicitly states, "the following, in particular, shall be considered unpatentable: (a) processes for cloning human beings; (b) processes for modifying the germ line genetic identity of human beings; (c) uses of human embryos for industrial or commercial purposes."[58] Thus, while the Directive certainly demonstrates the European Union's enthusiasm toward gene-editing patents, it outright suppresses innovation in the germline editing space and prohibits the patenting of embryos, which are needed for hereditary genome editing research. Of course, as EU directives are not binding legislation, the manner in which member states have incorporated 98/44/EC into national law impacts how each nation conducts research in the space. Importantly, as of 2007, all 28 current EU member states had implemented the Biotechnology Directive into national legislation.[59]

1.2.2.3 Intellectual Property Barriers in the EU

While the Biotechnology Directive laid out a clear groundwork for the patent system in Europe for biomedical inventions, intellectual property enforcement has been questionable in this regard and therefore has had the effect of hindering innovation. In the realm of somatic gene editing for the development of novel therapies, which generally raises no ethical problems, both the Broad Institute and the University of California Berkeley have had tremendous difficulty in having their inventions protected in Europe.[60] In the United States, both Doudna and Zhang's patents were granted, with the former covering the use of CRISPR–Cas9 in unicellular organisms and the latter demonstrating applications in multicellular organisms, including humans. However, both patent applications in Europe have faced tremendous opposition due to rather semantic issues.[61] Specifically, the Berkeley patent was questioned for containing too much similarity with that of Broad, and Broad's patent was not allowed as an inventor on the original application was not present in the final submission. By disallowing patents based on such trivial matters, the European Patent Office is working against the biotechnology innovation framework that 98/44/EC so skillfully laid out. As the Broad Institute states, this interpretation by the patent office "is inconsistent with treaties designed to harmonize the international patent process, including that of the United States and Europe."[62]

In the context of germline genome editing, while the Biotechnology Directive has aptly prohibited patents on editing hereditary sequences and embryos in accordance with the Oviedo Convention, the Directive does not prohibit the patenting of stem cells more broadly. However, the Court of Justice of the European Union ruled in Case C-34/10, *Brüstle v. Greenpeace eV,* 2011 E.C.R. I-09821 that patents on embryonic stem cell lines were banned even when no human embryos were destroyed in the process.[63,64] Subsequently, European scientists were extremely worried that the EU would introduce even stricter restrictions on stem cell research and that the decision would "drive development of stem cell therapies outside Europe."[65] Given the rise of revolutionary stem cell therapies that have been shown to address a wide range of diseases, performing gene editing on stem cells is crucial from a clinical perspective. Clearly, the EU's strict patent policy does not help Europe compete with the rest of the world in biomedical innovation, and decreased competition can result in limited European access to important biotechnologies.

1.2.2.4 Implementation by Select European Countries

While Europe as a bloc is generally seen as conservative with respect to engaging in ethically debatable research, the actual implementation of the Biotechnology Directive is nuanced and varies greatly by member state. Each state has unique policies on germline genome editing and, by extension, embryonic stem cell research. This differentiation makes it difficult to generalize European policy, and the lack of harmonization will be of great importance when germline or embryonic treatments soon enter the market, as important cross-border discussions will be needed.[66] Below, analysis is provided of the four leading European powerhouses in terms of both publications and patents in the biotechnology field.[67,68] These are the United Kingdom, Germany, France, and Switzerland, with the final of course never having been in the EU. With the UK's recent departure from the EU, the country's regulatory framework is subject to change.

1.2.2.4.1 United Kingdom

The United Kingdom has yet to sign or ratify the Oviedo Convention and has viewed the document as being too restrictive on important biotechnology research. Namely, the UK allows for hESC research, one of the few countries in the world to do so.[69] In fact, it was the British 1984 Warnock Committee that endorsed the 14-day rule, allowing for research on human embryos until 14 days postfertilization. Despite a prohibition on patenting both embryonic stem cells and germline editing in the UK, the 1990 Human Fertilization and Embryology Act has allowed for early-stage stem cell research, making the nation a leader in the space.[70] Since embryonic stem cell research is a prerequisite for germline editing discoveries, other countries in Europe may want to reconsider their stringent human embryo standards if they hope to remain competitive when germline therapies for genetic diseases begin to be approved worldwide.

1.2.2.4.2 Germany

Interestingly, Germany has also neither signed nor ratified the Oviedo Convention, but on the grounds that it is far too permissive, with strong public support for the preservation of human dignity. Germany has a hard-line policy against embryonic research and "bans the importation, utilization, and derivation of stem cells in the country" via its Embryo Protection Act of 1990.[71] Consequently, there is no ongoing gene-editing research in embryos in the country.

1.2.2.4.3 France

While France has both signed and ratified the Oviedo Convention, it has instituted reservations that limit the extent of the restrictions.[72] Although France initially had a comprehensive ban on experimentation with embryos, recent legislation as of 2014 permits both embryonic and hESC research.[73] Unlike the German case, "French

President François Hollande promised to transform life sciences and push Parliament to reform hESC research policies," with the explicit purpose of making France more competitive in creating gene-edited stem cell therapies on a global scale.[74] The French National Agency for the Safety of Medicine and Health Products has been authorizing clinical research protocols involving stem cell therapies since.[75]

1.2.2.4.4 Switzerland

Although Switzerland is not in the EU and therefore has not implemented the provisions of the Biotechnology Directive, it has both signed and ratified the Oviedo Convention as part of the Council of Europe. Like France, however, the nation only ratified the Convention with critical reservations, and like the UK, the Swiss government permits early-stage embryonic research.[76] Notably, by not being bound to the European Union's strict patent policy against stem cells, as per both the Biotechnology Directive and the subsequent *Brüstle* decision, Switzerland has done incredibly well in the gene-editing cell therapy space. For one, the largest gene-editing company in the world by market cap—USD 2 billion CRISPR Therapeutics—is based in Switzerland. Their lead drug candidate CTX001, a gene-edited stem cell therapy for β-thalassemia and sickle cell anemia, is one of the most advanced pipeline products undergoing clinical trials.[77] Furthermore, the Swiss pharmaceutical giant Novartis is leading the global development of treatments that make use of gene editing, as the CAR-T product Kymriah for leukemia was the first ever CRISPR-edited therapy approved by the US FDA in 2017.[78] Additionally, Novartis has created substantive inventions improving the delivery of CRISPR products into human cells and tissues using lipid nanoparticle systems.[79] Finally, the French company Genethon partnered with the Swiss CRISPR Therapeutics to "accelerate gene therapy research programs" after France loosened its policy on gene editing.[80] Overall, by being exempt from the EU's regulatory framework regarding gene editing, Switzerland has shown tremendous potential in this area.

1.2.3 China

The gene-editing space is no exception in the backdrop of China's rapid growth and recent dominance in technology-intensive sectors. With copious state initiatives for innovation—including state-owned enterprises, science and technology industrial parks, and additional STEM talent at illustrious research institutions and universities—alongside a massive market size, it is no surprise that China is progressing so quickly.[81,82] In the context of gene editing, the world-renowned biotechnologist André Choulika believes that China is winning today's version of the space race; as China is "investing USD 300 billion in gene editing technologies, and China's Natural Science Foundation has funded nearly 300 projects in the past four years," China's global rise in this industry has been prolific.[83] Although the country still remains behind the United States in terms of publications and companies in the space, it could soon overtake America as the CRISPR leader.[84]

Recently, China's intellectual property system has become more robust as it has integrated with WTO standards as a member of the TRIPS agreement.[85] Its policy on gene patents is rather nuanced; like the United States, known DNA fragments are not patentable as they fall under the category of mere scientific discovery, but like the EU, if a DNA sequence is isolated for the first time, it can be patented.[86] Unfortunately, law enforcement in China is rather questionable. Importantly, while China has banned the use of gene editing on embryos since 2003, limited oversight by the government makes it easy for researchers to cross the line, despite the Chinese Academy's view that germline gene editing should proceed very cautiously.[87] In November 2018, Chinese scientist He Jiankui announced "that he had created the world's first genetically edited babies," for which he employed CRISPR to alter the gene CCR5 and render the girls HIV-resistant.[88] This move led to serious uproar in the international community, as Jiankui not only violated Chinese law by forging ethical reviews but also ignored global protocols restricting such advanced gene-editing research on embryos.[89] This incident urgently called for international regulation of gene editing, as very little is known about research and regulation in China due to a lack of transparency.

1.3 GLOBAL HARMONIZATION AND POLICY RECOMMENDATIONS

The most recent statement on gene-editing regulation at an international level is from the Second International Summit on Gene Editing that took place in November 2018, hosted by the Academy of Sciences of Hong Kong alongside the US Academies of Sciences and Medicine as well as the UK Royal Society.[90] This summit once again reaffirmed the global community's worries about germline gene editing with scientific, religious, and ethical arguments. Since Jiankui's CRISPR twins incident, a World Health Organization (WHO) panel was also created in December 2018 with the goal of establishing a global policy on this subject.[91] Additionally, a group of leading scientists in March 2019 demanded adopting a moratorium on heritable gene editing until further research was conducted on its safety and morality.[92]

While there is no regulatory body that has global authority over sovereign nations per se, an organization such as the WHO panel can work to establish an international treaty on this subject. In particular, there is a tension between progressing too rapidly in embryonic gene-editing research, which raises serious ethical questions, and discouraging biomedical innovation, which can prevent needed treatments from coming to fruition. In an interview with Dr. Rumiana Yotova—a faculty member at the University of Cambridge's Centre for Law, Medicine, and Life Sciences—she stated that international law has a role to play in "institutionalizing a mechanism of oversight in gene editing." Importantly, she noted that there is a serious lack of transparency between different countries in this context. Responding to the Jiankui incident, Dr. Yotova believes that there is difficulty in knowing what R&D efforts are taking place in China both because there is a language barrier and because Western journals are sometimes reluctant to publish Chinese work given fears around unsubstantiated data. A solution to this problem, she thinks, would be a simple centralized database to which countries could submit approvals or licenses to undertake gene-editing research. On the flip side, she recognizes that too much regulation is not beneficial, as it could smother innovation in gene-editing technologies that improve human health. In my opinion, the following recommendations should be taken up by global standards bodies:

1. *A moratorium on advanced germline gene editing, but support for basic embryonic research:* He Jiankui's experiment on embryos sent shockwaves through scientific communities across the globe, and his research acutely demonstrates the need for some international body to regulate gene-editing research. Jiankui's experimentation was at odds with Chinese legislation, but clearly, this legislation was nonbinding and poorly enforced.[93] If a body such as the WHO were to hold the Chinese government responsible for such R&D, it is likely that the morally questionable experiment would not have occurred. Simultaneously, it is important that early-stage embryonic research is still funded to facilitate the development of lifesaving stem cell therapies. These *ex vivo* gene-edited treatments can only enter the market if their safety and efficacy are tested in hESCs.[94] While the WHO panel cannot legislate on stem cells, it should encourage member states to follow the liberal hESC research policies adopted by the US and the UK. To prevent experimentation on embryos from going too far, an open-source centralized database should exist that lists all research approvals for such research. The WHO should mandate signatory countries to submit all licenses regarding basic germline gene-editing research to this database to increase transparency.
2. *Continued emphasis on R&D for somatic gene therapy:* In light of the recent controversies regarding gene editing, it is easy to forget the power of CRISPR and its incredible potential in medical applications for ameliorating human health. The use of CRISPR on somatic cells that are not heritable circumvents many of the ethical oppositions to germline gene editing. Furthermore, the tremendous clinical potential of somatic gene therapies has been demonstrated scientifically, with examples in diseases including Huntington's disease, cystic fibrosis, and β-thalassemia. Additionally, CRISPR-edited CAR-T cells can be employed to treat a variety of cancers, and Novartis already has an approved oncology product for leukemia called Kymriah.[95]
3. *Encouraging innovation globally to increase competition and reduce problems of access, potentially through harmonization of IP regimes:* While strongly state-centered policy in China has clearly been effective in turning China into an innovation powerhouse, such policies are deemed protectionist by many and would be difficult to support. Alternatively, the WHO can push for innovation systems such as that of the United States, which encourages private-sector innovation through its harmonized R&D policy and robust IP system. On the other hand, the European Union should be criticized for its overregulation. As previously mentioned, French biotechnologist Choulika states, "The West's emphasis on regulation – and overregulation – is leaving us far behind countries whose governments are more open to the idea of unrestricted or less-restricted research."[96] Critically, the current system of restricting all research on embryonic stem cells and preventing it from being patented in the EU effectively makes the United States the only country

> innovating in this area. This decreases competition and will present a serious access barrier as high R&D costs will translate to high prices around the world. To prevent limited access to lifesaving biotechnologies, innovation should be spurred globally.

As Dr. Yotova believes, "the benefit of having a harmonized policy would be a level playing field in terms of innovation and a certainty for scientists regarding ethics, avoiding a race to the bottom by commercial providers." It is time for global agencies to take this matter into their own hands.

NOTES

1 *See generally* Eric S. Lander, *The Heroes of CRISPR*, 164(1–2) Cell 18, 18–28 (2016).
2 *See* Martin Jinek et al., *A Programmable Dual-RNA–Guided DNA Endonuclease in Adaptive Bacterial Immunity*. 337(6096) Science 816, 816–821 (2012).
3 Jim Yeadon, *Pros and Cons of ZNFs, TALENs, and CRISPR/Cas*, The Jackson Laboratory (Mar. 4, 2014), www.jax.org/news-and-insights/jax-blog/2014/march/pros-and-cons-of-znfs-talens-and-crispr-cas.
4 Diego Germini et al., *A Comparison of Techniques to Evaluate the Effectiveness of Genome Editing*, 36(2) Trends in Biotechnology 147, 147–159 (2018).
5 *See generally* United States National Library of Medicine, *What are Genome Editing and CRISPR–Cas9?*, Genetics Home Reference, https://ghr.nlm.nih.gov/primer/genomicresearch/genomeediting.
6 *See generally* Patrick D. Hsu et al., *Development and Applications of CRISPR–Cas9 for Genome Engineering*, 157(6) Cell 1262, 1262–1278 (2014).
7 *See, e.g.*, Nina Duensing et al., *Novel Features and Considerations for ERA and Regulation of Crops Produced by Genome Editing*, 6 Frontiers in Bioengineering and Biotechnology (2018).
8 *See, e.g.*, Matthew C. Nisbet, *The Gene-Editing Conversation*, American Scientist (2018), www.americanscientist.org/article/the-gene-editing-conversation.
9 David Baltimore et al., *A prudent path forward for genomic engineering and germline gene modification*, 348(6230) Science 36, 36–38 (2015).
10 *See* Adam P. Cribbs & Sumeth M. W. Perera, *Science and Bioethics of CRISPR–Cas9 Gene Editing: An Analysis Toward Separating Facts and Fiction*, 90(4) The Yale Journal of Biology and Medicine 625, 625–634 (2017).
11 *See generally* Mary T. Bergman, *Perspectives on Gene Editing*, The Harvard Gazette (Jan. 9, 2019), https://news.harvard.edu/gazette/story/2019/01/perspectives-on-gene-editing/.
12 Baltimore, *supra* note 9.
13 David Baltimore et al., *On Human Gene Editing: International Summit Statement*, The National Academies of Sciences, Engineering, and Medicine (Dec. 3, 2015), www.nationalacademies.org/news/2015/12/on-human-gene-editing-international-summit-statement.
14 *See id.*
15 *See* Baltimore, *supra* note 9; *see also* Baltimore, *supra* note 13.
16 *See* Michael Kosicki et al., *Repair of double-strand breaks induced by CRISPR–Cas9 leads to large deletions and complex rearrangements*. 36(8) Nature Biotechnology 765, 765–771 (2018).
17 Feng Zhang, *Questions and Answers about CRISPR*, The Broad Institute, www.broadinstitute.org/what-broad/areas-focus/project-spotlight/questions-and-answersabout-crispr.
18 *See* David Cyranoski, *Japan set to allow gene editing in human embryos*, Nature (Oct. 3, 2018), www.nature.com/articles/d41586-018-06847-7.
19 André den Exter, *Embryonic stem cell patents at European top court*, 24(3) European Journal of Human Genetics 311 (2016).
20 *See, e.g.*, Tiffany Jeung, *Countries Already Drafting Rulebooks for Ethically Editing Human Embryo DNA*, Inverse (2018), www.inverse.com/article/49725-governments-regulate-human-embryo-gene-editing.

21 *See CRISPR*. Elsevier (Dec. 10, 2016), www.elsevier.com/research-intelligence/campaigns/crispr.
22 *See generally* Philip Shapira & Jan Youtie, *The Innovation System and Innovation Policy in the United States*, Competing for Global Innovation Leadership: Innovation Systems and Policies in the USA, EU and Asia 5–29 (Rainer Frietsch & Margot Schüller eds., 2010).
23 *How much does your country invest in R&D?* UNESCO Institute for Statistics (2017), http://uis.unesco.org/apps/visualizations/research-and-development-spending/.
24 *See* Cyranoski, *supra* note 18.
25 *See generally* Food & Drug Administration, Fast Track, Breakthrough Therapy, Accelerated Approval, Priority Review (2018), www.fda.gov/patients/learn-about-drug-and-device-approvals/fast-track-breakthrough-therapy-accelerated-approval-priority-review.
26 *See* Rumiana Yotova, *The Regulation of Genome Editing and Human Reproduction Under International Law, EU Law and Comparative Law*, The Nuffield Council on Bioethics 45 (2017), www.nuffieldbioethics.org/wp-content/uploads/Report-regulation-GEHR-for-web.pdf.
27 See, e.g., Gene-edited cells fight cancer, 539 Nature 332, 332–333 (2016).
28 Bernd Zetsche et al., *Cpf1 is a Single RNA-Guided Endonuclease of a Class 2 CRISPR–Cas System*, 163(3) Cell 759, 759–771 (2015).
29 Pranam Chatterjee et al., *Minimal PAM specificity of a highly similar SpCas9 ortholog*. 4(10) Science Advances (2018).
30 Neil Davey et al., *Ananda Mohan 'Al' Chakrabarty 1938–2020*, Nature Biotechnology (2020), www.nature.com/articles/s41587-020-00785-4.
31 Mateo Aboy et al., *Myriad's impact on gene patents*, 34(11) Nature Biotechnology 1119, 1119–1123 (2016).
32 *See* Susan Finston et al., *Mayo, Myriad, America Invents Act and BPCIA: how has the United States biopharmaceutical market been affected?*, 5(3) Pharmaceutical Patent Analyst 159, 159–167 (2016).
33 *See* Neil Davey et al., *The CRISPR Controversy: Implications for Science, Law, and Policy in the US and EU*, 6(1) The GNLU Law Review (2019).
34 Broad Communications, *For Journalists: Statements and Background on the CRISPR Patent Process*, The Broad Institute (Sept. 27, 2020), www.broadinstitute.org/crispr/journalists-statement-and-background-crispr-patent-process.
35 *Supra* note 21; *supra* note 23.
36 European Commission, Improving Knowledge Transfer Between Research Institutions and Industry Across Europe 7 (2007), http://ec.europa.eu/invest-in-research/pdf/download_en/knowledge_transfe_07.pdf.
37 European Commission, OECD – Share of countries in biotechnology patents (2016), https://ec.europa.eu/knowledge4policy/dataset/beooecd-oecd-biotechnologypatentsshare_en.
38 Commission of the European Committees, Communication from the Commission on the Precautionary Principle (2000), http://eurlex.europa.eu/LexUriServ/LexUriServ.do?uri=COM:2000:0001:FIN:EN:PDF.
39 *See* Giovanni Tagliabue, *The EU legislation on "GMOs" between nonsense and protectionism: An ongoing Schumpeterian chain of public choices*, 8(1) GM Crops & Food 57, 57–73 (2017).
40 *See* Cyranoski, *supra* note 18.
41 CRISPR Therapeutics, *CRISPR Therapeutics and Vertex Announce Progress in Clinical Development Programs for the Investigational CRISPR/Cas9 Gene-Editing Therapy CTX001* (2019), http://ir.crisprtx.com/node/8596/pdf.
42 Intellia Therapeutics, *Intellia Therapeutics and Novartis Expand Cell Therapy Collaboration to Pursue CRISPR/Cas9-based Genome Editing in Additional Stem Cell Population* (2018), https://ir.intelliatx.com/news-releases/news-release-details/intellia-therapeutics-and-novartis-expand-cell-therapy.
43 Directive 98/44/EC of the European Parliament and of the Council of 6 July 1998 on the legal protection of biotechnological inventions, 1998 O.J. (L 213) 13, 13–21.
44 Kelly Servick, *Broad Institute takes a hit in European CRISPR patent struggle*, Science Magazine (2018), www.sciencemag.org/news/2018/01/broad-institute-takes-hit-european-crispr-patent-struggle.
45 *See generally* European Commission, What is Horizon 2020?, https://ec.europa.eu/programmes/horizon2020/en/what-horizon-2020.

46 Jeff Kipling, *The European Landscape for Human Genome Editing*, UK Academy of Medical Sciences (2016), https://acmedsci.ac.uk/file-download/41517-573f212e2b52a.pdf.
47 Council of Europe, *Convention for the Protection of Human Rights and Dignity of the Human Being with regard to the Application of Biology and Medicine: Convention on Human Rights and Biomedicine*, European Treaty Series (1997), https://rm.coe.int/168007cf98.
48 *See generally* Council of Europe, *Details of Treaty No. 164*, European Treaty Series (1997), www.coe.int/en/web/conventions/full-list/-/conventions/treaty/164.
49 *Supra* note 47.
50 *Id.*
51 *Supra* note 48.
52 *See* Yotova, *supra* note 26.
53 *Supra* note 43.
54 *Id.*
55 *See supra* note 38.
56 *Supra* note 43.
57 *See id.*
58 *Id.*
59 European Commission, State of Play of the Implementation of Directive 98/44/EC (2007).
60 *See* Servick, *supra* note 44.
61 Broad Communications, *supra* note 34.
62 *Id.*
63 Alison Abbott, *Europe rules against stem-cell patents*, 471(7338) Nature 280 (2011).
64 *See* Sonya Davey et al., *Interfacing of Science, Medicine and Law: The Stem Cell Patent Controversy in the United States and the European Union*, Frontiers in Cell and Developmental Biology (2015).
65 *European Court bans stem cell patents*, EuroStemCell (2012), www.eurostemcell.org/european-court-bans-stem-cell-patents.
66 Ana Nordberg et al., *Cutting edges and weaving threads in the gene editing (Я)evolution: reconciling scientific progress with legal, ethical, and social concerns*, 5(1) Journal of Law and the Biosciences 35, 35–83 (2018).
67 *Country/territory outputs*, Nature Index (2020), www.natureindex.com/country-outputs/generate/Life%20Sciences/Europe/All/n_article.
68 *Granted patents per field of technology and per country of residence for each individual year from 2010 until 2019*, European Patent Office (2019), www.epo.org/about-us/annual-reports-statistics/statistics.html.
69 *See* Insoo Hyun et al., *Embryology policy: Revisit the 14-day rule*, 533(7602) Nature 169, 169–171 (2016).
70 *See generally* Human Tissue Authority, *Stem Cell Science in the UK: Key facts about research, regulation and funders*, GOV.UK (2008), www.hta.gov.uk/sites/default/files/stem_cell_pack_200806170144.pdf.
71 Kipling, *supra* note 46.
72 *See supra* note 48.
73 Kipling, *supra* note 46.
74 Katherine Drabiak-Syed, *New President, New Human Embryonic Stem Cell Research Policy: Comparative International Perspectives and Embryonic Stem Cell Research Laws in France*, 32(6) Biotechnology Law Report 349, 349–356 (2013).
75 *See* Kipling, *supra* note 46.
76 *See id.*
77 CRISPR Therapeutics, *supra* note 41.
78 Novartis, *Novartis receives first ever FDA approval for a CAR-T-cell therapy, Kymriah™ (tisagenlecleucel, CTL019), for children and young adults with B-cell ALL that is refractory or has relapsed at least twice* (Aug. 30, 2017), https://novartis.gcs-web.com/novartis-receives-fda-approval-for-KymriahTM.
79 Andrew Hong, *CRISPR in personalized medicine: Industry perspectives in gene editing*, 42(8) Seminars in Perinatology 501, 501–507 (2018).

80 Kipling, *supra* note 46.

81 *See generally* Karen J. Lin et al., *State-owned enterprises in China: A review of 40 years of research and practice*, 13(1) China Journal of Accounting Research 31, 31–55 (2020).

82 *See generally* Albert G. Hu, *Technology parks and regional economic growth in China*, 36(1) Research Policy 76, 76–87 (2007).

83 André Choulika, *The West is losing the gene editing race. It needs to catch up*, STAT News (Oct. 29, 2018), www.statnews.com/2018/10/29/west-is-losing-gene-editing-race/.

84 *See supra* note 21.

85 *See, e.g.,* Thomas E. Volper, *TRIPS Enforcement in China: A Case for Judicial Transparency*, 33(1) Brooklyn Journal of International Law 309, 309–310 (2007).

86 Li Du, *Patenting human genes: Chinese academic articles' portrayal of gene patents*, 19(1) BMC Medical Ethics 29 (2018).

87 *See* Baltimore, *supra* note 13.

88 Gina Kolata et al., *Chinese Scientist Claims to use CRISPR to Make First Genetically Edited Babies*. New York Times (Nov. 26, 2018), www.nytimes.com/2018/11/26/health/gene-editing-babies-China.html.

89 *See* Kerry Grens, *Scientist Forged Ethical Reviews for CRISPRed Babies*, The Scientist (Jan. 21, 2019), www.the-scientist.com/news-opinion/scientist-forged-ethical-reviews-for-crispred-babies-65346.

90 National Academies of Sciences, Engineering, and Medicine, *Second International Summit on Human Genome Editing: Continuing the Global Discussion: Proceedings of a Workshop—in Brief*, The National Academies Press (2019).

91 *See generally* World Health Organization, Human Genome Editing: WHO Establishing Expert Panel to Develop Global Standards for Governance and Oversight of Human Genome Editing (Dec. 14, 2018), www.who.int/ethics/topics/human-genome-editing/en/.

92 Eric S. Lander et al., *Adopt a moratorium on heritable genome editing*, Nature (Mar. 13, 2019), www.nature.com/articles/d41586-019-00726-5.

93 *See, e.g.*, Achim Rosemann et al., *The regulatory and legal situation of human embryo, gamete and germ line gene editing research and clinical applications in the People's Republic of China*. Nuffield Council on Bioethics (2017), www.nuffieldbioethics.org/wp-content/uploads/Background-paper-GEHR.pdf.

94 *See generally* Emmanuelle Tuerlings, *WHO Expert Advisory Committee on Developing Global Standards for Governance and Oversight of Human Genome Editing*, World Health Organization 11 (2019), www.who.int/ethics/topics/human-genome-editing/WHO-Commissioned-Governance-1-paper-March-19.pdf.

95 Novartis, *supra* note 78.

96 Choulika, *supra* note 83.

2 IP Flexibilization for Accelerating the Achievement of the UN 2030 Agenda

Lessons from AIDS Programs in Brazil, India, Thailand, and Indonesia

Cristina Possas

2.1 INTRODUCTION

Intellectual property (IP) flexibility has been a key tool for accelerating the achievement of SDGs in the United Nations 2030 agenda for HIV/AIDS, as indicated in the evidence provided here. International and national IP initiatives supporting free and universal access to HIV treatment based on voluntary licensing, compulsory licensing and favoring generic competition of antiretroviral (ARV) drugs contributed to significantly lower ARV prices and have averted almost 20.8 million AIDS-related deaths worldwide in the past three decades. By 2022, AIDS-related deaths have been reduced by 69% since the epidemic peak in 2004, and 71% of people living with HIV in 2022 had suppressed viral loads, living long, healthy lives with zero risk of transmitting HIV [1].

The recent literature presented here provides evidence supporting the fact that IP flexibilization has played a significant role in this remarkable global decline in HIV/AIDS cases and deaths. We examine this impact from the experiences and lessons of four internationally recognized National AIDS Programs in Brazil, India, Thailand, and Indonesia. IP flexibilities resulted from the evolution and important changes in the international legal framework. Trade-Related Aspects of Intellectual Property Rights (TRIPS), an agreement by the World Trade Organization (WTO) member states [2] came into effect on January 1, 1995. TRIPS is considered the most comprehensive international IP agreement and has played a major role in the HIV/AIDS epidemic in resolving trade disputes over accelerating access of the poorest populations to essential ARV medicines worldwide. Later, the Declaration on the TRIPS Agreement and Public Health was adopted by the WTO Ministerial Meeting in Doha, Qatar, November

DOI: 10.1201/9781003405665-2

14, 2001 [3]. The Doha Declaration, internationally recognized as a major milestone, proposed a balanced interpretation of the system. This declaration reaffirmed the flexibility of TRIPS member states in patent rights, supporting WTO members' rights to protect public health and promote access to medicines for all. Moreover, the Doha Declaration was the first international trade, not health ministers, to recognize the gravity of global public health issues in the scenario of the HIV/AIDS epidemic and placed TRIPS into broader national and international contexts [4].

A report coordinated by the WTO in collaboration with the WIPO and WHO evidence this new public health perspective in international trade, examining the intersections between public health, intellectual property and trade [5]. Compulsory licensing and other patent flexibilities under the Doha Declaration have emerged as important tools to ensure access to needed patented medicines, particularly in critical health crises, and had, as presented here, a significant impact during the HIV/AIDS epidemic. Nevertheless, despite all these legal advances from TRIPS and the Doha Declaration, controversies and disputes remain in the pharmaceutical sector and related stakeholders on the utility of compulsory licensing as an instrument to reduce prices and improve access to expensive patented drugs.

Patent holders and pharmaceutical companies argue that patent policies that undermine IP protections, such as government seizure of patent protections to let competitors use patented inventions, threaten incentives to develop new treatments and respond to epidemic emergencies. We examine these controversies and disputes in light of the experience of four countries, Brazil, India, Thailand, and Indonesia, for the adoption of compulsory licensing for ARV drugs. A discussion is provided on the long-term feasibility of this legal and trade instrument, examining future prospects for the sustainability of National AIDS Programs in these countries.

2.2 CONCEPTUAL FRAMEWORK

The complexity of this IP debate and of the legal, trade and health policy issues involved in global access to ARV drugs requires an adequate conceptual framework and comprehensive interdisciplinary analysis. The results and impact of the synergy among these diverse fields and policy interventions have been the subject of diverse analyses in the literature.

TRIPS is a multilateral trade agreement by WTO member states to ensure fair and equitable protection of IP rights. In his seminal article, a leading author [6] notes that the right of governments to grant compulsory licenses on virtually any ground – including public interest, abuse, or anti-competitive conduct, or for non-commercial government use, among others – issued stronger and clearer from the TRIPS Agreement than had previously been the case under the Paris Convention for the Protection of Industrial Property in 1883.

A conceptual and legislative history pertaining to compulsory licensing of patented pharmaceuticals from the TRIPS Agreement from 1994 to the 2003 waiver can be found in the literature [6–8], including the later proposed amendment of article 31. This evolution in the legal framework article enables poor countries to obtain needed medicines from other countries that possess manufacturing capacity. Reichman evaluates controversial uses of the relevant legislative framework as viewed from different critical perspectives [6].

Clarifying how developing countries seeking access to essential medicines can collaborate in ways that would avoid undermining incentives to innovation and other social costs attributed to compulsory licensing, the author provides an important theoretical contribution to this topic. Reichman ends by defending the legality of recent measures taken to promote public health in developing countries and by reminding developed countries that unilateral retaliation against such measures is demonstrably illegal under WTO foundational law and jurisprudence.

Beyer provides a conceptual basis for developing socially responsible intellectual property licensing policies [7], examining nonexclusive licensing initiatives in the pharmaceutical sector and the role of governments among developed and developing countries in these initiatives.

Abbott and Reichman [8] examine the evolution in the legal framework and the Doha Round's public health legacy, discussing global and local strategies for the production and diffusion of patented medicines under the amended TRIPS.

Another study [9] provides evidence indicating how power in international politics is increasingly exercised by private interests rather than governments. In her compelling study of the influence of private interests in international legislation and local policies, Sell argues that TRIPS resulted from lobbying by powerful multinational corporations who wished to adequate international law to protect their markets.

Critical issues related to the conceptual framework and legal evolution in IP regime shifting in the TRIPS agreement are discussed by Helfer [10], highlighting how this shift resulted in new dynamics of international intellectual property lawmaking. Consequent to this IP regime shift, many legal flexibilities emerged in several countries, as presented by Coriat and several authors in a historical book on the political economy of HIV/AIDS in developing countries [11], indicating advances and constraints in their policies. Finally, Imtiazuddin and Mohiuddin [12] provide an innovative IP approach from a market perspective: the notion of socially responsible licensing, underlining the need to conceive and implement model partnerships for underserved markets. These contributions to the conceptual, legal, and global policy IP framework laid the foundations of a comprehensive and holistic perspective on the legislative transition from the Paris Convention in 1883 to TRIPS and the Doha Declaration, permeated by the new international demands from developing countries for expeditious access to ARV drugs.

2.3 IP FLEXIBILITIES AND LICENSING AGREEMENTS

Licensing agreements for IP can be voluntary or compulsory depending on the licensor's disposition. Voluntary licensing is the most common legal instrument to authorize use and make patented or copyrighted rights for commercial purposes. Nevertheless, there are occasions when licensors are not willing to authorize patent use to a potential licensee. When this occurs, governments may intervene, particularly in health crises and epidemics, and enforce compulsory licensing under certain conditions, without the licensor's consent [4, 13].

Compulsory licensing is a legal instrument by which a government licenses companies or individuals other than the patent owner to use the rights of the patent to make, use, sell, or import a product under patent without the permission of the patent owner. Nevertheless, compulsory licensing operation is complex, and in many developing countries, in addition to the difficulties of negotiating with pharmaceutical companies, there is another major constraint hindering its implementation: insufficient or absent local manufacturing capacity, particularly in the poorest countries.

Paragraph 6 of the Doha Declaration [3] alerted to this issue and called on the TRIPS Council "to find an expeditious solution to the recognized problem that WTO members with insufficient or no manufacturing capacities in the pharmaceutical sector could face difficulties in making effective use of compulsory licensing under the TRIPS Agreement."

The Doha Declaration affirmed that TRIPS "should be interpreted and implemented in a manner supportive of WTO members' right to protect public health and, in particular, to promote access to medicines for all." It affirmed the right of nations to use the exceptions of TRIPS, such as the compulsory licensing provision discussed above, to meet public health concerns, specifically stating that "public health crises, including those related to HIV/AIDS, tuberculosis, malaria and other epidemics which can represent a national emergency".

A waiver to the obligations of an exporting member under Art. 31(f) was agreed upon by the WTO General Council on August 30, 2003, with respect to the grant of a compulsory license and export to an eligible importing WTO member(s) under certain terms. In a meeting of the World Trade Organization Council for TRIPS on November 8–9, 2016, the intergovernmental South Centre held a side event to discuss experiences in the implementation and effective functioning of the system [14]. The meeting also discussed the effectiveness of a waiver to international trade rules intended to ease shipments of affordable medicines to low-income countries.

The flexibilities of Paragraph 6 of the Doha and the feasibility of the implementation of compulsory licensing in developing countries, due to the complexity of the process, have been highly controversial and were debated in this WTO meeting. In the literature on the topic, some authors discuss this feasibility and whether this paragraph has been a relief or just a regulatory ritualism [15].

Twenty countries have approximately either issued or publicly declared the intention to issue a compulsory license for one or more pharmaceutical products since the founding of the WTO [16]. In some cases, governments did not end up issuing a compulsory license. A mere public announcement or discussion of potentially issuing a compulsory license for a drug has sometimes led the patent holder to offer a discount or a voluntary license for the drug [17].

In the 2000s, Brazil, Ecuador, Ghana, Indonesia, Malaysia, Mozambique, Thailand, Rwanda, Zambia, and Zimbabwe each issued compulsory licenses for one or more antiretroviral drugs to respond to their HIV patients who could not afford antiretroviral therapy [18, 19]. While most countries issued licenses for a specific patented drug, Ghana and Zimbabwe issued categorical compulsory licenses on all antiretroviral drugs [16].

2.4 LEGAL AND TRADE DISPUTES

The main reason for legal and trade disputes on compulsory licensing is that pharmaceutical companies invest a significant number of resources in the long-term process of discovery and development of a new drug, from 10 to 20 years, researching, innovating, and gaining regulatory approval [20]. The long process for developing new ARV drugs since the onset of the AIDS epidemic illustrates this scenario well.

The Tufts University Center for the Study of Drug Development (CSDD) has estimated the cost of producing a new drug to be USD 2.6 billion based on information provided by 10 pharmaceutical companies on 106 randomly selected drugs [21]. This estimate is based on an average out-of-pocket cost of USD 1.4 billion and an estimate of USD 1.2 billion in returns that investors forego on that money during the 10-plus years a drug candidate spends in development [22].

Later, in a new study, which used the same methodology as earlier R&D cost assessments, the authors found that the cost of developing and gaining marketing approval for a new drug has increased at an annual rate of 8.5% since 2003. When adding post-approval R&D costs, which include long-term safety monitoring, the full life-cycle cost of an approved drug increased to US USD 2.9 billion [23].

Considering possible controversies on these estimated drug development costs, a recent study concluded that R&D costs per new medicine vary between USD 944 million and USD 2,826 million (adjusted to 2019 prices), with evidence indicating an increase in these costs over time [24]. The authors note that any cost estimate needs to be interpreted in context and is specific to the underlying methodology, sample, and time period of the study. According to The Pharmaceutical and Research Manufacturers of America – PhRMA [25], it takes 10–15 years on average to develop a new medicine from initial discovery through to regulatory approval. Figure 2.1 presents a flow chart for the development of a medicine, indicating the complexity of this process from discovery to launch/post-registration.

The industry group Biotechnology Innovation Organization – BIO, corroborated this information, supported by a follow-up study on 9,704 drug development programs over the course of a decade from January 2011 to November 2020. They found that,

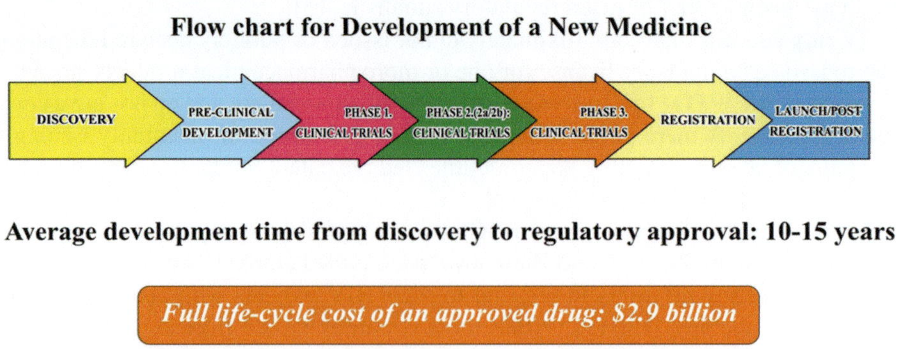

FIGURE 2.1 Overview of the process for the development of new medicines. (Elaborated by the author based on PhRMA [25] and BIO [26].)

on average, it took 10.5 years for a drug to get from Phase I to regulatory approval [26]. Patent holders argue that this high complexity and long-life cycle of successfully creating a new drug justify high prices as a consequence of needed long-term investments and innovation incentives, such as patent protection.

To examine these arguments from patent holders justifying the need for strong patent protection, a detailed study provided evidence to clarify whether stronger intellectual property rights increase incentives to innovate [27]. This study has tested the impact of ever more rigorous IPR systems on innovation through an index of economic complexity of 94 countries from 1965 to 2005. Its results confirm that stronger intellectual property systems engender higher levels of economic complexity. Nevertheless, the authors concluded that only countries with an initial above-average level of development and complexity could enjoy this effect.

The use of patent flexibilities and compulsory licensing has been a complex process for developing countries. A report from the UN High Level Panel on Access to Medicines [28] (UN Secretary General 2016) denounced obstacles to the use of TRIPS and alerted to many threats of retaliation from governments and corporations who have followed the use of compulsory licensing by middle-income countries, such as Thailand and Brazil.

This intimidation scenario has weakened the bargaining position of many WTO members. One such example is Thailand's 2006 decision to import generic versions of Efavirenz from India. According to this UN report, this decision was met with hostility from the manufacturer Merck and the United States Government, which questioned the legality of compulsory licensing in Thailand to rescind its decision.

Thailand's subsequent decision to issue two compulsory licenses in 2007 for lopinavir/ritonavir and clopidogrel also resulted in retaliatory measures, according to this report. In response, Abbott withdrew from the Thai market all medications waiting for registration and refused to register any new pharmaceutical products in the country, thereby denying patients access to the heat-resistant form of lopinavir/ritonavir for which no generic equivalent existed. The European Trade Commissioner wrote to the Thai Government criticizing its use of compulsory licenses as "detrimental" to medical innovation negotiations [28].

Brazil was also subject to potential trade retaliations by the United States in 2005, with implications to the successful Brazilian AIDS Program. The final resolution between Brazil and Abbott was reached after threats of commercial retaliations were made by the U.S. Government in response to Brazil's plan to locally produce a generic version of Kaletra. The commercial retaliation most readily available to the U.S. Government at the time of the dispute was the removal of Brazil from the Generalized System of Preferences. U.S. Government officials explained that tampering with AIDS drug patents threatens intellectual property rights protection, which ensures the search for new drugs [29].

2.5 COMPULSORY LICENSING: IMPACT ON PRICES AND ACCESS

To examine whether compulsory licensing is an effective mechanism to lower drug prices, a systematic review [30] verified a mean price reduction between

66.2 and 73.9% for the 24 compulsory licensing events in countries for which price data are available. They concluded, based on the retrieval and analysis of 51 observations of pre- and post-compulsory licensing prices, that a compulsory licensing event is likely to significantly reduce the price of a patented drug, albeit with some caveats.

These results indicate that compulsory licensing has indeed been an effective mechanism for price reduction, increasing availability and access, particularly in the poorest developing countries. The conclusion is that licensing options, price control and access policies should be approached in a holistic way, considering that they have an enormous impact on global market entry [31].

A World Bank report [32] corroborated this positive impact of compulsory licenses on prices, and access theses indicated the importance of TRIPS flexibility as a mechanism to sustain public HIV/AIDS programs. The authors estimated that Thailand could reduce the costs of second-line therapy by 90% through the issue of compulsory license, which would result in future cost savings of USD 3.2 billion through 2025.

These studies indicate that compulsory licensing should definitively remain a legal alternative for assuring universal access and free access to ARV drugs, particularly in the poorest developing countries. A recent study provides an extensive compilation of compulsory licensing attempts [33].

Another author [34], in a detailed study, found that compulsory licensing turns out to be very abstract and left undefined, especially regarding the conditions for local operation and manufacturing requirements. His study focuses on India and Brazil's compulsory licensing regimes, and the conclusion is that India has wisely exploited all the advantages of TRIPS, such as the 10-year transition. According to the author, unlike India, Brazil failed to adopt TRIPS flexibilities. To compensate for the country's missed opportunity, Brazil has turned compulsory licensing into a tool for price bargaining to ensure public health. Thus, while India has benefited from a proactive role in the private sector, Brazil's position is mainly based on its leaders' political efforts [35]. Nevertheless, although the authors are correct in their evaluation of the performance of both countries, further studies are necessary to examine the complex underlying legal, institutional, and political conditions of India and Brazil to understand why their achievements in the private and public sectors have been so different.

Finally, another study on compulsory licensing from a legal perspective [36] noted that TRIPS has been in existence for a relatively short time as far as international treaties are concerned. The Berne Convention for the Protection of Literary and Artistic Works and the Paris Convention for the Protection of Industrial Property, the other two major international intellectual property treaties, have existed for over a century (since the 1880s). The author underlines that international treaties have been subject to numerous revisions, occurring approximately every 20 years, so it is too early for a final conclusion on TRIPS, probably subject to future revisions. In his final considerations, the author concludes that whether TRIPS thus far has been a success or failure is not clear and that its message has been mixed. He notes that while there have been many positive developments, negotiations for further advancement have stalled.

2.6 LESSONS: BRAZIL, INDIA, THAILAND, AND INDONESIA

Here, we examine lessons from the experiences of four countries – Brazil, India, Thailand, and Indonesia – that are internationally recognized for successful National AIDS Programs. We present their diverse characteristics and peculiarities, examining events of voluntary licensing, compulsory licensing, and technology transfer of ARV drugs in a scenario of exponential increase in global and local demands for these drugs.

2.6.1 Brazil

The Brazilian policy of universal and free access to antiretroviral drugs for people living with HIV/AIDS is internationally recognized as a success in the implementation of a pioneer public health policy. Brazil introduced this universal coverage policy into its AIDS Program in the Ministry of Health, created in 1987, after the promulgation of Law 9, 313, in November 1996, which guaranteed the free provision of antiretroviral drugs by the Unified Health System (SUS). This pioneer policy made it possible to control the epidemic in the country [19], with, from 1996 to 2002, a drop of 70% in mortality and 80% in hospital admissions and maintaining the seroprevalence rate at 0.6% of HIV infection [37, 38].

Funds from the World Bank played an important role in this successful Brazilian policy. After a request from the Brazilian Government, these funds provided technical and financial assistance to support prevention, diagnosis, treatment, care, surveillance, research, monitoring and evaluation, and institutional development [39, 40]. In addition, Brazil has received international attention and recognition for its use of compulsory licensing in price negotiations with pharmaceutical companies. Starting in 2001, the country's strategy began to achieve its desired outcomes as patent holders reduced prices drastically. The successful Brazilian AIDS Program, including these pioneer threats to issue compulsory licenses, was named "the Brazilian model" and strongly recommended in other countries [19].

The international attention and recognition also resulted from the fact that Brazil challenged conventional belief because most of the international agencies supporting global HIV policies were against developing countries implementing therapeutic schemes for a treatment that often exceeded USD 10,000 per patient per year [41, 42]. Thus, the Brazilian Governmental strategy was not only crucial for the country´s HIV/AIDS patients but also, from a broader perspective, for contributing to the political economy of access to ARV drugs in developing countries [11].

Nevertheless, despite the global recognition of this Brazilian model, with significant cost reductions achieved through voluntary licensing negotiations with multinational pharmaceutical companies and stringent government measures such as the compulsory licensing of Efavirenz in 2007 and later the opposition procedure to the examination of the Tenofovir patent, it was not capable of containing very high public expenditures on innovative third-line drugs. These drugs, protected by patents, were consumed by only 3% of patients in the country. This high public expenditure, which started to increase exponentially in 2007 gradually constrained universal and free access by HIV/AIDS patients in the country [19, 37].

Consequently, HIV/AIDS patients from many Brazilian States started a strong judicialization process, requiring with success in Tribunals immediate free access to these third-line ARV drugs, the National Unified System-SUS, guaranteed by the 1988 Constitution and by Law 9, 313, November 1996 [43]. From a global perspective, another study provides an interesting overview of the judicialization debate, examining to what extent access to essential medicines, as part of the right to health, should be enforceable through the courts [44].

The previous Brazilian Government's price negotiation strategy threatening pharmaceutical companies with the possibility of a compulsory license, which had significantly contributed to reducing the prices of these antiretroviral drugs in the country over the past two decades, is a policy that now faces constraints from the new global scenario of IP flexibilization: the extensive global use of voluntary licensing and an exponential increase in ARV generic production in many countries, particularly in India, a leading manufacturer. The successful Brazilian AIDS Program experience provides insights into the potential synergies between industrial and health policies for access to ARV drugs in middle-income countries and on the role of local generic manufacturers and imports [45].

Compulsory licenses for ARV drugs in Brazil were issued based on a previous legal framework, the Brazilian Industrial Property Law No. 9. 279 of May 14, 1996, bringing national IP legislation into compliance with TRIPS standards. This law had a significant impact on the local pharmaceutical industry, as it introduced the patentability of previously nonpatentable chemical products and processes. Law 9,279/96 took effect in May 1997, and consequently, Brazil waived the 10-year adaptation period to comply with TRIPS. The enforcement of the Brazilian Patent Law created constraints in the local production of ARV drugs because only those molecules that were in the market before 1996 could be copied. Two other decisions aggravated this scenario: the inclusion of a provision for "pipeline patents" of pharmaceutical inventions and the exclusion of parallel imports [46]. On the other hand, the Brazilian Intellectual Property Law included some flexibilities allowed under TRIPS to undermine the potential negative impact of patents on access to medicines: compulsory licensing and experimental use of patented inventions. Nevertheless, recently, in the context of the COVID-19 pandemic, important changes were made in article 71 of this law. On September 3, 2021, the Brazilian President promulgated Law No. 14,200, September 2, 2021, amending the Industrial Property Law, Law No. 9,279 of May 14, 1996.

The new law provides for the compulsory licensing of patents or patent applications in cases of national or international emergency, public interest, or nationwide public calamity. Law No. 14,200 amended article 71 of Law No. 9,279, which determines that in cases of national or international emergency or public interest declared by law or an act of the federal executive branch, or a state of nationwide public calamity recognized by the National Congress, a compulsory, ex officio, temporary, and nonexclusive license may be granted "for the exploitation of the patent or patent application, without prejudice to the rights of the respective holder, provided that its holder or its licensee does not meet this need."

The main difference between the former legal text and the current text is the possibility of compulsory licensing of patent applications that are still under examination. The new law provides for an exception to compulsory licensing, in cases where the competent authority, established by the Executive Branch, deems that the patent holders of these technologies and products have made objective commitments capable of ensuring that domestic demand will be met in conditions of volume, price, and term compatibility with emergency needs.

Despite all the success and advances in the country´s model of universal and free access to ARV policy, Brazil now faces many challenges to end the HIV/AIDS epidemic by 2030: an adult prevalence rate of 1%, higher than India (0.2%), 1,170,000 people living with HIV/AIDS and 16,000 annual deaths, according to UNAIDS 2023 estimates [1].

2.6.2 India

In its 1970 Patent Law, India excluded drugs from product patent protection. This law was one of the reasons that the Indian generic drug industry was able to evolve and later become the main global pharmaceutical generics manufacturer and a leading ARV generics supplier to many developing countries.

The Indian pharmaceutical industry has come after the TRIPS agreement and the Doha Declaration with important implications for national IP policy [47,48]. Since 2010, with the support of the Medicines Patent Pool [49,50], the country became a leading player in the global generic market for antiretroviral drugs for HIV/AIDS treatment. India is the largest provider of generic medicines worldwide and a key generic ARV provider for the poorest developing countries. It is estimated that Indian companies are currently supplying approximately 60–65% of the total global supply of generics for HIV/AIDS treatment [51]. India has been an important source of affordable generic antiretroviral drugs for developing countries in Africa, Brazil, and other parts of the world. In a remarkable diversification of products, Indian generic manufacturers have been able to develop a wide range of generic drugs to treat HIV/AIDS. Considering India's important role as a global supplier, an extensive review of the legal and political situation in India [52] indicates some of the most crucial changes in Indian patent law and procedures that were considered likely to shape the production of local ARV drugs.

The Indian National AIDS Program, conducted with strong political will and a universal access perspective, greatly benefited from this possibility of immediately providing ARV to the country´s poorest HIV/AIDS patients [53]. According to a Johns Hopkins' study, potentially USD 843 billion is saved each year in India by using generic substitutions [54]. While this does not consider other countries who have benefitted from India's section 92(A), these numbers are a reminder of the enormous global impact of an increase in access to ARV drugs. However, it should be noted that since the enactment of the Indian Patent Laws and the country´s compliance with TRIPS, the application of rules regarding compulsory licensing in the country has remained very low and of an exceptional nature.

In 2009, The Indian Patent Office (Controller General of Patents, Designs and Trademarks) rejected applications from two U.S. pharmaceutical companies on grounds of "evergreening", raising hopes of a further lowering of prices of two life-saving drugs used by HIV/AIDS patients in the country. U.S. drug companies, Gilead Sciences and Tibotec Pharmaceuticals, manufacturers of Tenofovir and Darunavir, respectively, lost their cases for patents in India; Mumbai-based generic drug manufacturer, Cipla, filed cases against both companies. An extensive review of developments in compulsory licensing of pharmaceutical patents in India [55] indicates the complexity of the legal and implementation issues faced by the India Patent Office and Government officials.

The first compulsory license in India was issued only in 2012, when the Indian Patent Office granted for the first time a compulsory license to Natco, a domestic generic firm, to manufacture Nexavar, a patented medicine for liver and kidney cancer owned by German firm, Bayer. While this historical decision seemed to announce a new era of compulsory licensing in the region, the Indian Patent Office unexpectedly turned down two subsequent applications made by private companies. In 2013, BDR Pharmaceuticals International Pvt Ltd. filed for a compulsory license for a cancer drug SPRYCEL manufactured by Bristol Myers Squibb. In 2015, Lee Pharma filed for a compulsory license of a diabetes management drug Saxagliptin manufactured by AstraZeneca. Both applications were rejected by the Indian Patent Office on the basis that neither BDR Pharmaceuticals International Pvt Ltd. nor Lee Pharma had provided sufficiently justified reasons for such a grant. In the case of BDR Pharmaceuticals International Pvt Ltd., the Indian Patent Office argued that the applicants for compulsory license did not make a sufficient attempt to obtain a license from the patent holder and did not have the capacity to adequately manufacture the invention to meet public needs. On the other hand, Lee Pharma failed to demonstrate how the public would benefit from a generic version of the drug in view of already available drugs in the market in the same price range. Only later, in 2017, the Indian Government finally approved three compulsory licensing applications for ARV drugs manufactured by Merck & Co. Inc. Abbott and Boehringer for the AIDS medicines Efavirenz, Lopinavir, and Ritonavir, respectively [1, 54].

Indian Patent Law requires that a number of criteria should be taken into consideration when deciding whether a compulsory license should be granted to a third party, the applicant for the compulsory license. Under Indian Patent Laws, compulsory licensing can be granted after 3 years of obtaining a patent. The Indian Patent Office might grant a compulsory license to a third party only if the use of the patented product is not satisfying public requirements, the patented product is not accessible to the public at a reasonable price, or the patentee has not worked the patented product in India. Compulsory licenses in the country will only be imposed when an innovation that could be greatly beneficial to the public interest is not being used – or at least not sufficiently – by the patent owner, considering, for instance, if the third party has already approached the patent owner to obtain a license, whether the third party has capabilities to meet public interest by manufacturing the patented product, or the actual type of invention and its benefits for the public.

Several provisions have been put in place by the Indian Government to minimize the misuse of patent rights and provide a legal framework to the Indian Patent Office. For instance, under Indian Patent Laws, compulsory licensing can be granted after three years of obtaining a patent. Moreover, the Indian Patent Office might grant a compulsory license only if the use of the patented product is not satisfying public requirements, the patented product is not accessible to the public at a reasonable price, or the patentee has not worked the patented product in India. In other words, compulsory licenses will only be imposed when an innovation that could be greatly beneficial to the public interest is not being used – or at least not sufficiently – by the patent owner.

The intellectual property rights regime of the country has been modified by a number of legislations since 1995. For India, the WTO's TRIPS agreement became binding from 2005 onward as the country had a ten-year transition period to make the domestic legislation compatible with TRIPS. Here, India has received an additional five-year transition period because of not having previously a product patent regime in critical sectors such as pharmaceuticals. In this transition, existing laws were amended, and new legislation was introduced during this period.

Different amendments to the various existing Acts – Patent Amendment Act (2005) and Copyright Amendment Act (2010) have been made to strengthen the domestic legal framework and fulfill harmonization with the WTO's TRIPS agreement: Patent Amendment Act 1999, 2002, and 2005. Additional modifications are also expected in the near future to harmonize the country's IPR regime in the post-TRIPS scenario to achieve the UN 2030 Agenda SDGs.

India has, due to its successful AIDS Program and leadership in generic production of ARV, a relatively low adult prevalence of HIV/AIDS (0.2%) when compared to Brazil (1%) and many other countries. Nevertheless, the country faces important challenges to end the HIV/AIDS epidemic by 2030: 2,100,000 people living with HIV/AIDS and 69,000 annual deaths from HIV/AIDS [1].

2.6.3 Thailand

The Thai Government provides access to antiretroviral drugs for HIV/AIDS patients under the publicly funded universal health system, entitled to medicines on the National Essential Drug lists. Similar to Brazil, the Thai public health policy also includes the 2003 National HIV/AIDS program with universal coverage for ARV treatment for HIV/AIDS patients free of charge [38].

In late 2006 and early 2007, the Thai Government authorized compulsory pharmaceutical licenses for two ARV drugs: Kaletra (Abbott) and Efavirenz (Merck). Under compulsory licensing, the low-cost manufactured drugs could only supply the local market and be distributed to the poor, with important policy implications [56, 57]. For example, under the compulsory licensing of a five-year contract for Efavirenz, the drug would be manufactured within Thailand with a 50% cost reduction (approximately USD 28 million per year). In return, the patent owner, Merck, would receive 1% of sales in Thailand [58].

Despite declarations from the WTO, this policy led to a series of hostile reactions. From the end of 2006 to the beginning of 2008, the Minister of Public Health in Thailand granted seven compulsory licenses to allow patients' access to generic treatments against AIDS, cancer, or cardiovascular accidents. Placing it in its historical and social context, an excellent article focused on the conditions that led the Thai Minister of Health to make this decision [59] and on the collective action of a variety of actors that have enabled this development.

The history of intellectual property disputes between Thailand and the United States dates back at least to the 1980s. The question is at the heart of tensions marked by repeated use of threats of economic sanctions against Thailand through the use of provision 301 of American law on trade [59]. In 1989, Thailand was placed on the Priority Watch List. It was placed again on the Priority Foreign Country List in 1991 [60, 61]. These pressures resulted in an amendment to the Thai Patent Act in 1992, intended to establish more restrictive levels of protection. Following further pressure, the law was revised again in 1998.

Generic first-line ARV drugs are produced by Thai Government laboratories. The Government covers the costs of these HIV/AIDS treatment programs largely through domestic funds. Thailand's public health budget expenditures significantly increased as well as the costs for the national HIV/AIDS program. In 2019, the public health budget was approximately 2.1% of the federal budget. In March 2020, the Thai Government announced a federal budget increase of 5.9%. The establishment by the Thai Government of the universal access program for HIV/AIDS treatments triggered the demand for cheaper medicines. The number of HIV/AIDS treatment recipients rose from 20,000 in 2003 to approximately 120,000 in 2007. Moreover, the Government anticipated a rising demand for second-line ARV treatments in the public sector (because of drug resistance). For example, in 2007, it was estimated that the need for (second-line) lopinavir- and ritonavir-based patented ARV would grow from 13,000 to 50,000 patients in the near future.

Between 2004 and 2005, the Thai Government had unsuccessfully tried to directly negotiate price reduction for ARVs for public health programs from patent-holding pharmaceutical companies. This included official communications and several negotiation meetings, including debates on the impacts of the FTA agreements between the US and Thailand [62]. These discussions were conducted by a specifically established *ad hoc* working group to address the issues caused by high-priced essential drugs. The price cuts proposed by the firms were considered insufficient since the generic versions remained much cheaper than the original products. The failed negotiations led the working group to suggest the introduction of compulsory licensing as a measure to strengthen the position of negotiation.

The Thai Government established subcommittees that determined the conditions for the implementation of government use orders and the criteria for the selection of medicines. The drugs were chosen according to factors such as inadequate access, unaffordable costs, and patent protection. Between 2006 and 2008, Thailand issued seven government use orders, including two for Efavirenz and Lopinavir/Ritonavir, for the importation or production of generic versions for noncommercial use in the public health sector. Eventually, the authorizations were used for the import of generic drugs from Indian producers. The Government-use license on Efavirenz in 2006 was set up until December 31, 2011. The distribution of the drug had been limited to 200,000 patients per year. In January 2007, the government announced the government use of the second ARV (Lopinavir and Ritonavir). Thailand has, despite its successful AIDS Program and IP strategies, a relatively high adult prevalence of HIV/AIDS (1,16%) when compared to Brazil (1%), India (0.2%), and many other countries. The country faces important challenges to end the HIV/AIDS epidemic by

2030: 2,100,000 people living with HIV/AIDS and 69,000 annual deaths from HIV/AIDS [1].

2.6.4 Indonesia

The Indonesian AIDS Program created in 1987 has been successful and recognized as an international reference. Nevertheless, challenges persist to an effective HIV/AIDS response: while the number of new HIV infections in Indonesia decreased by 3.6% in 2021, the country remains one of the countries with the highest number of new infections in Asia and the Pacific region [1]. HIV cases in Indonesia are now the fastest growing in Asia with the increasing number of HIV transmissions among the heterosexual population. This epidemic has spread throughout Indonesia and has been reported from 34 provinces and 308 (61 percent) of 504 districts/cities [63, 1].

To accelerate its response, the Indonesian Ministry of Health, along with its policy of free and universal access to ARV drugs, is now strongly investing in a pre-exposure prophylaxis (PrEP) strategy that has proven highly effective at preventing HIV infection. Making it available to more vulnerable groups of people has recently become a key priority.

Indonesia was included in 2011, with India within the territory of the license granted by Gilead to the Medicines Patent Pool. This measure received substantial criticism because of its exclusion of the vast majority of middle-income countries, including all of Latin America, most of Asia, and all of North Africa and Eastern Europe/Central Asia.

In 2012, the Indonesian Government issued compulsory licensing for seven HIV/AIDS-related drugs [64]. This compulsory licensing strategy allowed the federal government to supply Indonesia's local health services demand with Gilead's Tenofovir + Emtricitabine and Tenofovir + Emtricitabine + Efavirenz, a strategy favored later by adequate provisions from the Medicine Patent Pool's voluntary license with Gilead.

These positive impacts of IP flexibilization in Indonesia illustrate well the importance of negotiating voluntary license terms that allow rejection of licensing rights on certain ARV drugs while retaining rights to others. This example indicates the importance of a strategy that assures voluntary licensees remain free to serve as licensees of compulsory licenses, ascertaining that they are able to choose licenses that already have access to significant markets, benefiting from lower generic prices.

Indonesia has strict local compulsory licensing legislation. Under the Indonesian Patent Law 13 in 2016, the invention of the patent must have been implemented within three years of the date of grant. If the invention has not been implemented before this date, the invention may be subject to a compulsory license by a third party. While this requirement is relatively strict compared to other countries, its goal was to support technology transfer and local investments.

Despite its successful AIDS Program, Indonesia faces many challenges to end the HIV/AIDS epidemic by 2030. In addition to remaining one of the countries with the highest number of new infections in Asia and the Pacific region, Indonesia has a relatively high adult prevalence of 0.4%, higher than that of India (0.2%) yet lower than that of Brazil (1.0%), with 640,443 people living with HIV/AIDS and 38,000 annual deaths from HIV/AIDS [1, 63].

2.7 LESSONS: POSITIVE IMPACTS AND DIFFICULTIES

The experiences of the four developing countries presented here evidence the positive impacts of IP flexibilities – voluntary licensing, compulsory licensing, and patent pools for generic production on their National AIDS Programs, accelerating response to the HIV/AIDS epidemic. They also brought to light the diverse interpretations from different stakeholders with respect to the implementation of TRIPS provisions.

Moreover, they also evidence the difficulties in local conditions in these countries to issue and conduct the licensing process and overcome operational limitations of local manufacturers for technology transfer. Another critical issue in these countries refers to the limited operational capacity of their IP offices for the implementation of compulsory licensing.

Despite these difficulties and constraints, remarkable progress has been made in these four countries. Progress and gaps in their National AIDS Programs should be considered in light of more favorable scenarios today, with more medicines on the market, significant increase in generic competition and in the number of potential ARV drug exporters, providing a broader scope for application and revision of the Doha paragraph 6 system.

2.8 THE MEDICINE PATENT POOL: GENERIC COMPETITION AND VOLUNTARY LICENSING

The Medicines Patent Pool (MPP), an innovative patent negotiation tool, was founded in 2010 by UNITAID, its main funder [49]. MPP has since played a major role in the HIV/AIDS epidemic, providing, through voluntary licensing negotiations with leading pharmaceutical companies, access to affordable ARV drugs for poor populations in developing countries [50].

In a patent pool, patent rights are aggregated among multiple patent holders. Pooled patents are made available to member and nonmember licensees, and the pool allocates a portion of the licensing fees it collects to each member in proportion to each patent's value.

Extraordinary progress has been made since the creation of the UNITAID-funded Medicines Patent Pool in 2010 and funding mechanisms of other international organizations: globally, the number of people on ARV treatment worldwide rose almost fourfold, from USD 7.7 million in 2010 to USD 29.8 million in 2022, as indicated in a recent UN Report [1]. IP Flexibilities and the Medicines Patent Pool certainly played in synergy an important role in these treatment advances.

The MPP has favored since its creation in 2010 a remarkable global expansion of generic manufacturers for ARV drugs, led by Indian manufacturers to date. MPP has signed licensing contracts with leading pharmaceutical companies covering 14 ARV drugs and related combinations and 1 HIV technology platform, as indicated in Table 2.1.

TABLE 2.1
Pharmaceutical Companies and HIV/AIDS Drugs Licensed in MPP Since 2010

Pharmaceutical Companies	HIV/AIDS Drugs	Generic Manufacturers
Abbvie	Lopinavir, ritonavir (adults) lopinavir, ritonavir (pediatric)	Cipla, Aurobindo, Desano, Hetero, Adcock Ingram, Arene Lifesciences, Lupin, Sun Pharma
Boehringer Ingelheim	Nevirapine (nonassert)	Cipla, Aurobindo, Mylan, Strides, Hetero, Laurus
Bristol-Myers Squibb	Atazanavir	Aurobindo, Cipla, Desano, Emcure, Mylan
Gilead	Bictegravir, Cobicistat, Elvitegravir, Emtricitabine, Tenofovir Alafenamide, Tenofovir Disoproxil, Fumarate	Adcock Ingram, Arene Lifesciences, Aurobindo, Biochem, Desano, Emcure, Langhua, Laurus Labs, Lupin, MacLeods, Micro Labs, Natco
Johnson & Johnson	Darunavir	Cipla, Mylan, Hetero Labs, Aurobindo, Synthon
Janssen	Darunavir (pediatric; nonassert)	Cipla, Mylan, Hetero Labs, Emcure
Merck Sharp and Dome – MSD	Raltegravir (pediatric)	Lupin
ViiV Healthcare	Abacavir (pediatric)	Aurobindo
	Dolutegravir (pediatric) Dolutegravir (adults) Dolutegravir (adults, for AZ, BY, KZ, MY)	Adcock Ingram, Arene Lifesciences, Celitrion, Cipla, Desano, Emcure, Hetero, Langhua, Lupin, MacLeods, Mangalam, Micro Labs, Mylan, Strides, Sun Pharma

Source: Elaborated by the author from different sources: WTO WIPO WHO [5], UNITAID [49]; The Medicines Patent Pool [50]; Urias and Ramani [30]; Stirner and Thangaraj [18].

Finally, for regulatory procedures, to ensure the availability of quality-assured products in LMICs, these generic medicines are developed and filed with regulatory authorities such as the U.S. Food and Drug Administration (USFDA) and the World Health Organization prequalification (WHO-PQ).

2.9 UN 2030 UNFINISHED AGENDA FOR HIV/AIDS: CHALLENGES

Despite all the advances promoted by international organizations, governments, stakeholders and from 2010 by the Medicines Patent Pool, universal and free access to quality antiretroviral therapy remains one of the main challenges for achieving the

UN 2030 SDGs and finally ending the HIV/AIDS epidemic. UN estimates in a recent report [1] evidence this unfinished agenda:

- 39.0 million people globally living with HIV and 29.8 million people accessing HIV;
- The persistence of a gap of 9.2 million people living with HIV without access to antiretroviral therapy.
- Nearly 1.3 million people in 2022 will become newly infected with HIV;
- 630,000 people die in 2022 from AIDS-related illnesses.

Progress has been greater in countries and regions with political determination and strong financial investments, such as in eastern and southern Africa, where new HIV infections have been reduced by 57% since 2010 [1].

According to this report, some African countries – Botswana, Eswatini, Rwanda, the United Republic of Tanzania, and Zimbabwe – have already achieved the "95-95-95" targets, as indicated in this UNAIDS report. This means that 95% of the people who are living with HIV know their HIV status, 95% of the people who know that they are living with HIV are on lifesaving antiretroviral treatment, and 95% of people who are on treatment are virally suppressed. A further 16 other countries, eight of them in sub-Saharan Africa, the region that accounts for 65% of all people living with HIV, are also close to doing so.

Almost one-quarter (23%) of new HIV infections were in Asia and the Pacific, where new infections are rising alarmingly in some countries. Sharp increases in new infections are continuing in eastern Europe and central Asia (a rise of 49% since 2010) and in the Middle East and North Africa (a rise of 61% since 2010). Global efforts to scale up combination prevention programs in this region, such as PrEP, as recently rolled out in Indonesia, will be key to changing this scenario.

Another important issue is sustaining adequate funding for increasing access to these preventive and therapeutic ARV strategies. It should be stressed that funding for HIV/AIDS and ARV access policies has also sharply declined in 2022 from both international and domestic sources, falling back to the same level as in 2013. Funding amounted to USD 20.8 billion in 2022, far short of the USD 29.3 billion needed by 2025 [1].

This funding decline is now affecting the world's most important multilateral HIV/AIDS initiative, the Global Fund to Fight AIDS, Tuberculosis and Malaria, which works interdependently with The President's Emergency Plan for AIDS Relief (PEPFAR), a bilateral program that represents the world's largest investment in delivering HIV treatment and prevention. The Global Fund operates in 78 countries where there is no US PEPFAR presence, drawing lessons from PEPFAR experience, expertise, and technical assistance. Models of innovation at the ground level developed by the Global Fund represent essential preparation for future epidemics and epidemics [65].

In conclusion, if there is strong global political mobilization and more effective and focused funding mechanisms, increasing global investments in a sustainable response and access to ARV treatment, there is a possibility that by 2030, the HIV/AIDS epidemic could finally end.

2.10 IP INTO EPIDEMIC PREPAREDNESS: NEED FOR INTEGRATED POLICIES

Epidemics such as HIV/AIDS often require exceptional trade and public health measures, which explains the terms of the WTO Doha Declaration in 2001 in its search for flexibilities. The TRIPS agreement refers to national emergencies or circumstances of extreme urgency in connection with compulsory licensing. A historical perspective on the four AIDS Programs examined here indicates the urgent need for a holistic approach to IP flexibilization and that a strict and narrow view of the TRIPS agreement became insufficient and even impossible.

The Doha Declaration placed TRIPS into broader national and international public health contexts. Nevertheless, many gaps persist in the current IP system, and a new socially responsible R&D-based IP model to support innovation, with pools, innovation prizes and other incentives [66] and knowledge-based and mission-oriented approaches [67,68] is necessary to integrate different IP and public health policies in a new framework, as indicated in Figure 2.2.

An integrated IP framework should thus be urgently conceived for a more expeditious response to the unfinished HIV/AIDS agenda. Finally, to achieve the UN 2030 Agenda SDGs for all, a synergy among all global players to assure an integrated approach to IP and sustained funding mechanisms to achieve ending HIV/AIDS epidemic, as indicated in Figure 2.3.

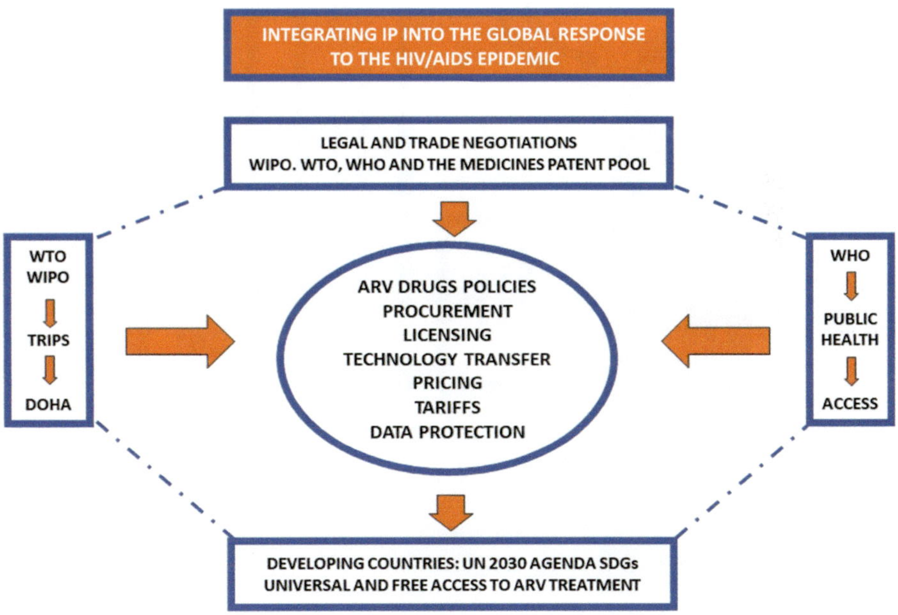

FIGURE 2.2 Integrating IP into the global response to the HIV/AIDS pandemic. (Elaborated by the author.)

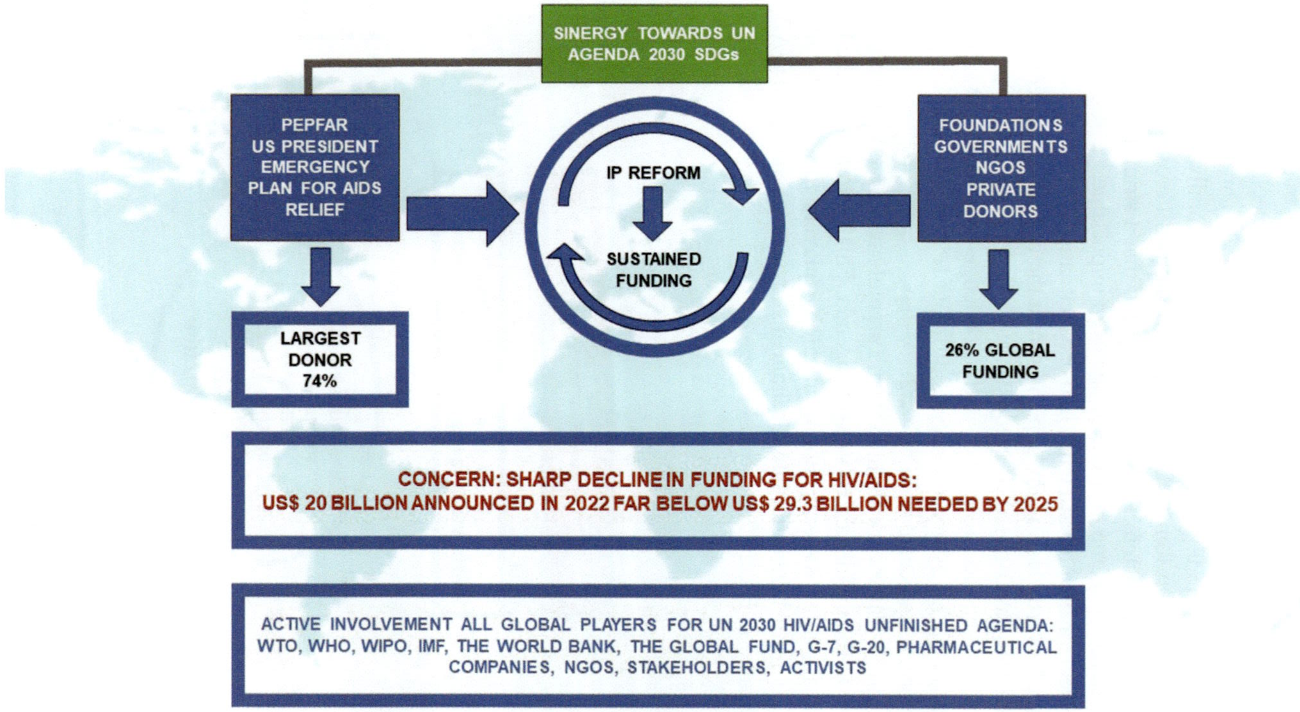

FIGURE 2.3 Synergetic approach to IP and sustained funding mechanisms to achieve ending the HIV/AIDS epidemic. (Elaborated by the author.)

2.11 FINAL CONSIDERATIONS

The third Sustainable Development Goal (SDG-3) of the UN 2030 Agenda has a target to end the HIV/AIDS epidemic by 2030. This goal would be achieved when the number of new HIV infections and AIDS-related deaths decline by 90% between 2010 and 2030. Despite extraordinary advances in the last two decades, many challenges persist.

AIDS vaccines and immunotherapies remain, despite extraordinary research breakthroughs in the last two decades, a distant target. The reason is that HIV is a complex and elusive virus, with very high variability and mutability, hiding its DNA cells inside immune cells supposed to destroy it.

Universal and free access to ARV, particularly in the poorest and more vulnerable developing countries, thus remains the best global strategy for ending HIV/AIDS. Generic competition of licensed ARV drugs supported by IP flexibilization through direct contract or through the UNITAID-funded Medicine Patent Pool has proven to be crucial, contributing to a sharp decline in ARV prices and to scale-up ARV treatments, with an enormous impact on access to treatment in the four countries discussed here – Brazil, India, Thailand, and Indonesia.

Compulsory licensing has been, as indicated by evidence provided here, an important exceptional measure in these countries, particularly as a negotiation strategy to lower prices in emergencies. Nevertheless, as indicated in our analysis, globally, there have been thus far few instances of compulsory licenses being issued. Moreover, many of the poorest countries lacking local capacity for drug development do not benefit from this alternative and depend on the import of generic drugs, mainly from India, a leading producer.

A reform of the global IP system, supported by an expansion in global funding for epidemic response, is thus urgently needed in the current HIV/AIDS epidemic scenario. A new socially responsible R&D-based IP model to support innovation, with innovation prizes and other incentives and knowledge-based and mission-oriented approaches, is urgently needed.

International IP treaties such as the Paris Convention for the Protection of Industrial Property in 1883 have existed over a century, with important institutional redesign implications and have historically been subject to revisions, in approximately every 20 years. Consequently, it is possibly too early for a final conclusion on TRIPS, probably subject to future revisions.

From this perspective, WTO member countries should revise the legal foundations and paragraph 6 of the TRIPS Agreement to conceive a reasonable strategy that enables fast and more expeditious exports to the poorest countries of medicines produced under compulsory licensing in developing nations. A permanent revision of the TRIPS Agreement should be put in place to implement this IP reform.

Finally, this IP reform aiming to accelerate achievement of the SDG Goals of the UN 2030 Agenda for ending HIV/AIDS epidemic should be designed and implemented in negotiations on the needed exceptionality with all global players: WTO (World Trade Organization), World Intellectual Property Organization (WIPO), World Health Organization (WHO), International Monetary Fund (IMF), the World Bank, the Global Fund to Fight AIDS, Tuberculosis and Malaria, G-7, G-20, Governments, pharmaceutical companies, stakeholders, public and private foundations, nongovernmental organizations, and activists.

ACKNOWLEDGMENTS

The author thanks Bio-Manguinhos Institute, Oswaldo Cruz Foundation, the Brazilian National AIDS Program and ANRS (Agence Nationale de Recherches sur le VIH/SIDA et les Hépatites Virales) for their support. The author also thanks Bernard Larouzé, Benjamin Coriat, James Love, and Norman Hearst for all the support and collaboration in the innovation and intellectual property debate in different moments of her trajectory as Director of the Scientific and Technological Development Unit in the Brazilian AIDS Program from 2003 to 2012.

REFERENCES

[1] UNAIDS 2023. The Joint Unite57-8-d Nations Programme on HIV/AIDS 2023. In The Path that Ends AIDS: UNAIDS global AIDS. Geneva.

[2] World Trade Organization, WTO 1994. Agreement on Trade-Related Aspects of Intellectual Property Rights (TRIPS), April 15, 1994, Marrakesh Annex 1C, 1869 U.N.T.S. 299, 33 I.L.M. 1197 (1994) [hereinafter TRIPS Agreement].

[3] World Trade Organization WTO 2001. Doha Declaration on the TRIPS Agreement and Public Health, November 20, 2001, WT/MIN (01)/DEC/2.

[4] Kampf R 2016. Comment: In: Nightingale A. WTO 'Paragraph 6' System for Affordable Medicine: Time For Change? Intellectual Property Watch, November 14, 2016.

[5] World Trade Organization WTO. World Intellectual Property Organization WIPO and World Health Organization WHO 2020. Promoting Access to Medical Technologies and Innovation: Intersections between Public Health, 2nd edition, Intellectual Property and Trade, Geneva.

[6] Reichman JH 2009. Compulsory licensing of patented pharmaceutical inventions: Evaluating the options. *Journal of Law, Medicine and Ethics* 37(2): 247–263.

[7] Beyer P 2012. Developing Socially Responsible Intellectual Property Licensing Policies: Non-Exclusive Licensing Initiatives in the Pharmaceutical Sector. In de Werra J (ed.), *Intellectual Property in the Pharmaceutical Industry*, Zürich, Schulthess Verlag.

[8] Abbott FM, Reichman JH 2007. The Doha round's public health legacy: Strategies for the production and diffusion of patented medicines under the amended TRIPS provisions. *Journal of International Economic Law* 10(4): 921–987.

[9] Sell SK 2003. Private Power, Public Law: The Globalization of Intellectual Property Rights. Cambridge University Press, Cambridge, UK.

[10] Helfer LR 2004. Regime shifting: The TRIPS agreement and new dynamics of international intellectual property lawmaking. *Yale Journal of International Law* 29(1): 1–83.

[11] Coriat B 2008. Introduction. In Coriat B (ed.), *The Political Economy of HIV/AIDS in Developing Countries: TRIPS, Health Systems and Free Access*. Edward Elgar Publishing, Cheltenham, UK.

[12] Imtiazuddin O, Mohiuddin M 2007. Socially responsible licensing: Model partnerships for underserved markets. *Acumen Fund Concepts* 2007:1–6.

[13] Singh HB, Jha A, Keswani C 2016. Intellectual Property Issues in Biotechnology. CABI, Wallingford, UK, p. 304.

[14] Nightingale A 2016. WTO 'Paragraph 6' System for Affordable Medicine: Time for Change? Intellectual Property Watch, November 14, 2016.

[15] Riaz S, Abbas MZ 2018. WTO 'Paragraph 6' system for affordable access to medicines: Relief or regulatory ritualism? *Journal of World Intellectual Property* 21(1–2): 32–51.

[16] Beall R, Kuhn R 2012. Trends in compulsory licensing of pharmaceuticals since the Doha Declaration: A database analysis. *PLOS Medicine* 9(1): e1001154.

[17] Wong H 2020 June. The case for compulsory licensing during COVID-19. *Journal of Global Health* 10(1).

[18] Stirner B, Thangaraj H 2013. Learning from practice: Compulsory licensing cases and access to medicines. *Pharmaceutical Patent Analyst* 2(2): 195–213.

[19] Possas C 2008. Compulsory licensing in the real world: The case of ARV drugs in Brazil. In Coriat B (ed.), *The Political Economy of HIV/AIDS in Developing Countries: TRIPS, Health Systems and Free Access*. Edward Elgar Publishing, Cheltenham, UK.

[20] Agarwal D 2019. Flexibilities Under the Indian Patents Act for Use of Pharmaceutical Inventions. In Singh HB, Keswani C, Singh SP (eds.), *Intellectual Property Issues in Microbiology*, 367–382. Springer Nature, London, UK.

[21] DiMasi JA, Grabowski HG, Hansen RW 2014. Innovation in the Pharmaceutical Industry: NEW Estimates of R&D Costs. Boston: Tufts Center for the Study of Drug Development, November 18.

[22] DiMasi JA, Grabowski HG, Hansen RW 2015. The cost of drug development. *New England Journal of Medicine* 372(20): 1972.

[23] DiMasi JA, Grabowski HG, Hansen RW 2016. Innovation in the pharmaceutical industry: New estimates of RandD costs. *Journal of Health Economics* 47: 20–33.

[24] Simoens S and Huys I 2021. R&D costs of new medicines: A landscape analysis. *Frontier in Medicine* 8: 760762.

[25] PhRMA. The Pharmaceutical Research and Manufacturers of America 2022. Industry Profile 2022 Report, Washington DC.

[26] BIO, Biotechnology Innovation Organization 2021. Clinical Development Success Rates and Contributing Factors 2011–2020. Report.

[27] Sweet CM, Maggio D 2015. Do stronger intellectual property rights increase innovation? *World Development* 66: 665–677.

[28] UN Secretary General 2016. *UN Report on the High-Level Panel on Access to Medicines: Promoting Innovation and Access to Health Technologies*, September 2016.

[29] Bjornberg J 2006. Brazil's recent threat on Abbott's patent: Resolution or retaliation. *Northwestern Journal International Law & Business* 27: 199.

[30] Urias E, Ramani SV. 2020. Access to medicines after TRIPS: Is compulsory licensing an effective mechanism to lower drug prices? A review of the existing evidence. *Journal of International Business Policy* 3(4): 367–384.

[31] Lanjouw JO 2005. *Patents, Price Control, and Access to New Drugs: How Policy Affects Global Market Entry*, NBER Working Paper No. 11321.

[32] Revenga A, Over M, Masaki E, Peerapatanapokin W, Gold J, Tangcharoensathien V. Thanprasertsuk S 2006. *The Economics of Effective AIDS Treatment: Evaluating Policy Options for Thailand*. World Bank, Washington DC, USA. 1–270.

[33] Son KB, Lee TJ 2018. Compulsory licensing of pharmaceuticals reconsidered: Current situation and implications for access to medicines. *Global Public Health* 13(10): 1430–1440.

[34] Le VA 2022. The Indian Case Study of Compulsory Licensing. In Le VA (ed.), *Compulsory Patent Licensing and Access to Medicines: A Silver Bullet Approach to Public Health*? Palgrave Macmillan, Cham.

[35] Nguyen SL 2022. Compulsory patent licensing and access to medicines: A silver bullet approach to public health? by Van Anh LE, London, New York Shanghai: Palgrave Macmillan 2022. *Asian Journal of International Law* 12(2): 415–416.

[36] Harris D 2010. TRIPS after fifteen years: Success or failure, as measured by compulsory licensing. *J Intellectual Property Law* 18: 387.

[37] Possas C, Scapini R, Simão M 2013. Soutenabilité de la politique brésilienne d'accès universel et gratuit aux médicaments antirétroviraux: conquêtes et défis. In Possas C and Larouzé B (eds.), *Accès aux antirétroviraux dans les pays du Sud: propriété intellectuelle et politiques publiques. Collection Sciences Sociales et SIDA, Paris: Agence Nationale de Recherches sur le VIH/SIDA et les Hépatites Virales (ANRS)*, Paris, France.

[38] Ford N, Wilson D, Costa CG, Lotrowska M, Kijtiwatchakul K 2007. Sustaining access to antiretroviral therapy in the less-developed world: Lessons from Brazil and Thailand. *AIDS* 21(Suppl 4): S21–S29.

[39] Possas C, Buchalla CM Hearst N 2011. Building AIDS research capacity in Brazil introduction. *JAIDS* 57(Suppl 3): S125–S128.

[40] Urias EM 2015. *Improving Access to HIV/AIDS Treatment in Brazil: When Are Compulsory Licenses Effective in Price Negotiations*? Datawyse/Universitaire Pers Maastricht, Maastricht, the Netherlands.

[41] Nunn A, Fonseca EM, Bastos FI, Gruskin S, Salomon JA 2007. Evolution of antiretroviral drug costs in Brazil in the context of free and universal access to AIDS treatment. *PLoS Medicine* 4(11): e305.

[42] Nunn A 2009. *The Politics and History of AIDS Treatment in Brazil*. Springer, New York.

[43] Ferraz OLM 2009. The right to health in the Courts of Brazil: Worsening Health Inequities? *Health and Human Rights* 11(2): 33–45.

[44] Hogerzeil HV, Samson M, Casanovas JV, Rahmani-Ocora L 2006. Is access to essential medicines as part of the fulfilment of the right to health enforceable through the courts? *Lancet* 368: 305–11.

[45] Urias E 2019. The potential synergies between industrial and health policies for access to medicines: Insights from the Brazilian policy of universal access to HIV/AIDS treatment. *Innovation and Development*, 9: 2, 245–260.

[46] Orsi, F, Hasenclever L, Fialho B, Tigre P, Coriat B 2003. Intellectual property rights, anti-AIDS policy and generic drugs. lessons from the brazilian public health program. In J. P. Moatti JP, Barnett T, Coriat B, Souteyrand Y, Dumoulin J, Flori YA (eds.), *Economics of AIDS and Access to HIV-AIDS Care in Developing Countries: Issues and Challenges*, 110–135. Agence Nationale de Recherches sur le VIH/SIDA et les Hépatites Virales (ANRS), Paris.

[47] Mueller JM 2007. The Tiger awakens: The tumultuous transformation of india's patent system and the rise of indian pharmaceutical innovation. *U. Pitt. L. Rev*. 491: 543.

[48] Waning B, Diedrichsen E, Moon S 2010. A lifeline to treatment: The role of Indian generic manufacturers in supplying antiretroviral medicines to developing countries. *JAIDS* 13(1): 1–9.

[49] UNITAID. The Medicines Patent Pool 2014. Patents and licenses on antiretrovirals: A snapshot. Geneva. https://medicinespatentpool.org/uploads/2017/07/ARV-Snapshot_April2014Rev.pdf (accessed June 6, 2024).

[50] The Medicines Patent Pool 2023. Our Strategy 2023–2025 Greater Access to Medicines and Health Technologies for Those Who Need Them. Annual Report 2022.

[51] IBEF, India Brand Equity Foundation, IBEF 2023. The Emergence of India as a Global Startup Hub. October 16 2023. www.ibef.org/research/case-study/the-emergence-of-india-as-a-global-startup-hub (accessed: October 28 2023).

[52] Sweet C, Das K 2013. Institutional and Procedural Challenges to Generic Production in India: Antiretrovirals in Focus. In Possas C and Larouzé B (eds.), Accès aux antirétroviraux dans les pays du Sud: propriété intellectuelle et politiques publiques. Collection Sciences Sociales et SIDA. Agence Nationale de Recherches sur le VIH/SIDA et les Hépatites Virales (ANRS), Paris.

[53] Guennif S, M'Fuka C 2003. Impact of Intellectual Property Rights on AIDS Public Health Policy in Thailand. In: JP Moatti, B Coriat, Y Souteyrand, T Barnett, J Dumoulin, YA Flori (eds.), Economics of AIDS and access to HIV/AIDS care in developing countries: Issues and challenges, 137–149. Agence Nationale de Recherches sur le VIH/SIDA et les Hépatites Virales (ANRS), Paris.

[54] Gandhi B 2019. India's compulsory license model: Increased pharmaceutical access and innovation coexis. *Brigham Young University Prelaw Review* 33: Article 5.

[55] Konde V 2016. Recent developments in compulsory licensing of pharmaceutical patents in India, 22. *Journal of Commercial Biotechnology* 64: 64–65.

[56] Yamabhai I, Mohara A, Tantivess S, Chaisiri K, Teerawattananon Y 2011. Government use licenses in Thailand: An assessment of the health and economic impacts. *Globalization and Health* 7(28): 11–15.

[57] Wibulpolprasert S Chokevivat V, Oh C, Yamabhai I 2011. Government use licenses in Thailand: The power of evidence, civil movement and political leadership. *Globalization and Health* 7(32): 18–23.

[58] Yang D 2012. Compulsory licensing: For better or for worse, the done deal lies in the balance. *Journal of Intellectual Property Rights* 17: 76–81.

[59] Krikorian G 2013. Conditions d´usage des licenses obligatoires: l´action du gouvernement thailandais. In Possas C et Larouzé B (eds.), *Accès aux antirétroviraux dans les pays du Sud: propriété intellectuelle et politiques publiques. Collection Sciences Sociales et SIDA*. Agence Nationale de Recherches sur le VIH/SIDA et les Hépatites Virales (ANRS), Paris.

[60] Sallstrom LR 1994. US withdrawal of Thailand's GSP benefits: Real or imagined? *TDRI Quarterly Review* 9(3): 15–22.

[61] Markandya S 2001. Timeline of trade disputes involving Thailand and access to medicines. Consumer Project on Technology, 2001.

[62] Limpananont J 2005. Impact of US-Thailand FTA on Access to Medicines in Thailand. In Bamford CC, A Bamford A (eds.), Free Trade Agreement: Impact in Thailand. FTA Watch, Bangkok, Thailand, 61.

[63] ILO, International Labor Organization 2019. Report, HIV/AIDS Programme in the World of Work in Indonesia, Project IDN/19/50/UNA, ILO/UNAIDS, Geneva.

[64] Velásquez G 2012. Indonesia Government Granted 7 compulsory licenses to promote access to HIV related medicines. Editorial *Vitae* 19 (3) Medellín Seot./Dec. 2012.

[65] AVAC 2022. Global Fund and PEPFARS´s Essential Collaboration, September 15 https://avac.org/blog/global-fund-and-pepfars-essential-collaboration/ (accessed on 28 October 2022).

[66] Love J, Hubbard T 2004. Comment – Make Drugs Affordable: Replace TRIPs-Plus by R&D-Plus International Centre for Trade and Sustainable Development number 6 June 2004 3–4.

[67] Mazzucato M, Penna C 2015. *Mission-Oriented Finance for Innovation: New Ideas for Investment-Led Growth* Rowman & Littlefield Publishers, Lanham, Maryland, US.

[68] Mazzucato M, Li HL (2021). A market shaping approach for the biopharmaceutical industry: Governing innovation toward the public interest. *Journal of Law, Medicine & Ethics* 49(1): 39–49.

3 Cancer Drug Patent Disputes

Balancing Innovation, Access, and Ethics

Homa Fatma, Sadaf R. Khan, and Hifzur R. Siddique

3.1 INTRODUCTION

Although we have made significant progress in understanding the molecular basis of cancer, it remains a complex and evolving disease that continuously interacts with its microenvironment. The global number of new cancer cases was reported to be 19 million, with a mortality of 10 million [1]. It is estimated that globally, the number of new cancer cases will rise to 21.6 million per year, and the number of cancer-related deaths will increase to 16.5 million annually. More than two-thirds of global cancer cases and deaths occur in low- and middle-income countries (LMICs), and in the future, this number is expected to increase over the next 10 years [2]. Approximately every year, 0.4 million children develop cancer. The prevalence of specific cancers varies across nations [3]. The growing global burden of cancer necessitates continuously developing new and more effective treatments. Cancer is not a single type of disease but comprises many different types of cancer with unique characteristics, genetic mutations, and biological pathways [4]. Progress in imaging and tissue study techniques has allowed us to categorize cancer diagnosis more accurately into distinct stages and grades, each indicating different treatment outcomes [5]. This complexity requires the development of a wide array of drugs explicitly tailored to cancer. Advancements in translational research with the development of targeted and personalized medicine have transformed the cancer treatment landscape, enabling researchers to develop drugs based on the molecular and genetic characteristics of individual tumors. The process of drug development is extensive and intricate, demanding substantial financial commitments and considerable commercial uncertainties [6]. This process requires strict regulations and specific permissions from a designated state authority to distribute a drug. These permissions are only approved by creators when they prove a medication's safety and efficacy, requiring lengthy and expensive clinical trials [7].

DOI: 10.1201/9781003405665-3

In such a scenario, patents play a pivotal role in drug development by helping inventors, pharmaceutical companies, and institutes protect their intellectual property rights and bar counterfeits. This protection fosters innovation and encourages collaboration and investment in research and development by ensuring that companies can profit from their discoveries [8]. It provides investors and companies rights to their invention for a fixed period, mostly 20 years, the minimum duration approved by the TRIPS (Trade-Related Aspects of Intellectual Property Rights). The protection period ranges from the date of filing, and they can recoup their investment by generating profits through the commercialization of their proprietary products [9]. In the legal framework, cancer drug patent disputes have become increasingly common among the pharmaceutical industries, which invest heavily in oncology research and development (R&D). In R&D investment, oncology remains the leading area, with a forecast spending of USD 82 billion, which is almost three times that of any other area. This spending would result in 152 novel FDA (Food and Drug Administration) approvals and, by 2026, estimated drug sales of USD 311 billion [10]. The disputes arise due to disagreement over the validity, infringement, or scope of intellectual property rights related to cancer drugs. Patent law's ambiguities, particularly in the fast-evolving field of life sciences, contribute to these disputes: high drug prices and limited access to patents. To make cancer drugs more affordable and accessible to the public, Government and nonprofit organizations have challenged patent rights [11]. Therefore, more strategic change is needed for genuine innovation to take place and to make drugs more accessible. A thorough examination of drug prices and patent methodology is needed, which will utilize public funds. Furthermore, the Government should prioritize public health by the pharmaceutical industry as a solution component rather than a sole answer [12].

3.2 THE PATENT SYSTEM AND CANCER DRUG INNOVATION

In the field of cancer drug discovery, innovation is greatly aided by the patent system. The approach encourages substantial research and investment in ground-breaking medicines by giving innovators exclusive rights. This exclusivity guarantees pharmaceutical businesses a window of time within which they can recoup their R&D expenses and turn a profit. As a result, the complexity of the patent system is inextricably related to the rise in new and better cancer treatments, highlighting its crucial role in advancing oncology.

3.2.1 Patent System and Drug Development

A patent represents an agreement between the investor or the patent seeker and the Government. Exclusive rights are provided to the applicant for a particular duration in return for revealing the invention's specifics. A patent is one of the facets of intellectual property that empowers the patent holder with the legal rights to prevent others from producing, utilizing, and marketing the said innovation for a set period. The patent guidelines ensure that the details of the invention are shared so that they contribute to paving the way for more technological and economic growth [13]. There are three criteria for an inventor to receive the rights: (i) the invention should be novel, which means there is no prior work to be found; (ii) the idea should have a unique perspective and creativity and should stand out as not being an easily attainable discovery; (iii) the useful idea should be inventive, novel, and have industrial application [14].

An international patent agreement known as the Patent Cooperation Treaty (PCT) makes it easier to obtain patent protection for innovations in several different nations by establishing a standardized method for filing and processing patent applications. The World Intellectual Property Organization (WIPO) oversees the PCT System [15]. The PCT System has gained widespread acceptance and is now a crucial instrument for inventors, businesses, universities, and research institutes looking to protect their ideas globally. Before the PCT System was established, applicants were needed to submit separate patent applications in each nation where they sought to protect their innovation within a year of submitting a first patent application on a particular subject matter, known as the "*Paris route*," based on the Paris Convention for the Protection of Industrial Property. The PCT innovated within patent space by offering an alternative that provided a centralized application process, and it became WIPO's largest international intellectual property (IP) filing system [13].

3.2.2 How the Patent System Drives Innovation in Cancer Drugs

The patent system should pave the path to encourage innovation for economic growth and development. This aim is accomplished by rewarding the patent owners with a period of exclusive rights for their innovative efforts and an incentive to keep them involved and encourage others to innovate [16]. These incentives and opportunities to make a profit encourage innovation by creating market competitiveness and maintaining it by preventing market foreclosure and preserving market access. Firms are put under pressure to innovate by the current competition since patent laws grant them exclusive rights and the opportunity to profit. To foster innovation and ensure competitive exploitation, intellectual property rights and competition are crucial. Through innovation, it encourages a productive market and long-term, dynamic competition [17]. Patent applications require investors to disclose the details of their work, as patent disclosures stimulate knowledge sharing and promote further innovation, as other researchers can build on existing inventions [18]. According to The International Federation of Pharmaceutical Manufacturers and Associations (IFPMA), a robust IP system contributes to continuous sustainable innovation, which yields the majority of today's and tomorrow's breakthroughs and facilitates' ongoing advancements in pharmaceuticals and vaccines. It aids in providing patients with the broadest range of treatments with significant health results. Due to IP, this industry has created new medications to address the world's changing healthcare requirements. This business has created many new drugs in the past 20 years, concentrating on treating diabetes, cardiovascular disease, and cancer [19]. Since 2010, the number of published PCT applications has grown by 5% per year. With medical technology as one of the top fields that grew fastest, pharmaceuticals were in the top five in 2021 [20].

The fields of biotechnology, medical technology, organic fine chemistry, and pharmaceuticals have seen a steep increase in the proportion of applications filed between April and July 2020. This increase is seen predominantly in the pharmaceutical field, whose share rose in the 4 months between April and July 2020 by 1.3 to

1.9% [21]. The legal system temporarily prevents competition but has a limited duration of time. However, the law anticipates that the competition to enter leads to price reduction and further fosters innovation [22].

3.2.3 Access to Cancer Drugs and Patent Protection

Access to cancer drugs is intricately linked to patent protection, with the patent system frequently raising drug prices due to its monopoly grant, which can hinder accessibility for many patients. To mitigate these challenges, mechanisms such as compulsory licensing and parallel importation are employed by Governments to ensure more affordable access to crucial drugs. Additionally, the availability of generic and biosimilar medications following the patent's expiration offers a cost-effective substitute, ensuring that the same therapeutic advantages are provided at a small portion of the cost of the original medicines.

3.2.3.1 Patent-Induced Price Inflation and Its Impediments to the Accessibility of Cancer Drugs

Due to the monopoly granted by patents, companies charge a lot for their medications in an effort to make a profit and recuperate their research and development expenses. After a patent expires, drug prices drastically decline. This price reduction's scope varies significantly between products and countries [23]. This exclusivity period, typically 20 years from the filing date, allows companies to set higher prices for their patented drugs without the pressure of competition. This mechanism is particularly evident in oncology, where novel therapies are often introduced at steep prices. From 1995 to 2013, there was an increase of an average of USD 8500 annually in the launch price of cancer drugs. There has been an upward trend in the launch price of products, as manufacturers set the price slightly above the existing therapies [24]. Imatinib was launched in 2001 and was priced at USD 2200 per month, based on the price of interferon, the recommended course of medication at the time [25]. In India, sorafenib and imatinib were priced significantly higher due to patents, which later paved the way for compulsory licensing and generic versions [26,27].

In LMICs, anticancer medications are less affordable based on the individual patient's income approach compared to high-income countries, as a higher percentage of deaths from cancer occur in LMICs because of the lack of resources for treating cancer. According to a WHO technical report, nations with lower national incomes had less access to anticancer medications for more expensive medicines such as targeted therapies. Based on reports, in lower-middle-income countries, only 32.0% of cancer treatment drugs were accessible to patients who could bear their costs. This figure was 57.7% in low-income countries [28]. In India, the largest LMIC, it is estimated that patients bear almost 68% of their healthcare expenditure, a steep contrast to the world average of 18% [29].

3.2.3.2 Compulsory Licensing and Parallel Importation

Patents can, paradoxically, create barriers to accessing these very innovations, particularly regarding essential medicines. These problems are addressed by introducing compulsory licensing and parallel importation. To ensure the use of a patented invention, the Government authority or a court issues a compulsory license so that the patent holder's consent is not needed. This is not an arbitrary provision; it is invoked under specific circumstances. Many WTO members utilize this flexibility within the TRIPS Agreement. However, these rules are initially restricted to the local level or consumption unless anti-competitive practices exist. Nations possess the right to determine their stance on the exhaustion of IP rights, permitting parallel imports. Parallel importation allows countries to import a drug to a nation where it would be sold at a more affordable price. Canada issued 613 compulsory licenses to produce or bring in generic drugs between 1969 and 1992, a move influenced by the North American Free Trade Agreement, resulting in some of the lowest prices in the pharmaceutical realm [30]. In 2012, India was granted its first compulsory licensing for sorafenib, and in 2007, compulsory licensing was granted for various cancer drugs, such as docetaxel, letrozole, and erlotinib, in Thailand [31]. In 2016, Colombia was denied compulsory licensing but reduced the price of imatinib [32].

Parallel importation covers the importation of a noncounterfeit drug across borders without the permission of the product owner. It is also a gray product implicated in international trade and intellectual property issues. Many regulatory rules can be imposed on the parallel importation of medicinal products, which interferes with the patent rights of trademark owners or proprietors. In low- and middle-income countries, healthcare becomes more out of range for patients, as the patent system makes the drug more costly. Parallel importation is one such legitimate option that makes obtaining patented medicines at a lower cost possible. The concept of international exhaustion implies that once a patent owner makes the first sale across nations, their right is considered "used up" or "exhausted" [33]. Such a provision permits nations to use parallel trade or importation as a means to access medications. These parallel imports allow the country to take advantage of various prices set by IP holders, who set the cost of drugs based on the economic strength of a particular region [34]. By using this flexibility, the strategy of parallel importation capitalizes on price discrepancies across markets. If any drug is under patent protection and sold at a lower cost in one country than another, the latter can resort to parallel importation from a cheaper country to circumvent the monopoly price [35]. Challenges arise in the parallel trade of patented drugs, as IP owners obstruct such importation, which makes parallel trading more complex. Producers often market these products under different brand names across different countries to sort out this impediment [34].

3.2.3.3 Generic and Biosimilar Cancer Drugs

Drugs with expired patent protection led to the development of generic drugs, which may be manufactured by companies other than the original developer. These medications are identical to the dose, recommended usage, effects, side effects, route of administration, safety profile, and strength of generic medicines to those of their original drug. Within the first year after a drug's generic counterpart is released, its price is roughly cut in half. The 1984 Hatch-Waxman Act promoted the production of generic medications to balance patient access to care and pharmacological innovation. The Act has helped the market for generic medications grow for more than 30 years [36]. The latest surge in the adoption of generic medicines is driven mainly by the intention to cut healthcare prices, as generic medicines come at 20 to 90 percent lower than their original patented drugs [37], which has revolutionized the affordability and accessibility of treatments. Recently, a Hyderabad-based pharmaceutical company launched a generic version of palbociclib in tablet form under the brand name Palborest, a breast cancer drug. Tamoxifen, patented initially as Nolvadex, became a go-to for breast cancer drugs [38]. Anastrozole paved the way for generic versions after its patent expired in 2010. Gemcitabine was introduced as a drug for NSCLC, breast cancer, and pancreatic cancer after its patent expired in 2011; it allowed generics to offer similar treatment but at a more affordable price [39]. In 2012, the generic drug erlotinib was approved in India [31]. Biosimilar drugs are similar to biological drugs in terms of safety, efficacy against disease, and biological activity. Unlike generic drugs, biosimilar drugs are not exact replicas of brand-name drugs. For a biosimilar to be approved, proper trial and testing are conducted [40].

In 2021, generic and biosimilar medicines saved approximately USD 17.9 billion in cancer treatment; this amount for the past decade was USD 114 billion. The National Cancer Institute (NCI) highlights that the number of U.S. cancer diagnoses per year is estimated to be 1.8 million, and there are more than 6 million deaths annually from cancer. In the pharmaceutical field, the rise of biosimilars is a significant example of how fostering competition can lead to reduced drug prices. In 2021 alone, new biosimilars saved USD 3.5 billion for cancer patients. It is estimated that biosimilars can cut spending by nearly half, and this reduction trend will persist in upcoming years. A breakdown of savings from generic treatment includes USD 3.1 billion in blood cancer, USD 204 million in myeloma, USD 7.8 billion in breast cancer, USD 102.1 million in lung cancer, and USD 3.9 billion in colorectal cancer. In 2019, the U.S. healthcare system had a massive savings of USD 313 billion, accounting for USD 2.2 trillion savings in the overall decade [41]. Nationally, using generics, the Medicare program saved USD 1,053 per participant, or USD 96.1 billion in 2019. On the other hand, Medicaid programs saved USD 770 per participant. U.S. biosimilars contributed to USD 2.2 billion savings in 2019 and approximately USD 4.5 billion over the past 10 years, based on data from IQVIA [42]. Cipla's launch of drugs with reduced prices brought global media attention, pointing to the dominant position of major pharmaceutical corporations, especially during health emergencies. This highlights the importance of generic drug competition as a means to lower medication prices [43].

3.2.4 Ethical Issues in Cancer Drug Patent Disputes

Ethical dilemmas permeate cancer drug patent disputes, balancing rightful pharmaceutical profits with public health. Patient advocacy underscores the moral obligation to ensure equitable access to crucial treatments, highlighting the intricate interplay between patents, profits, and health rights (Table 3.1).

TABLE 3.1
Notable Cancer Drug Patent Disputes and Their Outcomes

S.No.	Drugs	Company involved	Nature of dispute	Year	Outcome	Refs.
1	Herceptin	*Genentech v Pfizer*	Biosimilar version	Case filed in 2017	Settled in 2019	[44]
2	Keytruda	*Merck v Bristol Myers Squibb*	Use of anti-PD-1	2014	Settled in 2017 in favor of BMS	[45]
3	Revlimid	*Celgene v Sun Pharma industry*	Generic version	2018	Settled in 2021 in favor of Celgene	[46]
4	Enbrel	*Amgen v Sandoz*	Generic version	2019	Settled in favor of Amgen	[47]
6	Sprycel	*BMS v Dr Reddy*	Generic version		Settled	[48]
7	Rituxan	*Roche v Pfizer*	Biosimilar version		Settled in 2019	[49]
8	Avastin	*Genentech v Samsung Bioepis*	Biosimilar Drug	2020	Settled in 2022	[50]
9	Thalomid	*Celgene v Lannet Co. Inc*	Infringement of Patent	2019	Settled	[51]
10	Zytiga	*Janssen Pharmaceutical v Generic Manufacturers*	Infringement of Patent	2016	It settled in 2018, and the generic version was approved	[52]
11	Tarceva	*Cipla v Roche*	Generic drug	2014	It settled in 2016 in favor of Roche; however, it allowed a generic version for sale	[53]
12	Xtandi	*Pfizer v Indian Patent Office*	Patent Dispute	2016	Patent rejected in India in 2016	[54]
13	Glivec	*Novartis v Union of India*	Patent Dispute	2006	The patent was rejected in India	[55]
14	Opdivo	*Bristol Myers Squibb v Astra Zenca*	Infringement of Patents	2022	Settled in favor of BMS in 2023	[56]
15	Yervoy	*Bristol Myers Squibb v Astra Zenca*	Infringement of Patents	2023	It settled in favor of BMS in 2023	[57]
16	Tarceva	*Glenmark v Roche*	Generic Drug		Settled in favor of Roche	[53]

3.2.4.1 Balancing Profit and Public Health

Balancing profit and public health are complex and often contentious, especially in the pharmaceutical industry. To pursue ground-breaking treatments, millions of dollars are invested in research on new drugs by pharmaceutical companies, which they expect to recoup through the sales of their patented products. They argue that the profits generated from successful drugs fund future innovations and compensate for the vast number of experimental drugs that never make it to market. These companies emphasize that without the potential for substantial returns, the incentive to invest in research could diminish, leading to fewer therapeutic advancements. While all this is understandable from a business standpoint, it becomes an ethical dilemma when life-saving drugs, such as those for cancer, are priced. In the healthcare space, these incentives have assets as well as liabilities. It has resulted in many exceptional success stories; however, many undesirable outcomes, such as the development of identical (but inferior) products (also known as "me too drug"), are also consequences of it. An emphasis on medicines for diseases prevalent in underdeveloped and developing countries rather than those predominant in developed and wealthy nations and reduced focus on researching medications that might not offer substantial financial gains [22].

3.2.4.2 Ethical Challenges in Patenting Medicine

The act of patenting life-saving medications raises ethical concerns; when a company patents a drug, it essentially has a monopoly on its production and distribution, sometimes making it inaccessible to those in dire need. The overpriced launch of drugs for human immunodeficiency virus/AIDS treatment, which caused an enormous outcry across countries for human immunodeficiency virus/AIDS treatment in the late 1990s, is a reminder of this conflict [43]. Another ethical challenge is that patent law creates incentives to pursue a potentially riskier dosage instead of the safer range of doses advised by the FDA. This issue is further complicated if the patent is granted specifically for these doses, then the drug sponsor has little motivation to explore the potential of milder doses. To make matters worse, after securing a patent that covers solely a higher dose, the drug sponsor lacks the incentive to explore the possibility of a lower dose, as any data supporting the efficacy and safety of a lower dose would diminish the worth of the patent. Considering that the USPTO (United States Patent and Trademark Office) had already dismissed the lower doses as not being innovative, any data in favor of a lower dose would jeopardize the commercial value of a patented high drug dose [22]. Although the patent system is intended to promote innovation, a balance must be maintained between encouraging research and development and the moral need to save lives when it becomes a barrier to accessing important medicines.

3.2.4.3 Patient Advocacy and Access to Essential Medication

Patients, particularly those in lower-income countries, often rely on generic or low-cost alternatives, but when patents block their production, it poses a significant challenge. Numerous patient advocacy organizations focus on particular illnesses or more general healthcare issues, impacting everything from funding for research to public policy and patient support. The essence of patient advocacy is prioritizing patients' needs and rights, ensuring that they remain central in the healthcare journey. Incentive structures where expensive medications receive preferential treatments result in patients being directed toward higher-cost options. Pharmaceutical firms craft financial incentives targeting doctors, insurance, hospitals, and even patient advocacy groups. Although this system benefits various entities, it burdens taxpayers and society [58]. A patient advocacy group campaigned for access to affordable versions of trastuzumab, a breast cancer drug in South Africa. Such advocacy reduces the ethical tension between IP rights and public health [59]. In India, the high price of sorafenib, a kidney cancer drug patented by Bayer, led to the country issuing its first-ever compulsory license in 2012, allowing a local firm to produce a generic version at low cost [27]. This dilemma poses an ethical conundrum: how to guarantee just pharmaceutical investment returns without undermining the larger social duty to guarantee that everyone, regardless of economic circumstances, has access to the pharmaceuticals they require. To achieve this balance, regulatory measures such as tiered pricing, mandatory licensing, and promoting the manufacturing of generic medications are used. The argument rages on as players try to balance commercial interests with the broader objective of achieving global health equity, but the equilibrium is still elusive.

3.2.5 Strategies to Balance Innovation and Access

Balancing innovation with access in the pharmaceutical landscape necessitates strategic approaches. Tiered and differential pricing methods have been devised, allowing drug prices to vary based on regional economic conditions, thus enabling wider accessibility while upholding profitability and innovation incentives. Furthermore, the rise of public–private partnerships and nonprofit initiatives in drug development symbolizes a collective pursuit to spearhead innovative drug discoveries and guarantee that people in need have equitable access to these medical developments.

3.2.5.1 Tiered Pricing and Differential Pricing Strategies

The pharmaceutical industry has embraced complex pricing systems, notably tiered and differential pricing schemes, intending to increase worldwide accessibility to life-saving medications while retaining commercial viability. Even though these initiatives take different techniques, they all share the same goal of eliminating healthcare disparities between different income groups and geographic areas. The concept of differential pricing is based on the economic idea of price discrimination, in which a manufacturer seeking to maximize profits sells its goods to different consumers at different prices, with the price variations reflecting the various levels of each consumer's willingness and ability to pay for the goods [60].

3.2.5.2 Public Private Partnerships and Nonprofit Drug Development

In recent decades, due to complex issues surrounding cancer research and the unquestionable significance of accelerating access to life-saving treatments, public–private partnerships and the involvement of nonprofit organizations have emerged as significant players in the oncology drug development landscape. PPP in cancer research is a prime example of how public sector policies, emphasizing overall public health outcomes, may coexist with the private sector's cutting-edge research skills and business tactics. The Cancer Moonshot effort, introduced in the U.S. in 2016, is one such example. This ambitious project aimed to improve patient access and involvement, advance early diagnosis and treatment methods and speed up cancer research [61].

Similarly, nonprofit entities are making substantial roads in cancer drug development, such as the Multiple Myeloma Research Foundation (MMRF), which was founded in 1998; it is one of the largest nonprofit organizations in the world that is solely focused on accelerating a cure for multiple myeloma disease, expecting that increased research funding would lead to new medication options and cure of the disease. The MMRF played a crucial role in escalating the development of new myeloma therapies by pioneering collaborative research models and promoting open-access data sharing [62]. Another such nonprofit platform, which was established in 2008, is "Stand Up to Cancer," raised over USD 700 million to date; over the years, it has brought together over 2000 of the best scientists from the leading institutions to collaborate toward ending cancer's reign. Its research has contributed to nine FDA approvals for new cancer therapies and many clinical trials. It aims to increase diverse enrollment in cancer clinical trials and fund innovative research [63]. As drug development endeavors surge in cost, these nonprofit organizations in cancer research ensure that funding restraints do not obstruct crucial research.

3.2.6 The World Trade Organization and TRIPS Agreement

The World Trade Organization, which was established in 1995, is a chief international body governing global trade rules between nations. The TRIPS Agreement, which came into effect on January 1, 1995, is the most comprehensive multilateral agreement on intellectual property. The crucial component of its framework is the Agreement on TRIPS, which aims to harmonize IP laws worldwide. The TRIPS Agreement outlines the details of the criteria of IP protection that every member is provided with. It specifies the subject need to be protected, the rights and exceptions that come along with and the minimum duration of protection [64]. The TRIPS agreement mandates that member nations provide a minimum of 20 years of patent protection to innovators, including pharmaceutical companies. While this provision is crucial for incentivizing research and development, it has become a subject of international debate due to its implications for healthcare access and affordability, particularly in LMICs. The World Trade Organization (WTO), which was established in 1995, has significantly shaped international trade norms, including regulations surrounding intellectual property rights (IPRs). Among its various agreements, TRIPS is particularly noteworthy because it seeks to establish a peaceful system of IP legislation across its member nations.

In November 2001, WTO members, during the Fourth Ministerial Conference in Doha, Qatar, established the Doha Declaration on the TRIPS Agreement and

Public Health. It aimed to clarify uncertainties about public health concerns in the Agreement on TRIPS about public health concerns. The Doha Declaration is especially significant because it confirms the leeway of TRIPS member states to bypass patent rights for vital conditions. Originally, the TRIPS framework of rules mainly confined compulsory licenses to local markets, with exceptions related to curbing non-noncompetitive actions.

The Doha Declaration identified an issue: countries without a strong pharmaceutical industry were disadvantaged after the restriction on compulsory licensing. In response, the TRIPS Agreement was revised and introduced a new license option, especially to enable the production and export of essential medicines to impoverished countries [65]. The TRIPS Agreement's new article 31 legally allows the exportation and production of low-cost generic medicines to countries that lack the infrastructure to manufacture these on their own [66]. Specifically, the Doha Declaration affirmed the right of member states to issue compulsory licenses, allowing them to bypass patents in urgent public health situations by manufacturing generic versions of patented medicines or importing them from generic manufacturers elsewhere. This can be critical for countries struggling to improve drug accessibility, particularly for expensive treatments.

The relationship between patent rules and access to cancer treatments frequently illustrates the conflict between fostering pharmaceutical innovation and maintaining public health. Depending on their economic resources, healthcare systems, and public health agendas, several nations have managed to strike this equilibrium. The WTO's inability to achieve unanimous agreement on advancing additional trade pacts has led to the rise of bilateral and multilateral frameworks such as preferential trade agreements (PTAs), free trade agreements (FTAs), closer economic relation trade agreements (CERs), and economic partnership agreements (EPAs), which is more evident in the Asia-Pacific region [67].

Brazil, known for its assertive approach toward pharmaceutical patents, requires the granting of patents for pharmaceutical products and processes depending on the prior consent of the Brazilian Health Regulatory Agency (ANVISA – Agência Nacional de Vigilância Sanitária) [68].

The Australian Government established the Pharmaceutical Benefits Scheme (PBS), which subsidizes the cost of many medicines for Australians [69]. The Government has added more medicines to the PBS as of October 1, 2022, including pembrolizumab for the treatment of head and neck squamous cell carcinoma, in addition to other types of cancer and growth hormone deficiency in children [70]. A semi-independent organization was established by Parliament in 1987 under the Patent Act named the Patented Medicine Prices Review Board (PMPRB) in Canada. It operates independently from the Government, which safeguards the interests of consumers by overseeing that the patented drug prices are reasonable [71].

However, China revised its drug patent policies in 2021 by incorporating extensions for drug patent durations and a compensation mechanism to guard against untimely and unfair generic drug market entry. After these reforms, market analysts foresee the Chinese pharmaceutical market as more favorable for innovative research firms, both local and global [72].

While the guidelines provided by the WTO and TRIPS provide some background, they do not give a full explanation. Membership of the WTO intrinsically carries with it an affiliation to TRIPS—the Uruguay Round's extensive negotiations, where participants reached a collective and overarching agreement. In the 1994 Final Act, nations were mandated to adopt all the encompassed agreements. While adherence to TRIPS was imperative for all WTO members, the path to compliance offered considerable freedom. This adaptability is further highlighted by an extensive body of literature that not only details numerous TRIPS-compatible policies but also emphasizes the diverse strategies countries can employ to align with these standards [73].

3.3 CONCLUSION

In the multifaceted world of pharmaceuticals, the realm of cancer drugs sits at the intersection of innovation, accessibility, and ethics. Following recent patent conflicts, the fundamental conflict between intellectual property rights and patient access to critical medications has worsened, while patents serve as the bedrock of pharmaceutical innovation, incentivizing research and development, they can also inadvertently gatekeep critical therapies from those in dire need [74]. This chapter has delved into these complexities, shedding light on instances where patent rights clashed with humanitarian concerns. As global health priorities evolve, there is an urgent need for a balanced strategy that protects both the spirit of innovation and the requirements of ethical responsibility. It involves negotiating a complex web of moral commitments as well as understanding legal systems. To ensure that the upcoming generation of cancer therapies is cutting-edge, easily available, and morally sound, the task of moving forward is to build a more symbiotic connection between pharmaceutical corporations, legal networks, and the larger community [75].

ACKNOWLEDGMENTS

The authors are thankful to the Department of Zoology, Aligarh Muslim University, for providing the necessary facilities. HRS is thankful to the UGC-India [Grant no. F.30-377/2017(BSR)] and DST-SERB, India (Grant no. EMR/2017/001758), New Delhi, for providing financial help.

REFERENCES

[1] H. Sung et al., "Global cancer statistics 2020: GLOBOCAN estimates of incidence and mortality Worldwide for 36 Cancers in 185 countries," *CA. Cancer J. Clin.*, vol. 71, no. 3, pp. 209–249, May 2021, doi: 10.3322/caac.21660.

[2] K. Jenei et al., "Cancer medicines on the WHO model list of essential medicines: processes, challenges, and a way forward," *Lancet Glob. Health*, vol. 10, no. 12, pp. e1860–e1866, Dec. 2022, doi: 10.1016/S2214-109X(22)00376-X.

[3] "Cancer." www.who.int/news-room/fact-sheets/detail/cancer (accessed Aug. 13, 2023).

[4] "The Genetic Basis of Kidney Cancer: A Metabolic Disease | Nature Reviews Urology." www.nature.com/articles/nrurol.2010.47 (accessed May 31, 2023).

[5] R. J. Gilbertson, "Mapping cancer origins," *Cell*, vol. 145, no. 1, pp. 25–29, Apr. 2011, doi: 10.1016/j.cell.2011.03.019.
[6] "The Role of Competition in the Pharmaceutical Sector and its Benefits for Consumers".
[7] O. Gurgula, "Strategic patenting by pharmaceutical companies – should competition law intervene?," *IIC – Int. Rev. Intellect. Prop. Compet. Law*, vol. 51, no. 9, pp. 1062–1085, Nov. 2020, doi: 10.1007/s40319-020-00985-0.
[8] D. H. Hsu and R. H. Ziedonis, "Resources as dual sources of advantage: implications for valuing entrepreneurial-firm patents," *Strateg. Manag. J.*, vol. 34, no. 7, pp. 761–781, 2013.
[9] B. V. P. D. L. Potterie, D. Guellec, B. van Pottelsberghe, and G. Rassenfosse, "The Worldwide Count of Priority Patents: A New Indicator of Inventive Activity," Jan. 2012. www.academia.edu/3516490/The_worldwide_count_of_priority_patents_A_new_indicator_of_inventive_activity (accessed May 31, 2023).
[10] "Evaluate pharma world preview 2020, outlook to 2026," *Evaluate.com*, Jul. 16, 2020. www.evaluate.com/thought-leadership/pharma/evaluatepharma-world-preview-2020-outlook-2026 (accessed Jun. 01, 2023).
[11] "Promoting Value, Affordability, and Innovation in Cancer Drug Treatment – The Rising Cost of Cancer Drugs: Impact on Patients and Society." https://prescancerpanel.cancer.gov/report/drugvalue/Part1.html (accessed Jun. 02, 2023).
[12] O. Gurgula, "Drug Prices, Patents and Access to Life-Saving Medicines: Changes Are Urgently Needed in the COVID-19 Era (February 6, 2021)". Forthcoming in the European Intellectual Property Review (EIPR), Available at SSRN: https://ssrn.com/abstract=3780630 or http://dx.doi.org/10.2139/ssrn.3780630.
[13] "Intellectual Property: a Powerful Tool to Develop Biotech Research – Giugni – 2010 – Microbial BIOTECHNOLOGY – Wiley Online Library." https://ami-journals.onlinelibrary.wiley.com/doi/full/10.1111/j.1751-7915.2010.00172.x (accessed Aug. 17, 2023).
[14] E. Ullberg, "Patent system," in *Encyclopedia of creativity, invention, innovation and entrepreneurship*, E. G. Carayannis, Ed., New York, NY: Springer New York, 2013, pp. 1432–1433, doi: 10.1007/978-1-4614-3858-8_506.
[15] L. Jorgenson and C. Fink, "WIPO's contributions to international cooperation on intellectual property," *J. Int. Econ. Law*, vol. 26, no. 1, pp. 30–34, Mar. 2023, doi: 10.1093/jiel/jgac049.
[16] "The Interface Between Intellectual Property Rights and Competition Policy – Google Books." https://books.google.co.in/books?hl=en&lr=&id=4GV8t25cr5YC&oi=fnd&pg=PA1&ots=7-eImCoQTW&sig=eSrIZoNznlnbeBoGHAedM3uyAAk&redir_esc=y#v=onepage&q&f=false (accessed Aug. 04, 2023).
[17] IP and Antitrust: An Analysis of Antitrust Principles Applied to Intellectual Property Law, 3rd Edition. https://books.google.com/books/about/IP_and_Antitrust_An_Analysis_of_Antitrus.html?id=nTuxDQAAQBAJ (accessed Aug. 04, 2023).
[18] H. L. Williams, "How do patents affect research investments?," *Annu. Rev. Econ.*, vol. 9, no. 1, pp. 441–469, Aug. 2017, doi: 10.1146/annurev-economics-110216-100959.
[19] "Intellectual Property – IFPMA." www.ifpma.org/areas-of-work/fostering-innovation/intellectual-property/ (accessed Jun. 12, 2023).
[20] (:Unas), "Patent cooperation treaty yearly review 2023: The international patent system." *Unknown*. doi: 10.34667/TIND.48061.
[21] "Patent cooperation treaty yearly review 2021," 2021.
[22] R. C. Feldman, D. A. Hyman, W. N. Price, and M. J. Ratain, "Negative innovation: when patents are bad for patients," *Nat. Biotechnol.*, vol. 39, no. 8, pp. 914–916, Aug. 2021, doi: 10.1038/s41587-021-00999-0.

[23] G. T. Vondeling, Q. Cao, M. J. Postma, and M. H. Rozenbaum, "The impact of patent expiry on drug prices: a systematic literature review," *Appl. Health Econ. Health Policy.*, vol. 16, no. 5, pp. 653–660, Oct. 2018, doi: 10.1007/s40258-018-0406-6.

[24] D. H. Howard, P. B. Bach, E. R. Berndt, and R. M. Conti, "Pricing in the market for anticancer drugs," *J. Econ. Perspect.*, vol. 29, no. 1, pp. 139–162, Feb. 2015, doi: 10.1257/jep.29.1.139.

[25] Experts in Chronic Myeloid Leukemia, "The price of drugs for chronic myeloid leukemia (CML) is a reflection of the unsustainable prices of cancer drugs: from the perspective of a large group of CML experts," *Blood*, vol. 121, no. 22, pp. 4439–4442, May 2013, doi: 10.1182/blood-2013-03-490003.

[26] R. Gabble and J. Kohler, "To patent or not to patent? the case of Novartis' cancer drug Glivec in India," *Glob. Health*, vol. 10, no. 1, p. 3, 2014, doi: 10.1186/1744-8603-10-3.

[27] S. K. Rathod, "Compulsory licenses on pharmaceutical patents in India: a short article," *J. Generic Med. Bus. J. Generic Med. Sect.*, vol. 13, no. 3, pp. 108–113, Sep. 2017, doi: 10.1177/1741134317691804.

[28] P. Ocran Mattila, R. Ahmad, S. S. Hasan, and Z.-U.-D. Babar, "Availability, affordability, access, and pricing of anticancer medicines in low- and middle-income countries: a systematic review of literature," *Front. Public Health*, vol. 9, p. 628744, Apr. 2021, doi: 10.3389/fpubh.2021.628744.

[29] V. Radhakrishnan, "Drug pricing: a major barrier to access to cancer care in India," *Cancer Res. Stat. Treat.*, vol. 4, no. 2, p. 195, 2021, doi: 10.4103/crst.crst_66_21.

[30] J. H. Reichman, *Non-Voluntary Licensing of Patented Inventions: Historical Perspective, Legal Framework under TRIPS, and an Overview of the Practice in Canada and the United States of America.* International Centre for Trade and Sustainable Development, Geneva, Switzerland, 2003, doi: 10.7215/IP_IP_20030601D.

[31] C. L. F. B. Bognar, B. L. Bychkovsky, and G. D. L. Lopes, "Compulsory licenses for cancer drugs: does circumventing patent rights improve access to oncology medications?," *J. Glob. Oncol.*, vol. 2, no. 5, pp. 292–301, Oct. 2016, doi: 10.1200/JGO.2016.005363.

[32] "Colombia Cuts Price of Novartis Cancer Drug by 44 Percent," Reuters, Dec. 21, 2016. www.reuters.com/article/us-novartis-colombia-idUSKBN14A1FD (accessed Aug. 16, 2023).

[33] C. Ho, *Access to Medicine in the Global Economy: International Agreements on Patents and Related Rights*. Oxford University Press, UK, 2011.

[34] I. Linnosmaa, T. Karhunen, and I. Vohlonen, "Parallel Importation of pharmaceuticals in finland: effects on markets and expenditures," *Pharm. Dev. Regul.*, vol. 1, no. 1, pp. 67–74, Mar. 2003, doi: 10.1007/BF03257366.

[35] M. R. Kurpad, "The crack in the wall: parallel importation as a flexibility within the indian patent system to ensure access to medicine," *Indian J. Intellect. Prop. Law*, vol. 7, p. 29, 2015.

[36] X. Qin, P. Huckfeldt, J. Abraham, D. Yee, and B. A. Virnig, "Generic entry of aromatase inhibitors and pharmaceutical access: initiation of hormonal therapy, timeliness of initiation, and drug choice," *Res. Soc. Adm. Pharm.*, vol. 17, no. 9, pp. 1588–1595, Sep. 2021, doi: 10.1016/j.sapharm.2020.12.006.

[37] S. Dunne, B. Shannon, C. Dunne, and W. Cullen, "A review of the differences and similarities between generic drugs and their originator counterparts, including economic benefits associated with usage of generic medicines, using Ireland as a case

study," *BMC Pharmacol. Toxicol.*, vol. 14, no. 1, p. 1, Dec. 2013, doi: 10.1186/2050-6511-14-1.

[38] N. S. Blencowe, C. Reichl, J. Gahir, and I. Paterson, "The use of Nolvadex in the treatment of generic Tamoxifen-associated small joint arthralgia," *The Breast*, vol. 19, no. 3, pp. 243–245, Jun. 2010, doi: 10.1016/j.breast.2010.02.004.

[39] "Patent Expiration Offers Hope of Cheaper Cancer Drugs – ClinicalKey." www.clinicalkey.com/#!/content/playContent/1-s2.0-S1470204511704004?returnurl=https:%2F%2Flinkinghub.elsevier.com%2Fretrieve%2Fpii%2FS1470204511704004%3Fshowall%3Dtrue&referrer=https:%2F%2Fwww.thelancet.com%2F (accessed Aug. 17, 2023).

[40] R. Sayah, S. Awada, L. Ismaiil, and G. Hatem, "Assessment of the differences between generic and biosimilar drugs: a brief literature review," *J. Generic Med. Bus. J. Generic Med. Sect.*, vol. 19, no. 3, pp. 124–129, Feb. 2023, doi: 10.1177/17411343231157309.

[41] "2022 Report| Association for Accessible Medicines." https://accessiblemeds.org/resources/blog/2022-savings-report#map (accessed Jun. 12, 2023).

[42] "2020 Report | Association for Accessible Medicines." https://accessiblemeds.org/2020-Access-Savings-Report (accessed Jun. 02, 2023).

[43] E. 'T Hoen, J. Berger, A. Calmy, and S. Moon, "Driving a decade of change: HIV/AIDS, patents and access to medicines for all," *J. Int. AIDS Soc.*, vol. 14, no. 1, pp. 15–15, Jan. 2011, doi: 10.1186/1758-2652-14-15.

[44] GaBI Journal Editor, "Patent expiry dates for biologicals: 2016 update," *Generics Biosimilars Initiat. J.,* vol. 6, no. 1, pp. 27–30, Mar. 2017, doi: 10.5639/gabij.2017.0601.006.

[45] "Bristol Myers Squibb – Bristol-Myers Squibb and Ono Pharmaceutical Company Enter Settlement and License Agreement with Merck to Resolve PD-1 Antibody Patent Litigation." https://news.bms.com/news/details/2017/Bristol-Myers-Squibb-and-Ono-Pharmaceutical-Company-Enter-Settlement-and-License-Agreement-with-Merck-to-Resolve-PD-1-Antibody-Patent-Litigation/default.aspx (accessed Aug. 17, 2023).

[46] "Sun Pharma and Celegene Settle Patent Dispute Over Lenalidomide (Revlimid®) – Patent – India." www.mondaq.com/india/patent/1099008/sun-pharma-and-celegene-settle-patent-dispute-over-lenalidomide-revlimid%C2%AE (accessed Aug. 17, 2023).

[47] "Amgen's Enbrel Fends Off Biosimilar Threat—And Heads Toward Blockbuster Superstardom | Fierce Pharma." www.fiercepharma.com/pharma/amgen-catches-a-break-enbrel-biosimilar-threat-as-federal-court-upholds-patent-win (accessed Aug. 17, 2023).

[48] "Bristol-Myers Settles Two Patent Suits on Blockbuster Sprycel." https://news.bloomberglaw.com/ip-law/bristol-myers-settles-two-patent-suits-on-blockbuster-sprycel (accessed Aug. 17, 2023).

[49] "Roche Settles with Pfizer Over Rituximab Patent." www.centerforbiosimilars.com/view/roche-settles-with-pfizer-over-rituximab-patent (accessed Aug. 21, 2023).

[50] "Genentech and Samsung Bioepis Settle Bevacizumab Biosimilar Patent Dispute | Insights & Resources | Goodwin Procter." www.goodwinlaw.com/en/insights/blogs/2022/09/genentech-and-samsung-bioepis-settle-bevacizumab-b (accessed Aug. 17, 2023).

[51] "Celgene Wins in District Court Case, Invalidating Patent Asserted Against its Thalomid® and Revlimid® Brand Cancer Drugs | Experience | Jones Day." www.jonesday.com/en/practices/experience/2015/07/celgene-wins-in-district-court-case-invalidating-patent-asserted-against-its-thalomid-and-revlimid-brand-cancer-drugs (accessed Aug. 17, 2023).

[52] D. Miller, B. Ippolito, I. Hernandez, and B. Davies, "The costs of delayed generic drug entry: evidence from a controversial prostate cancer drug patent," *J. Gen. Intern. Med.*, vol. 37, no. 3, pp. 668–670, Feb. 2022, doi: 10.1007/s11606-021-07000-3.

[53] "Glenmark and Roche Settle Tarceva Patent Dispute | Fierce Pharma." www.fiercepharma.com/regulatory/glenmark-and-roche-settle-tarceva-patent-dispute (accessed Aug. 17, 2023).

[54] A. Panditrao, "Xtandi versus Indian patent laws: a persepective," *Int. J. Appl. Basic Med. Res.*, vol. 7, no. 1, p. 1, 2017, doi: 10.4103/2229-516X.198496.

[55] K. Jayaraman, "Gleevec loses Indian patent battle," *Nat. Biotechnol.*, vol. 31, no. 5, pp. 371–371, May 2013, doi: 10.1038/nbt0513-371b.

[56] "AstraZeneca Pays $510 M to BMS to Settle PD-L1, CTLA-4 Patent Disputes | BioSpace." www.biospace.com/article/astrazeneca-pays-510 m-to-bms-to-settle-pd-l1-ctla-4-patent-disputes/(accessed Aug. 17, 2023).

[57] "Bristol Myers Says AstraZeneca's Imjudo Infringes Yervoy Patent." https://ipwatchdog.com/2023/01/26/bristol-myers-says-astrazeneca-imjudo-infringes-yervoy-patent/id=155783/ (accessed Aug. 17, 2023).

[58] R. Feldman, "Perverse Incentives: Why Everyone Prefers High Drug Prices – Except for Those Who Pay the Bills," *Harv. J. Legis.*, vol. 57, p. 303, 2020.

[59] M. Mbali, "From AIDS to cancer: health activism, biotechnology and intellectual property in South Africa," *Soc. Dyn.*, vol. 46, no. 3, pp. 449–470, Sep. 2020, doi: 10.1080/02533952.2020.1853954.

[60] P. Yadav, "Differential Pricing for Pharmaceuticals: Review of Current Knowledge, New Findings and Ideas for Action," A study conducted for the U.K. Department for International Development (DFID), August 2010.

[61] "Cancer Moonshot Research Initiatives – NCI." www.cancer.gov/research/key-initiatives/moonshot-cancer-initiative/implementation (accessed Aug. 14, 2023).

[62] K. Giusti, "The multiple myeloma research foundation: educating the myeloma community on recent treatment innovations," *J. Oncol. Pract.*, vol. 1, no. 2, pp. 76–76, Jul. 2005, doi: 10.1200/jop.2005.1.2.76.

[63] "Stand Up To Cancer Raises More Than $143.18 Million In The U.S. In addition, Canada In Connection With Its Seventh Biennial Roadblock Telecast And Streaming Event." www.prnewswire.com/news-releases/stand-up-to-cancer-raises-more-than-143-18-million-in-the-us-and-canada-in-connection-with-its-seventh-biennial-roadblock-telecast-and-streaming-event-301360309.html (accessed Aug. 14, 2023).

[64] "WTO | Intellectual Property – Overview of TRIPS Agreement." www.wto.org/english/tratop_e/trips_e/intel2_e.htm (accessed Aug. 07, 2023).

[65] "WTO | Intellectual Property (TRIPS) – TRIPS and Public Health." www.wto.org/english/tratop_e/trips_e/pharmpatent_e.htm (accessed Aug. 08, 2023).

[66] "WTO | Intellectual Property (TRIPS) – Agreement Text – Institutional Arrangements Etc." www.wto.org/english/docs_e/legal_e/31bis_trips_annex_e.htm (accessed Aug. 08, 2023).

[67] K.-C. Liu, "Christoph Antons and Reto M. Hilty (eds.): intellectual property and free trade agreements in the Asia-Pacific region: Springer-Verlag, Berlin/Heidelberg 2015. viii + 438 pp. ISBN 978-3-642-30887-1. €106.99," *IIC – Int. Rev. Intellect. Prop. Compet. Law*, vol. 47, no. 4, pp. 511–514, Jun. 2016, doi: 10.1007/s40319-016-0483-8.

[68] ANVISA Agência Nacional de Vigilância Sanitária [Brazilian Health Regulatory Agency] v. Novartis International Pharmaceutical Law 9.279/1996 (Industrial Property Law), Arts. 18, 229-C; Law 8.080/1990, Arts. 6,19-M; Law 9.782/1999, Art. 7 item XXV, "'Penciclovir': Decision of the Superior Court of Justice of Brazil (Superior

Tribunal de Justiça) 5 August 2021 – Case No. 1.543.826 – RJ (2015/0173736-6)," *IIC – Int. Rev. Intellect. Prop. Compet. Law*, vol. 53, no. 8, pp. 1228–1236, Sep. 2022, doi: 10.1007/s40319-022-01234-2.
[69] "The Pharmaceutical Benefits Scheme: A Quick Guide – Parliament of Australia." www.aph.gov.au/About_Parliament/Parliamentary_Departments/Parliamentary_Library/pubs/rp/rp1516/Quick_Guides/PBS (accessed Aug. 21, 2023).
[70] "Australians Have Access to More Affordable and Lifechanging Medicines on PBS | Health Portfolio Ministers and Aged Care." www.health.gov.au/ministers/the-hon-mark-butler-mp/media/australians-have-access-to-more-affordable-and-lifechanging-medicines-on-pbs (accessed Aug. 21, 2023).
[71] "Patented Medicine Prices Review Board – Canada.ca." www.canada.ca/en/patented-medicine-prices-review.html (accessed Aug. 21, 2023).
[72] "Chinese Patent Reforms to Bolster Pharma Innovation." www.europeanpharmaceuticalreview.com/news/156869/chinese-patent-reforms-to-bolster-pharma-innovation/ (accessed Aug. 21, 2023).
[73] "Coalitions and Compliance: The Political Economy of Pharmaceutical Patents ... – Kenneth C. Shadlen – Google Books." https://books.google.co.uk/books?hl=en&lr=&id=FWIwDwAAQBAJ&oi=fnd&pg=PP1&ots=jc2n5i5a4q&sig=onT4PraSOMF3TK9_-fDYEe7GINk#v=onepage&q&f=false (accessed Aug. 15, 2023).
[74] "Preserving the Promise." https://books.google.com/books/about/Preserving_the_Promise.html?id=HJbkCwAAQBAJ (accessed Aug. 17, 2023).
[75] I. Olver, "Improving cancer treatment by addressing legislative and regulatory issues," *Public Health*, vol. 125, no. 12, pp. 876–881, Dec. 2011, doi: 10.1016/j.puhe.2011.03.016.

4 Patent Disputes and Access to COVID-19 Vaccines

Structural Heterogeneity and Lessons for Disease X Preparedness

Cristina Possas

4.1 INTRODUCTION

The COVID-19 pandemic began in 2020 and tragically resulted by November 22, 2023 in nearly 7 million deaths globally. As of November 19, 2023, a total of 13.5 billion vaccine doses had been administered worldwide [1]. Nevertheless, despite these extraordinary global efforts in vaccine development and immunization, significant gaps in vaccine coverage persist, particularly in the poorest developing nations. The pandemic exposed the global need for equitable and free access to vaccines, with public and private supply chains unsuccessfully striving to provide, under these emergency conditions, vaccines to the more vulnerable populations worldwide.

Notwithstanding, contrasting with the exponential increase in global vaccine demand in the peak of the pandemic, patent protection and the consequent high prices of these products persisted as one of the main obstacles to achieve the goal of equitable access to these products. Although vaccines are internationally recognized as the most cost-effective and low-cost public health technologies, they have become over the last three decades increasingly complex and multipatented commodities, often requiring patents for each of their diverse processes, from basic research to final products [2,3].

Challenges for vaccine preparedness and pandemic response emerged in a scenario of paradigmatic changes in pharmaceutical manufacturing technologies, in the context of Vaccinology 5.0. This scenario of disruptive technological changes in vaccinology and extensive vaccine patenting drastically impacted on prices and consequently on the possibility of access of the poorest populations to these essential products.

Global shortage of pharmaceutical components, such as nucleotides, enzymes, and lipids, and delays in vaccine production resulted in extreme concentration in few

 DOI: 10.1201/9781003405665-4

nations. Developed countries virtually monopolized access to COVID-19 vaccines in the peak of the pandemic, concentrating nearly 90% of vaccines and excluding the poorest nations [4]. By December 2022, G-20 countries had immunized nearly 80% of their populations, while the poorest countries had not achieved 25% of vaccination. In this scenario, a patent waiver would be ineffective since no patented product would be available in time for local manufacturing [1].

In this scenario, the international regulatory debate on IP rights reemerged, with resurgence of controversies on compulsory licensing and patent waiver on COVID-19 vaccines, which had for decades permeated the HIV/AIDS epidemic debate on antiretroviral drugs. The debate, supported by the Trade-Related Aspects of Intellectual Property Rights (TRIPS) Agreement [5] and the Doha Declaration [6], focused on whether the global intellectual property regime would be able to contribute to accelerate access to vaccines, especially for the poorest populations in developing countries. It also resulted in controversies on which revisions of the legal framework were necessary to achieve this goal.

In October 2020, India and South Africa proposed to the World Trade Organization (WTO) that in order to accelerate access to these products, there should be no trade sanctions on member countries that decide to implement compulsory licensing on vaccines and medicines against COVID-19. This recommendation was based on a long-standing historical debate since the peak of the HIV/AIDS epidemic on the detrimental consequences of the global patent regime introduced by the WTO Agreement on TRIPS on prices, affecting access to essential drugs and vaccines in developing countries. This recommendation to WTO from India and South Africa, supported by more than 100 nations and based on the exceptional circumstances caused by the COVID-19 pandemic, proposed a patent waiver with temporary interruption on imposition of patent rules, trade secrets, and pharmaceutical monopolies.

This recommendation aimed to provide to local developing countries manufacturers legal and technological conditions to accelerate global production capacity for COVID-19 vaccines and drugs. The main argument was that multinational manufacturers' production capacity had proved insufficient to respond to global demands in the case of a pandemic, even in developed nations. These constraints in leading multinational manufacturers' production capacity became more evident as the pandemic scenario aggravated, with the emergence of new COVID-19 variants, evidencing the imperative and urgent need to redesign vaccines for these variants. With the proposed IP measures, such as temporary patent waiver and compulsory licensing of vaccines, it was expected that, in the medium term, laboratories and institutes worldwide could increase their production capacity and contribute to expand the COVID-19 immunization of the global population. Nevertheless, some countries opposed this recommendation for patent waiver and compulsory licensing and the proposal was not approved, since the WTO decisions require consensus. These nations included developed countries with leading multinational pharmaceutical companies, such as the United States, the European Union, Switzerland, Norway, Canada, Japan, Australia.

At the WTO meeting, the Brazilian delegation aligned with developed nations and multinational pharmaceutical companies and opposed the patent waiver recommendation, contrasting with the country's previous positions for patent flexibilities during the HIV/AIDS epidemic. The country's delegation argued that a waiver of COVID-19 vaccine patents would hardly be a solution: due to the global shortage of pharmaceutical components, such as nucleotides, enzymes, and lipids, and vaccines provided by the leading multinational pharmaceutical companies, a patent waiver would not result in the short term in local vaccine production in developing countries. The delegation argued that the best strategy would be the direct licensing in technological transfer agreements from multinational pharmaceutical companies with local manufacturers.

The opponents of the patent waiver recommendation supported that the scarcity of the COVID-19 vaccines in the global pharmaceutical market did not derive exclusively from high prices consequent to patents, but from low manufacturing and vaccine production capacity, particularly in developing countries. Thus, implementing patent waiver of vaccines would be ineffective if these countries had no option to locally produce or import vaccines, considering that the main pharmaceutical components for vaccine manufacturing were in shortage worldwide. Later, on May 5, 2021, the U.S. Biden administration announced, contrasting with the previous US position on the topic, that it would support waiving intellectual property (IP) protections for COVID-19 vaccines under the World Trade Organization's Agreement on TRIPS. This announcement was received with great criticism from drug companies, patent holders, and also independent experts, who argued patent waiver would not be a feasible solution, due to the shortage of the main vaccine manufacturing components worldwide.

Contrasting with this US position, the European Union submitted a proposal in June 2021 with the recommendation that there should be no suspension of COVID-19 vaccine and drug patents, but individual negotiations of licensing agreements between countries and pharmaceutical companies. These negotiations should be supported by three basic strategies: facilitating trade and lifting export restrictions; expanding production and facilitating TRIPS flexibilities related to compulsory licensing.

A critical issue in this debate, as observed in a recent publication [7], is that temporary suspension of IP rights by a national government does not necessarily presuppose that the government will establish some mechanism to compensate owners for IP intellectual property that may be used during the period of suspension. Moreover, COVID-19 vaccines, particularly the mRNA ones, are complex products that require a wide range of processes in diverse development and manufacturing stages, most of them not available in a patent. The overlap of different IP rights from diverse patent holders and different levels of disclosure severely restricts the sharing of know-how and technology transfer needed for vaccine manufacturing. Thus, for a patent waiver to be effective it would be necessary strategies to ensure sustainable technology transfer agreements to qualified manufacturers in developing countries, for sharing know-how and trade secrets.

4.2 LICENSING IN TECHNOLOGY TRANSFER AGREEMENTS: LAC AND AFRICA

Consequent to this IP debate, an alternative for patent waiver has been licensing within the technology transfer agreements, with pharmaceutical companies licensing third-parties in developing countries. It was expected that this strategy would strengthen manufacturing capabilities in the poorest countries, particularly in Africa and Latin America and Caribbean (LAC) region. Notwithstanding, although it seemed feasible and emerged as a solution to support local manufacturers, this alternative proved insufficient, with very few agreements so far in place in these countries. As we will discuss here, this is not just an issue of insufficient manufacturing capacity in the poorest developing countries. The main necessary conditions are full technology transfer from the leading multinational companies and sustainable global funding to support local manufacturers for pandemic preparedness and response.

So far, 15 contracts have been signed with the Medicines Patent Pool [8] within the WHO mRNA technology transfer agreements from the following manufacturers/organizations: 1. Afrigen/Biovac; 2. BioVax; 3. Bio-Manguinhos; 4. Biofarma; 5. BiologicalE; 6. BioGeneric Pharma; 7. Biovaccines Nigeria; 8. Darnitsa; 9. Incepta Vaccine; 10. Institut Pasteur de Dakar; 11. Institut Pasteur deTunis; 12. Institut Torlak; 13. National Institutes of Health; 14. Polyvac; 15. Sinergium Biotech. In spite of the relevance and great success of this international initiative, the number of contracts is still very low. To reverse this scenario, it an exponential increase in global support will be necessary for technology transfer agreements and funding for the development of innovative vaccines for Disease X preparedness.

Technology transfer for innovative vaccine development from local manufacturers is feasible if global IP governance, sustainable funding to implement them, and support capacity building is in place. A recent report, based on data from European Medicines Agency, WHO, and FDA databases [9] identified 120 manufacturers worldwide from emerging and low-income countries with the technical requirements and quality standards to begin manufacturing mRNA vaccines, as indicated in Figure 4.1: 106 in Asia – 55 in India, 34 in China, nine in South Korea, three in Vietnam, two in Malaysia, one in Bangladesh; one in Thailand and one in Indonesia; eight in Africa – three in Egypt, two in Tunisia, one in Morocco; six in Latin America – three in Brazil, one in Argentina, one in Chile and one in Cuba.

These 120 companies were identified in this report from the European Medicines Agency's database of companies that have passed quality standards to import pharmaceutical products to the European Union: the World Health Organization's Pre-Qualification Programme and the US Food and Drug Administration's approval for export to the United States.

In Table 4.1 we provide a list of pharmaceutical companies and partners producing COVID-19 vaccines that could be potential global players in these technological transfer agreements, but very few of them, with the exception of AstraZeneca and the three mRNA vaccine companies included in the MPP database (Moderna, Pfizer/BioNtech, and Curevac) have so far implemented and completed technology transfer agreements with developing countries' manufacturers.

A good example of technology transfer to developing countries' manufacturers is in Africa. The leading vaccine development, production, and capacity building initiative for technology transfer is the "WHO Afrigen mRNA vaccine hub" created in June 2021 and based in South Africa [10]. In January 2022, Afrigen successfully reproduced Moderna's full vaccine [11], both the mRNA technology and final vaccine formulation. It is now experimenting with other formulations that are cheaper

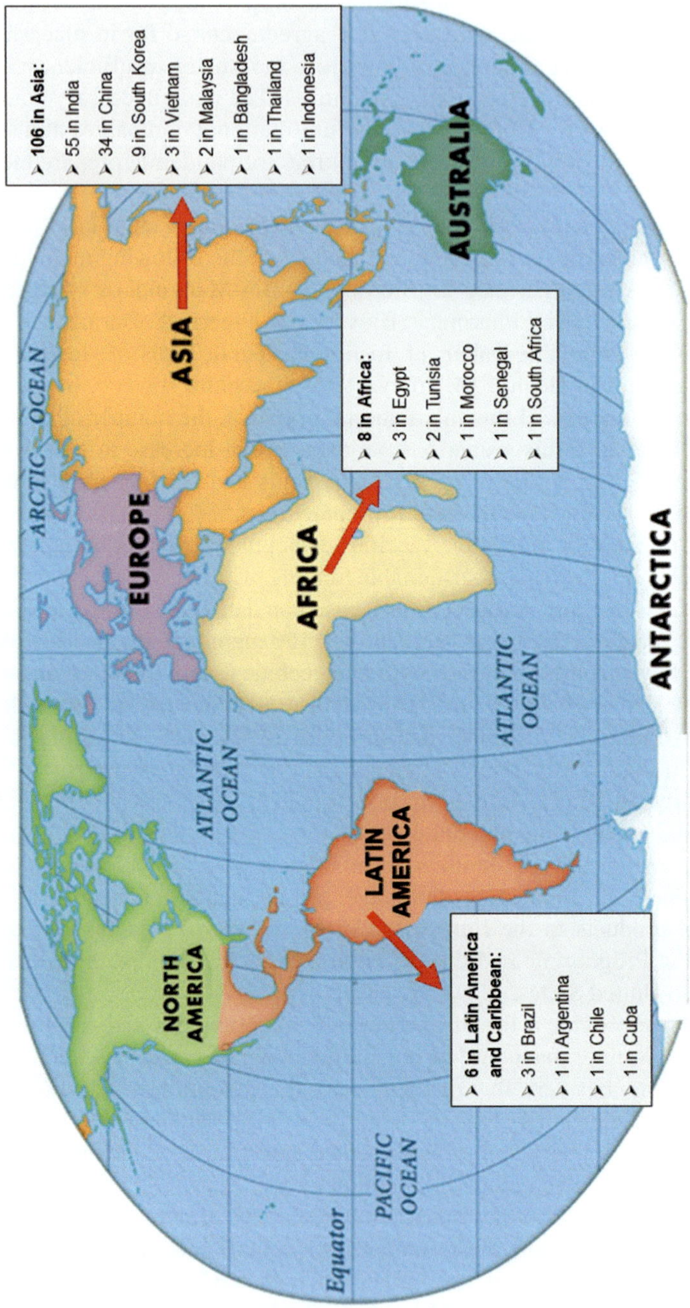

FIGURE 4.1 Local manufacturers with quality standards for mRNA vaccines. (Elaborated by the author based on data from European Medicines Agency, WHO and FDA databases [9].)

TABLE 4.1
Pharmaceutical Companies Producing COVID-19 Vaccines

Pharmaceutical Company/Partner	Vaccine Type
Pfizer/ BioNTech	mRNA
Sanofi/ GlaxoSmithKline	Protein subunit
Moderna	mRNA
Novavax	Protein subunit
AstraZeneca/ Oxford University	Non-replicating viral vector
Johnson & Johnson/Janssen	Non-replicating viral vector
CureVac	mRNA
Inovio	DNA
Covaxx/ Nebraska University	Protein subunit
SK Biosciences	Protein subunit
Biological E	Protein subunit
AnGes/Osaka University	DNA
Bharat Biotech	Inactivated virus
CanSino	Non-replicating viral vector
Gamaleya	Non-replicating viral vector
SII/ Max Planck Institute	Live attenuated virus
Sinopharm/ Beijing Institute	Inactivated virus
Sinovac Biotech	Inactivated virus

Source: Elaborated by the author based on enterprises' information.

to produce and do not need to be frozen in storage, an urgent need in many African countries [12]. Afrigen is now negotiating with biotech companies to help scale up manufacturing. Two South African companies, Afrigen Biologics and Biovac, have joined a consortium to develop and manufacture the mRNA vaccine, with guidance from the WHO, the Medicines Patent Pool, the South Africa Medical Research Council, and the Africa Centres for Disease Control and Prevention (Africa CDC). This initiative is part of the wider Partners for African Vaccine Manufacturing project coordinated by Africa CDC, launched in April 2021 [13]. The Afrigen mRNA hub seeks to scale up African production of lifesaving vaccines from 1% to 60% by 2040.

Notwithstanding its success, this Afrigen initiative has been under political pressures and threats of retaliation. The BMJ [12] obtained documents from the kENUP Foundation, hired by BioNTech, recommending mRNA Afrigen hub efforts should be terminated immediately, on the claim that it would infringe on patents. The Medicines Patent Pool, a UNITAID organization [8] responsible for intellectual property and licensing in Afrigen said that statements sustaining that the mRNA hub intends to infringe patents are "unfounded rumours" [14]. BioNTech later said in a statement that its plans to establish mRNA-based vaccine manufacturing on the African continent "will be done in close alignment with the WHO, the African Union, and the Africa CDC" [15].

In the LAC region, the successful technology transfer agreement of Oxford/AstraZeneca vaccine in Brazil is often cited as an example to be followed. AstraZeneca implemented one of the most important global strategies to increase access to this COVID-19 vaccine. Licensing was provided for local production in technology transfer agreements with Bio-Manguinhos/Fiocruz and with the Serum Institute in India. Another Brazilian manufacturer, the Butantan Institute in São Paulo, made a partnership with the Chinese laboratory Sinovac Biotech for a COVID-19 vaccine, Coronavac, while two private companies, Eurofarma and União Química, started technology transfer agreements for the BioNtech/Pfizer and SputnikV vaccines, respectively.

These successful technology transfer experiences in Brazil were followed by government policies aiming to accelerate vaccine development capacity to support such initiatives. Bio-Manguinhos/Fiocruz is implementing the Industrial Center of Biotechnology in Health (CIBS), in Santa Cruz, Rio de Janeiro, with the capacity to exponentially increase its production of vaccines and capacity to export vaccines. In the same direction, the Butantan Institute inaugurated a new plant in 2022, with the capacity to produce 100 million doses of vaccines per year, not only for COVID-19 but also for Zika, Dengue, and other neglected diseases

Notwithstanding, in spite of these successful experiences in South Africa and Brazil, technology transfer initiatives have so far been very limited and exposed to trade retaliations and political constraints. These IP and local capacity building constraints have particularly impacted on the poorest countries in Africa and Latin America and the Caribbean, with these regions most in need relying on a heavy dependence on imports.

Shadlen [16] provides a thorough analysis of Astrazeneca technology transfer initiatives in Latin America and the Caribbean (LAC) region and concludes that in spite of its success it has been less extensive than what might be expected from a region with many middle-income countries that have important pharmaceutical sectors. In COVID-19 pandemic, most of these LAC firms were applying the final steps to vaccine production, using imported vaccine manufacturing components. The author's conclusion is that with the exception of Cuba, which developed and produced its own vaccines, only two LAC manufacturers, Brazil's BioManguinhos/Fiocruz and Argentina's mAbxience, advanced significantly on manufacturing drug substance; and only BioManguinhos produced the full vaccine, drug substance plus fill-finish. Considering these constraints, the WHO proposed a "coordinated technology transfer," in which universities and manufacturers license their vaccines to other companies through a global mechanism (mRNA Technology Transfer Hub), which would also facilitate the training of staff at the recipient companies, and coordinate investments in infrastructure. The WHO considers this approach is more coherent and transparent than one-off tech-transfer deals such as that between AstraZeneca and Serum Institute [3].

A final consideration is that the benefits of technology transfer agreements have been so far very low. Under current global RD&I and manufacturing conditions these benefits might come only later, in two or three decades, after these developing countries' manufacturers succeed, supported by sustained global funding and incentives, to incorporate know-how and to strengthen their domestic capacity for production. Such a timeframe would be too late for Disease X Pandemic preparedness and response. Overcoming gaps in technology transfer and bottlenecks for supply chains remains thus the main challenge for local production capacity of COVID-19 vaccines in these developing nations.

4.3 COVID-19: GLOBAL IP-RELATED INITIATIVES

In this item, we provide an overview of the global governance attempts and IP-related initiatives to respond to the COVID-19 pandemic. A broad range of collaborative public and private initiatives, including voluntary licensing and purchasing mechanisms, have been launched since the start of the pandemic [10]. These initiatives, most of them implemented in an emergency scenario, aimed to develop vaccines, kits for diagnosis and medicines and simultaneously ensure equitable and free access to these new technologies. These efforts included substantial investments in product development collaborations and large multiple stakeholder R&D initiatives. The leading initiative is the COVAX Facility.

4.3.1 COVAX Facility: Vaccine Support of ACT Accelerator

The COVAX Facility is the main multilateral initiative conceived to accelerate the development of COVID-19 vaccines and ensuring they are equitably available in low- and middle-income countries. Coordinated by the Coalition for Epidemic Preparedness Innovations (CEPI), The Vaccine Alliance (GAVI), and the World Health Organization (WHO), alongside its key delivery partner the United Nations Children´s Fund (UNICEF), the COVAX Facility played a key role in the COVID-19 pandemic [10].

COVAX was implemented to accelerate the development and manufacture of COVID-19 vaccines in signatory countries and to guarantee fair and equitable access for every country in the world. COVAX is the vaccines' support of the Access to COVID-19 Tools (ACT) Accelerator, a pioneer global collaboration to accelerate the development, production, and equitable access to COVID-19 vaccines, kits for diagnosis, and medicines.

The COVAX Facility was conceived to provide governments with the opportunity to benefit from a large portfolio of COVID-19 candidate vaccines using a broad range of technology platforms, produced by more manufacturers across the world, with a bigger market to provide security of demand. Even where governments already have access to a portfolio of candidate vaccines, it acts as an insurance policy through economies of scale and portfolio diversification. Through the CEPI's portfolio, which will be transferred to the COVAX Facility, the Facility benefits from a growing number of COVID-19 vaccine candidates and manufacturing capacity.

The COVAX Facility has also contributed to a global scale-up of vaccines and therefore to controlling the pandemic. The Facility is reinforced by the Gavi COVAX Advance Market Commitment (AMC), which provides funding for vaccines for lower-income countries, creating scale and therefore making the Facility more attractive to manufacturers, creating the opportunity to tackle the pandemic around the world. This successful strategy promoted a centralization of purchases, with an implementation strategy that has significantly contributed to overcome the dichotomy between the private and public sectors, resulting in a sharp reduction in the costs of developing countries' immunization programs.

In their first thorough analysis of the COVAX Facility [17], the authors argue that the funding mechanism deployed by the COVAX pillar appears to be effective at fostering at-risk investments in R&D and the production of doses in advance of confirmation of clinical efficacy, with development of two vaccine candidates funded by the ACT Accelerator. Notwithstanding, the authors caution that the mechanism heavily favors pharmaceutical companies, for which up-front investment from the COVAX pillar represents a win-win situation and an opportunity to benefit regardless of whether their vaccine candidate ever goes on to gain regulatory approval.

The authors conclude arguing that, like any multilateral purchasing system, securing a sufficient degree of interest and participation is essential to its success. From their perspective, the COVAX Facility has so far failed to do this, primarily as a result of the prevalence of vaccine nationalism, whereby countries adopt policies that heavily prioritize their own public health needs at the expense of others.

4.3.2 WHO Initiatives

The WHO has conceived and implemented diverse initiatives to accelerate the response to the COVID-19 pandemic, involving licensing, patent pooling, and technology transfer of vaccines, kits for diagnosis and medicines [10], as indicated below.

4.3.2.1 mRNA Vaccines Hub: Afrigen and Other Countries

In February 2022, the WHO announced that Egypt, Kenya, Nigeria, Senegal, South Africa, and Tunisia would be the first countries on the African continent to receive the technology needed to produce mRNA vaccines through the Afrigen mRNA hub. As of August 2022, the list of recipients has been expanded to include Argentina, Bangladesh, Brazil, India, Indonesia, Pakistan, Serbia, Ukraine, and Vietnam.

4.3.2.2 WHO COVID-19 Technology Access Pool (WHO C-TAP)

C-TAP's priority has been to realize equitable global access to COVID-19 health technologies through pooling of knowledge, intellectual property and data. Its main strategy has been engagement with technology patent holders to encourage sharing of intellectual property, know-how or data to promote global equitable access of COVID-19 vaccines, therapeutics, diagnostics, or other medical devices. C-TAP provides pooling of COVID-19 health technologies or products to promote innovation or further development and facilitate scaling up production through multiple quality-assured manufacturers that currently have untapped capacity to scale up and diversify production. The US National Institutes of Health (NIH) has licensed COVID-19 technologies to the Medicines Patent Pool (MPP) for access through the World Health Organization's (WHO) COVID-19 Technology Access Pool (C-TAP).

The technologies include the stabilized spike protein used in currently available COVID-19 vaccines, research tools for vaccine, drug, and diagnostic development as well as early-stage vaccine candidates and diagnostics. While recent licenses indicate the potential of the mechanism to promote access and innovation, significant challenges remain in realizing the full potential of C-TAP.

4.3.2.3 Genetic Sequences of Viral Samples Shared Openly Worldwide

The WHO made a considerable effort to ensure in the COVID-19 pandemic that viral samples, clinical samples, and associated data would be rapidly shared for immediate public health purposes. Fair and equitable access to any medical products or innovations that are developed using these materials must be part of such sharing.

4.3.2.4 Access to COVID-19 Tools Accelerator (ACT-A)

In 2020, the WHO launched ACT-A, in collaboration with private sector partners and other stakeholders, to accelerate the development, production, and equitable global access to new COVID-19 health technologies. ACT-A is organized around three main pillars of work (diagnostics, therapeutics, vaccines). As of March 2023, COVAX, the vaccines' pillar of ACT-A has been allocated 1.94 billion COVID-19 vaccine doses since the beginning of the pandemic.

4.3.2.5 COVID-19 Vaccine Delivery Partnership

In January 2022, the WHO created in partnership with UNICEF and Gavi, the COVID-19 Vaccine Delivery Partnership to expedite vaccine delivery in 34 countries that were at or below 10 percent full vaccination coverage to progress towards national and global targets. This initiative prioritizes full vaccination for high-priority populations, such as older adults, health-care workers and persons with co-morbidities. based on the principles of equity, transparency, and ethics.

4.3.2.6 COVAX Manufacturing Task Force

The WHO established with partners the COVAX Manufacturing Task Force aiming to increase immediate vaccine supply (three to six months) for existing vaccines, ensuring that vaccines coming into market could be produced at maximum scale and not constrained by existing contracts. This strategy would enable low- and middle-income countries to acquire COVID-19 vaccine production technology and to establish sustainable outbreak response capacity for regional health security.

4.3.2.7 WHO Emergency Use Listing

The WHO Emergency Use Listing (EUL) procedure defines the process by which new or unlicensed products can be used during public health emergencies. The EUL provides a time-limited listing for unlicensed products in an emergency context when limited data are available and when products are not yet ready to apply for WHO prequalification. In this context, products are still undergoing development and are not yet licensed. The WHO assesses the quality, safety, and efficacy of the data generated during development and conduct a risk–benefit assessment to determine use outside of clinical trials.

4.3.3 WTO Initiatives

WTO promoted, since the onset of the COVID-19 pandemic, diverse fora for the debate by experts and stakeholders on intellectual property, vaccine supply chains, trade disputes, and regulatory transparency related to the COVID-19 pandemic. Three of them are a reference for this debate: the Technical Symposium on COVID-19 Vaccine Supply Chain and Regulatory Transparency in June 2021; the Webinar on Regulatory Cooperation during the COVID-19 Pandemic, in June 2021; the Technical Workshop on COVID-19 Vaccines R&D, Manufacturing and Distribution in February 2022, conceived to improve knowledge on the practical aspects of this topic.

4.3.4 WIPO Initiatives

4.3.4.1 WIPO COVID-19 Response Package

The COVID-19 Response Package launched in 2022 is a commitment from WIPO to support member states in ensuring that IP is used as a tool to strengthen economies, support communities, and drive sustainable recovery post-pandemic [10], including the following: bilateral technical assistance; access to resources and capacity building at the intersection of IP, health and trade including through the WHO–WIPO–WTO trilateral cooperation; access to IP dispute resolution and contract negotiation services, including through the WIPO Arbitration and Mediation Center, particularly for the life sciences sector; capacity building related to innovation and technology transfer resources, including new policy and decision-making tools to respond to COVID-19 issues through WIPO Technology and Innovation Support Centers (TISCs) and other technology transfer structures; new knowledge resources, publications, guidance documents and training materials including the Patent Landscape Report on COVID-19 Related Vaccines and Therapeutics, IP and studies on innovation

4.3.5 Government and Intergovernmental Initiatives

4.3.5.1 United States

The most important US initiative was the immediate and extraordinary Operation Warp Speed (OWS), which allowed a rapid global vaccine development and response to the COVID-19 pandemic in less than six months, due to previous coronavirus vaccine developments and strong funding mechanisms from the US government and international agencies. The OWS was conceived as a public–private partnership initiated by the US government to accelerate the development, manufacturing, and distribution of COVID-19 vaccines, therapeutics, and diagnostic tools. The US Congress has directed almost USD 10 billion to this effort through supplemental funding, including the CARES Act. Congress has also appropriated other flexible funding. The nearly USD 10 billion included more than USD 6.5 billion designated for countermeasure development through BARDA and USD 3 billion for National Institutes of Health (NIH) research. As of January 2021, five of the six OWS vaccine pharmaceutical companies had started commercial scale manufacturing and as of January 31, 2021, companies had already released 63.7 million vaccine doses. Later, on June 23, 2022, the US Congress enacted six emergency supplemental funding bills to address the COVID-19 pandemic, which provided approximately USD 19.03 billion for the global response, including health and humanitarian efforts.

4.3.5.2 European Commission

In May 2020, the European Commission launched Coronavirus Global Response and reached a total of €15.9 billion by the end of June 2020 with pledging events to fund collaborative development and universal deployment of, and access to, diagnostics, treatments, and vaccines against coronavirus. The European Commission has also instituted a temporary framework to allow state aid to go to COVID-related R&D if beneficiaries commit to grant non-exclusive licences under non-discriminatory market conditions to third parties in the European Economic Area.

4.3.6 International Partnerships and Organizations

4.3.6.1 Coalition for Epidemic Preparedness Innovations (CEPI)

CEPI is a Product Development Partnership (PDP) created during the 2014 Ebola virus outbreak by philanthropies and several governments. It initially implemented a USD 3.5 billion action plan to fund late-stage clinical trials of promising COVID-19 vaccine candidates from pharmaceutical companies that could offer doses to COVAX. CEPI requires manufacturers to provide equitable access to any vaccine developed through its funding. It further requires product developers to be willing to undertake technology transfer to enable production by a global network of manufacturers. CEPI has played a key role in COVAX's operation.

4.3.6.2 The World Bank

Vaccine development financing by the World Bank has been key for global COVID-19 response. In June 2021, the World Bank announced that it had expanded its COVID-19 vaccine financing from an initial USD 12 billion to USD 20 billion over a period of 18 months.

Figure 4.2 provides a global overview of the main organizations leading these extraordinary international initiatives, striving to operate and to respond to an urgent global demand in the peak of the COVID-19 pandemic, when no vaccines or drugs were available.

4.4 IP GOVERNANCE FOR DISEASE X PREPAREDNESS: AN URGENT ISSUE

Notwithstanding all these collaborative multilateral initiatives and efforts, implemented under emergency pandemic conditions, many gaps and bottlenecks persist. Providing incentives for COVID-19 vaccine development during the pandemic and at the same time ensuring a feasible IP framework for equitable access to them has proved to be a difficult task. The main issue is how to conceive and implement socially responsible IP licensing policies [18, 19] into a new global governance paradigm.

Although IP flexibilities are legitimate and clearly supported by international laws, their implementation has been politically challenging and complex amidst the

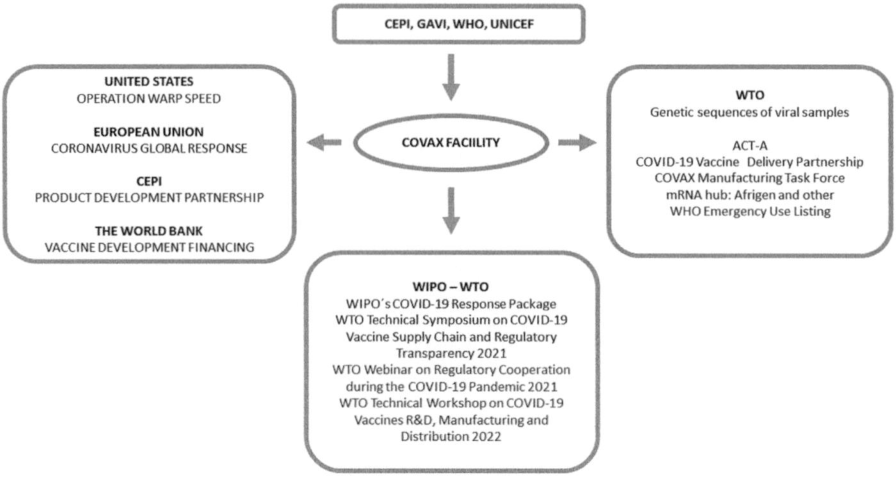

FIGURE 4.2 Global initiatives in COVID-19 pandemic. (Elaborated by the author.)

COVID-19 pandemic, particularly for developing countries. Threats of retaliation by developed nations and by leading pharmaceutical companies as well other trade, legal, and political constraints, urged their governments to require emergency debates at WTO, WIPO, and WHO. The debate on these controversial regulatory and implementation issues, in spite of their relevance, proved ineffective, with limited impact on local vaccine manufacturing and access.

The lesson is that the solution for patent flexibility and local manufacturing impasses in pandemics should not be searched under emergency conditions. It will be necessary for a future Disease X pandemic preparedness to urgently move from the search for emergency solutions to a sustainable new International Treaty for Pandemic Preparedness and Response. This Treaty should be supported by a Global IP Governance Pandemic Framework (Figure 4.3), favoring an extensive acceleration of technology transfer agreements and an exponential expansion of local manufacturing capacity in developing countries, supported by an international collaborative funding effort to align vaccine supply with global demand.

4.4.1 Disease X Pandemic: How Devastating Could It Be?

A new Disease X virus is probably replicating now in the more vulnerable regions of the globe and might result in a new pandemic, favored by the amplification of conditions, such as accelerated urbanization, climate change, and very high populational mobility. The World Health Organization (WHO) added Disease X in 2017 to a list of priority pathogens, including COVID-19 later to this list. There is a concern among international experts that a Disease X pandemic might result in 20 times more fatalities than COVID-19. In the worst scenario, this new disease could have a global impact similar to the devastating Spanish Flu of 1918–1919, with enormous social and economic losses.

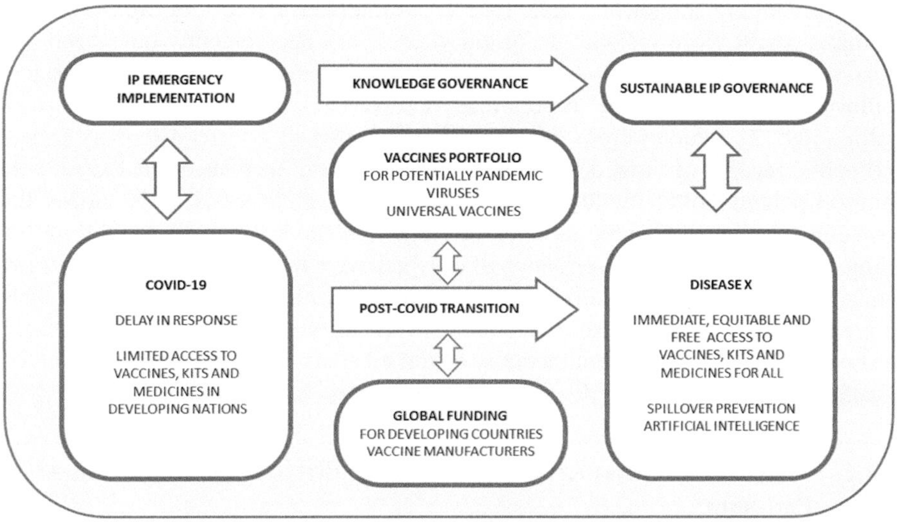

FIGURE 4.3 IP governance: from COVID-19 to Disease X. (Elaborated by the author.)

The COVID-19 pandemic benefited from a rapid global vaccine response, in less than six months, due to previous coronavirus vaccine developments and strong funding mechanisms from the US Government (Operation Warp Speed) and international agencies. The concern is that in the case of a future Disease X pandemic, different from the COVID-19 one, probably no previous vaccine developments will be available and therefore the global vaccine preparedness and response might possibly be much slower than the ones for COVID-19.

The world should urgently prepare for Disease X universal vaccination and create local manufacturing and immunization preparedness in developing countries to deliver the vaccine doses immediately, overcoming current technological and production platform bottlenecks [20]. A vaccine portfolio for potentially pandemic viruses should immediately be designed, considering that nearly 70% of these viruses are of zoonotic origin.

The social and economic resources generated by the COVID-19 vaccine response should urgently be redirected to the development of this vaccine portfolio and universal vaccines. Funding and access to vaccines have often been considered public health expenditures to reduce high mortality in epidemics and pandemics. The fact that universal access to vaccines enables an enormous global saving of resources and a safer re-opening of social and economic activities has unfortunately not been sufficiently recognized by global economic players and authorities as it should.

Scientists have identified 25 virus families, including thousands of viruses, many of them with pandemic potential [21]. Notwithstanding, there are millions of viruses yet to be discovered and with the potential to evolve into pandemics. The identification of these viruses and understanding their evolution is critical for strengthening global preparedness against future pandemics.

A risk ranking framework supported by an interactive web tool, SpillOver, that estimates a risk score for wildlife-origin viruses, has been recently developed [22] creating a comparative risk assessment of viruses with uncharacterized zoonotic spillover potential alongside those already known to be zoonotic. Using data from testing 509,721 samples from 74,635 animals as part of a virus discovery project and public records of virus detections around the world, they succeeded to rank the spillover potential of 887 wildlife viruses. In addition, there are over 250 viruses that have already jumped the species barrier from animals to humans. Nearly 1.7 million animal viruses are still undescribed, half of which have the potential to spill over into humans and could result in a future Disease X pandemic. Scientists should be able, in preparedness for a Disease X pandemic, to identify potential locations for spillover of these viruses in environmentally disturbed areas, supporting timely and more effective response.

4.4.2 Vaccine Development: Challenges and Bottlenecks for a Disease X Pandemic

Vaccine manufacturing capacity varies enormously across countries and regions, which will require from their governments and international agencies diverse governance strategies. There will be many bottlenecks to overcome to ensure global access in a future Disease X pandemic: production, affordability, allocation, and deployment [23]. It will also be necessary to address the shortcomings of current vaccines, with short duration of protection and adverse events. New technologies and approaches

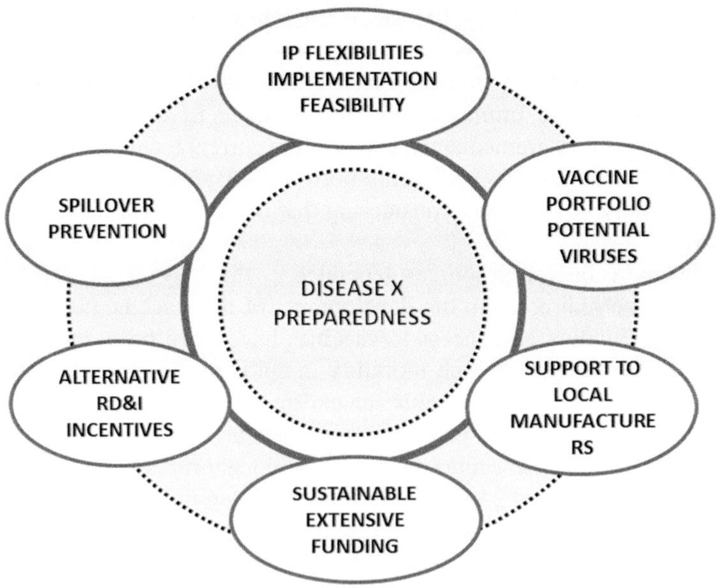

FIGURE 4.4 Disease X preparedness: equitable access to vaccines. (Elaborated by the author.)

to vaccine design are thus urgently needed, such as innovative universal vaccines offering protection to all viral variants, potentially leading to more effective vaccines in the future.

The global costs of an omission in vaccine preparedness could be extremely high for economics and societies. The COVID-19 pandemic, possibly less severe than a future Disease X one, resulted in the US alone having USD 16 trillion of global public health expenditure and direct and indirect social and economic costs [24, 25]. In Figure 4.4 we provide an overview of the main vaccine development policies to ensure global access to innovative vaccines for all.

4.5 CONCLUSION AND FUTURE TRENDS

The main question is: how IP lessons from the COVID-19 pandemic could be effectively incorporated into global governance and preparedness for a future Disease X? Our results indicate that the main challenge is conceiving and implementing socially responsible licensing and technology transfer policies [18,19].

TRIPS flexibilities and IP exceptionalities have proved insufficient and very difficult to implement in the COVID-19 pandemic. The patent disputes to increase global access to COVID-19 vaccines through temporary waiver and technology transfer initiatives with multinational pharmaceutical companies evidenced more complex technological and political conditions than the ones in the HIV/AIDS epidemic, when generic production of antiretroviral drugs proved feasible and gained scale.

Vaccines are, in contrast, multipatented and much more complex products, particularly the innovative mRNA vaccines, with IP detained by many pharmaceutical companies, requiring negotiations with a large number of patent holders in the different stages of the vaccine development chain. This explains the difficulties in the international IP debate and initiatives.

Even the immediate COVID-19 vaccine development response, in less than 6 months by the US Operation Warp Speed, was not sufficient to ensure rapid access to global immunization. G-20 countries were, in the peak of the COVID-19 pandemic, virtually monopolizing access to vaccines, by concentrating nearly 90% of innovation-intensive vaccines and excluding the poorest nations.

This social exclusion in immunization coverage is concerning, considering that a Disease X pandemic might result in a global disaster, with 20 times more fatalities than the COVID-19 pandemic. COVID-19 caused nearly 7 million deaths worldwide, but the majority of people infected with the virus recovered. Moreover, in the most severe Disease X scenario, the virus could be as infective as measles (90%) and as lethal as Ebola (from 25% to 90% in past outbreaks).

This unacceptable and unethical “vaccine apartheid” [26] evidences the strong inequality and structural heterogeneity [27] in access to vaccines in the COVID-19 pandemic, even with all international initiatives in place. This social exclusion from immunization was aggravated by the global shortage of vaccine manufacturing components, such as nucleotides, enzymes, and lipids from multinational pharmaceutical companies and by few vaccine producers participating at in the COVAX Facility, based on voluntary licensing mechanisms.

The social and economic costs of a global omission in vaccine preparedness could be extremely high. The COVID-19 pandemic, possibly less severe than a future Disease X one, resulted in the US alone having USD 16 trillion of global public health expenditure and direct and indirect social and economic costs. Sustainable and long-term global funding to reverse this scenario and collaborative international governance strategies supported by science diplomacy [28, 29] should be a priority.

It is therefore necessary to conceive new global governance strategies with a more flexible IP framework and alternative incentives for vaccine and drug innovation, which should be economically feasible and easy to implement. Incentives such as patent pools, prizes and awards [30–32] should be urgently put in place and strengthened, supported by economic fora, such as G-7, G-20, WTO, IMF, and the World Bank.

A new IP paradigm aiming global access to vaccines is thus urgently needed. Public health goods, such as vaccines and drugs, should no longer be subject to the same IP rules as other products in the global market [33]. Knowledge governance strategies should be supported by mission-oriented approaches [34, 35]. Extensive funding incentives for innovation and for supporting local manufacturers in developing nations should be urgently directed to overcome extreme inequalities in access to life-saving products in a Disease X pandemic.

REFERENCES

[1] World Health Organization 2023. WHO Coronavirus (COVID-19) Dashboard 2023. In *WHO Health Emergency Dashboard*, Geneva. https://data.who.int/dashboards/covid19/cases (accessed on April 7 2024)

[2] Possas C, de Souza Antunes A, Mendes F, Schumacher S, Homma A 2015. Access to new technologies in multipatented vaccines: challenges for Brazil. *Nat Biotech* 33:599–603.

[3] Irwin A, Nkengasong J 2021. What it will take to vaccinate the world against COVID-19. *Nature*. 592(7853):176–178.

[4] Our World in Data 2022. As of 28 Jan 2022. https://ourworldindata.org/grapher/cumulative-covid-vaccinations-continent?country=Africa~Asia~Europe~North+America~Oceania~South+America (accessed in 22 Nov 2023)

[5] World Trade Organization, WTO 1994. WTO Agreement on Trade-Related Aspects of Intellectual Property Rights (TRIPS), Apr 15, 1994, Marrakesh Annex 1C, 1869 UNTS. 299, 33 ILM. 1197 (1994) [hereinafter TRIPS Agreement].

[6] World Trade Organization, WTO 2001. WTO Ministerial Declaration of 14 Nov 2001, WTO Doc. WT/MIN (01)/DEC/1, 41 ILM 746 [hereinafter Doha Declaration].

[7] Abbott FM 2020. The TRIPS Agreement Article 73 Security Exceptions and the COVID-19 Pandemic, August 1, 2020. Research Paper 116, South Centre, Geneva.

[8] The Medicines Patent Pool, MPP 2023. Our Strategy 2023–2025 Greater access to medicines and health technologies for those who need them. Annual Report 2022.

[9] Médecins Sans Frontières, MSF 2021. Pharmaceutical firms across Asia, Africa and Latin America with potential to manufacture mRNA vaccines – Covid-19. Technical Report 16 December 2021.

[10] WTO, WIPO, WHO 2023. An Integrated Health, Trade and IP Approach to Respond to the COVID-19 Pandemic. Second update, Geneva, May. www.wipo.int/publications/en/details.jsp?id=4590 (accessed on April 20 2024).

[11] Moderna 2021. Moderna to build state-of-the-art mRNA facility in Africa to manufacture up to 500 million doses per year. 7 Oct 2021. https://investors.modernatx.com/news/news-details/2021/Moderna-to-Build-State-of-the-Art-mRNA-Facility-in-Africa-to-Manufacture-up-to-500-Million-Doses-Per-Year-10-07-2021/default.aspx (accessed in 10 Nov 2023)

[12] Davies M. 2022. Covid-19: WHO efforts to bring vaccine manufacturing to Africa are under-mined by the drug industry, documents show. *BMJ* 376:o304.

[13] Partnership for African Vaccine Manufacturing, PAVM 2021. Concept note. 7 Dec 2021. https://africacdc.org/wp-content/uploads/2021/11/ENGLISH_PAVM_Public-Stakeholder-Engagement_Concept-note_Final-version.pdf (accessed on 5 Oct 2023).

[14] Medicines Patent Pool, MPP 2021. MPP position statement on patents with regards to the mRNA vaccine technology transfer hub. 16 Nov 2021. https://medicinespatentpool.org/news-publications-post/mpp-position-statement-on-patents-with-regards-to-the-mrna-vaccine-technology-transfer-hub (accessed on 24 Nov 2023)

[15] BioNTech 2021. BioNTech plans to initiate the construction of an mRNA vaccine manufacturing facility in Africa in mid-2022. 26 Oct 2021. https://investors.biontech.de/news-releases/news-release-details/biontech-plans-initiate-construction-mrna-vaccine-manufacturing (accessed on 19 Nov 2023)

[16] Shadlen KC 2023. Technology Transfer for Production of COVID-19 Vaccines in Latin America. In Sun H, Sunder M (Eds.), Intellectual Property, COVID-19, and the Next Pandemic: Diagnosing Problems, Developing Cures, Cambridge University Press, Cambridge.

[17] Eccleston-Turner M, Upton H 2021. International collaboration to ensure equitable access to vaccines for COVID-19: the ACT-Accelerator and the COVAX facility. *Milbank Quarterly* 99(2):426–449.

[18] Beyer P 2012. Developing Socially Responsible Intellectual Property Licensing Policies: Non-Exclusive Licensing Initiatives in the Pharmaceutical Sector. In de Werra J (Ed.), Intellectual Property in the Pharmaceutical Industry, Zürich, Schulthess Verlag.

[19] Wong H 2020. The case for compulsory licensing during COVID-19. *J Glob Health* 10(1): 5–10.

[20] Possas C, de Souza Antunes AM, de Oliveira AM et al 2021. Vaccine innovation for pandemic preparedness: patent landscape, global sustainability, and circular bioeconomy in post-COVID-19 era. *Circ Econ Sust* 1: 1439–1461.

[21] World Health Organization, WHO 2022. WHO to Identify Pathogens that Could Cause Future Outbreaks and Pandemics. WHO News, Geneva, Nov 21 2022. www.who.int/news/item/21-11-2022-who-to-identify-pathogens-that-could-cause-future-outbreaks-and-pandemics (accessed on 24 Nov 2023)

[22] Grange ZL, Goldstein T, Johnson CK, Anthony S, Gilardi K, Daszak P et al. 2021. Ranking the risk of animal-to-human spillover for newly discovered viruses. *PNAS* 118: 14–20.

[23] Wouters OJ, Shadlen KC, Salcher-Konrad M, Pollard AJ, Larson HJ, Teerawattananon Y, Jit M 2021. Challenges in ensuring global access to COVID-19 vaccines: production, affordability, allocation, and deployment. *The Lancet* 397(10278): 1023–1034.

[24] Cutler DM, Summers LH 2020 The COVID-19 pandemic and the $16 trillion virus. *JAMA* 324(15):1495–149613.

[25] Possas C, Marques ET, Risi Jr JB, Homma A 2021. COVID-19 and future disease X in circular economy transition: redesigning pandemic preparedness to prevent a global di-saster. *Circ Econ Sustain* 1(4):1463–1478.

[26] Bajaj SS, Lwando M, Stanford F 2022. Vaccine apartheid: global cooperation and equity. *The Lancet* 399 (10334): 1452–1453.

[27] Possas C 1989. Epidemiologia e Sociedade: Heterogeneidade Estrutural e Saúde no Brasil. [Epidemiology and Society: Structural Heterogeneity and Health in Brazil]. HUCITEC SP, 271 pp.

[28] Lord KM, Turekian VC 2007. Time for a new era of science diplomacy. *Science*. 315(5813):769–770.

[29] Ruffini PB 2017. *Science and diplomacy. a new dimension ofinternational relations*. Springer, Berlin, p. 132 .

[30] Stiglitz J 2004 *Towards a Pro-Development and Balanced Intellectual Property Regime. Keynote address presented at the Ministerial Conference on Intellectual Property for Least Developed Countries*. World Intellectual Property Organization (WIPO).

[31] Love J 2006a. Drug development incentives to improve access to essential medicines. *Bull World Health Organ* 84(5): 20–25.

[32] Love J 2006b *Measures to Enhance Access to Medical Technologies, and New Methods of Stimulating Medical R&D*. Paper prepared for the WIPO Open Forum on the draft Substantive Patent Law Treaty (SPLT).

[33] Possas C, Martins RM, Homma A 2020. Vaccine Innovation and Global Sustainability. In Adenle A, Chertow M, Moors E Pannell D (Eds.), *Science Technology and Innovation for Sustainable Development Goals: Insights from Agriculture, Health, Environment and Energy*, Oxford University Press, Oxford, London, UK, Vol. 1, pp. 219–242.

[34] Mazzucato M, Li HL 2021. A market shaping approach for the biopharmaceutical industry: governing innovation towards the public interest. *J L Med Ethics* 49(1):39–49.

[35] Mazzucato M, Penna C (Eds.) 2015. *Mission-Oriented Finance for Innovation: New Ideas for Investment-Led Growth*. Rowman & Littlefield Publishers, Laham, Maryland.

5 Analysis of Factors Associated With Patent Quality in Tissue Engineering

Misael Linares-Bravo and Alejandro Barragán-Ocaña

5.1 INTRODUCTION

There are different stages within the process of technological development in the tissue engineering area that are necessary before clinical application. Its fundamental purpose is to create applications that allow for tissue regeneration or, where appropriate, tissue repair. Among these applications, the different aspects related to characterization, design, and conceptualization stand out, as well as their functional and preclinical evaluation, regulatory compliance, and adequate manufacturing controls [1,2]. In addition, three elements are essential for advancing tissue engineering: 1. cell source; 2. scaffolding development; and 3. growth factor as another critical element [3]. For example, to design functional tissue such as ligaments, it is necessary to create entire technological structures involving elements such as autologous cell management or highly specialized bioreactors [4]. This study area is relevant because advances in cell reprogramming make it possible for autologous cells to be considered for the design of applications aimed at tissue repair [5].

Thanks to the most recent advances, clinical-use products are already in the market, and many others will be introduced soon. However, first, it will be important to continue studying cell dynamics and construction materials in compliance with the most demanding technical and legal requirements for the development of biologics [6]. Although tissue engineering has been advancing and addressing different technical challenges, its final applications are still few. In addition, it is necessary to gradually migrate from the development stages to commercial stages that prove clinically effective and demonstrate their viability after a cost–benefit analysis [7,8]. Nevertheless, experimental inventions have a long path ahead compared to those reaching the market because profitability, quality, and compliance with complex regulations are necessary conditions for achieving this end [9,10]. Other relevant aspects of this field of knowledge are the need for collaboration between different

DOI: 10.1201/9781003405665-5

scientific disciplines, the obtainment of funding, and the progress in regulatory issues to face future challenges [11,12].

Tissue engineering has multiple application areas; among them, the following stand out: regeneration of bone tissue caused by trauma, aging, and diseases, although each case is different, and consequently, responses vary, and problems related to structure, mechanics, use of biomaterials, and the biology of the body itself must be considered [13,14]. Cardiac tissue applications represent another element of interest, although their progress depends on fulfilling different technical requirements. One consists of immunotolerant, autologous, or phenotypically stable cells that integrate cell populations, and those found natively in tissue should serve as a reference for matrix construction. In fact, tissue engineering has found an area of opportunity in cardiovascular tissue research focused on the replacement of blood vessels through various methodologies that contribute to the achievement of the biological functionality of these developments [15–17].

Undoubtedly, the potential of tissue engineering is relevant, and it will require considerable investment to introduce these inventions into the market and to make them available to all interested parties. For this reason, it is essential to protect these technologies through different intellectual property options, including patents. Derived from the above, we pose the following research questions: 1. What is the outlook of basic research in the area of tissue engineering? 2. What is its technological inventory as indicated by granted patents? 3. What quality factors are associated with this group of patents? Thus, the objectives of this study are, first, to carry out a network analysis based on bibliographic information from the Web of Science (WoS) to outline the main advances in basic research; second, to carry out a granted patent analysis to study different relevant indicators, such as the technological sectors and the main issues addressed in the area; and third, to appraise different factors associated with quality in these patents. A discussion of the analysis results, presented in the following sections of this chapter, provides reflections that shed light on the dynamics of tissue engineering, the potential of its technological inventory, and the different elements associated with quality.

5.2 TISSUE ENGINEERING AND PATENT ANALYSIS

Tissue engineering applications offer solutions oriented toward the generation of skin substitutes to benefit patients suffering from burns and other health conditions, although it is certainly essential to continue working on the cosmetic and functional challenges arising in the context of these developments [18]. Dentistry is a clinical area in which applications have been increasing; specifically, treatments for common ailments such as periodontal disease have been enhanced by the regeneration of oral tissues in very focused cases. In this area, the pending tasks are the reconstruction of complete teeth and the design of tissue replacement methods with greater reproducibility conditions [19]. The potential for applications is manifold, and the examples are numerous. Some areas under development with variable advances include genitourinary tissues [20] and bone and cartilage tissue [21–23]; craniofacial applications [24]; blood vessels [25]; liver regeneration [26]; and skeletal muscle tissue [27].

Concerning technological developments and tools to materialize tissue engineering applications, the advances, and challenges in bioinert and bioactive materials stand out [28,29], as well as hydrogels [30,31], bioreactors [32], bioink, and 3D printing [33]. To understand complexity, bibliometric and network approaches have been shown to help reveal basic research patterns and trends of knowledge production in areas such as stem cell research, in addition to other exciting fields such as clinical applications related to bone tissue regeneration [34,35]. For their part, patent studies are idoneous to understand the dynamics of an area of knowledge. Some examples of patent studies within tissue engineering, such as the case of hydrogels, reveal elements of interest, such as their market, production costs, and the progress of the techniques used in the area [36]. Another case of patent analysis that focuses on progress in this field of knowledge is studies on polymeric biomaterials useful in scaffold construction [37].

In addition, different studies in the area of biomedical applications showcase a wide array of tools to examine its progress, including coevolution studies (scientific articles-patents) that use networks, citations, actors, and other elements for the analysis [38]. This is relevant because the use of networks has demonstrated its worth in addressing issues that range from the social to the technical, thereby contributing to the management of their complexity [39]. 3D printing in the biomedical area is an interesting case; in this case, patent analysis has been used to identify and outline trends [40], and studies combining literature and patent reviews have served to sketch future research and development (R&D), for example, the search for potential biomedical applications based on brown algae [41]. Likewise, patent analysis work makes it possible to identify applications and the technological categories of the inventions, as has been done in the case of stem cell research for regenerative dentistry, revealing the limitations and advances in the field [42].

Additionally, patent analysis has been used to describe the evolution of technological applications and the historical vision of tissue engineering [43]. Other studies show that patent analyses and academic document analysis can draft the trajectories of different fields of knowledge in tissue engineering [44]. Likewise, other studies have shown how patent document analysis can shed light on innovation dynamics in cellulose nanomaterials, as well as the quantitative evolution of published documents and citations [45]. However, patent analysis goes beyond the status of knowledge in a specific technological area; for example, in fields as novel as 3D bioprinting, studying this form of intellectual property contextualizes its management system. Other conjunctural issues must also be studied, including the scope of technological management, especially intellectual property, and the role of the government and industrial dynamics, as happens in the field of biotechnology, as well as clear bioethical guidelines, to allow for this emerging area of knowledge to yield results [46]. Given its usefulness for the industrial and commercial sectors, patent documents are a pertinent information source representing a fundamental input for economic progress [47]. Additional elements, such as regulatory policies, are fundamental for the generation and market placement of innovative technology to create the necessary scaffolding for its operation through the articulation of laws and technical standards, among other elements [48]. In this context, technological convergence will be necessary to generate radical innovations; an example of this will be the challenges of tissue engineering at the nanometric level [49].

On the other hand, patent portfolios represent a valuable object of study to understand R&D and its potential commercial use, the characterization and quality of the technology generated by companies, its international and technological scope, and its eventual commercial use [50]. Due to the importance of patents as precursors of the innovation process, their quality becomes a central element for analysis because such quality tends to vary [51]. In this regard, when discussing factors associated with patent quality, different indicators come into play, such as the ratio between patent applications and granted patents, citations received, and technological and international scope [52,53], as well as the role of collaboration [54]. Thus, the quality of the information sources consulted is essential, in addition to other elements, such as the size of the patent families and the analysis of the patent claims [55].

5.3 METHODOLOGY

The methodological proposal consists, first, of carrying out a patent analysis for the area of tissue engineering to analyze the main indicators concerning jurisdictions, applicants, and technological classes, in addition to a network analysis to identify the leading technologies and other elements in this area of knowledge. A second objective is to evaluate the quality and strength of the patent portfolios associated with the leading companies and individuals. To carry out these objectives, we carried out a patent search in the Lens database [56] using the following query: [Title: ("tissue engineering") OR (Abstract: ("tissue engineering") OR Claims: ("tissue engineering")], in which two filters were applied to obtain, first, only the granted patents, and second, granted patents within the period from 2013 to 2022. A total of 1,301 granted patents were obtained, which were grouped into 913 simple patent families and 889 extended patent families. With this information, we obtained the first ten positions regarding indicators related to jurisdiction, applicants, and technological sectors, according to IPC classifications. The network analysis was based on text data; the criteria to carry out this exercise were the terms from the title and abstract fields, and the counting method was full counting with a minimum occurrence of ten.

Thus, 6,056 terms were identified, of which 262 met this criterion. Additionally, only 60% of the most relevant items were extracted, so only 157 words were used to create the network, resulting in a network comprising 12 clusters, 157 nodes, and 2,044 links with a total link strength of 19,701. In a second analysis level, based on the database obtained from Lens, we evaluated patent quality by considering the companies and individuals with the highest number of granted patents. Their patent portfolios were assessed using the following criteria: 1. validity: defined by the number of active patents divided by the number of granted patents; 2. international scope: defined by the average size of simple patent families; 3. technological scope: defined by the average of IPC technological classes in the analyzed portfolio; and 4. citation frequency: defined by the average number of citations received by all the patents in a portfolio. Subsequently, these indicators were added up for each of the 30 identified portfolios; this result was considered as the value of the quality of each portfolio. With this information, the scores were ordered in descending order, and quartiles were determined for this data group. For classification purposes, the highest scores were placed within the Q1 and Q2 quartiles and the lowest within the Q3 and Q4 quartiles, thus generating an adequate metric to provide a level and grouping criteria for the scores obtained.

It is important to point out that this indicator is only a starting point for a discussion around the concept of patent quality, and it cannot be interpreted as an absolute indicator since, in principle, all patents have undergone an examination. In other words, our purpose was to show how, based on the information obtained, organizations or individuals have made progress in the quality of their patents according to the previously described indicators; at the same time, this is also an indication of the success that an invention can have in the market. Subsequently, to determine their innovative potential, we calculated the general strength of the patent portfolio by multiplying the results of the patent portfolio quality based on the patenting activity, that is, the number of granted patents. With these two indicators, we carried out a cluster analysis using the furthest neighbor method with standardized data to identify the patterns of the 30 analyzed portfolios; four clusters were thus generated based on similarity. This test was important because the highest-quality portfolio is not necessarily the one with the highest number of patents; therefore, an organization or individual with a large number of high-quality patents will have a greater chance of generating a significant market share.

5.4 DISCUSSION AND ANALYSIS OF RESULTS

Concerning the jurisdictions where these technologies are being protected, the most dominant technology markets appear in the first two places. The case of China and its blunt incursion into the protection of cutting-edge technology also stands out, as well as other developing countries that show significant participation. However, when analyzing the principal applicants, it is possible to observe a strong presence of the academy, which could be explained by the somewhat experimental nature of this technological field. Derived from this, it will be necessary to study the role of the companies that venture into this technological field and, in the specific case of the academy, address its participation not only in terms of its contribution to the construction of technological progress in tissue engineering but also in terms of the different technological management mechanisms and regulatory policies intended for these inventions to reach the market for the benefit of end users. Among the most important points are those referring to technology transfer, where the context of each economy must be considered, especially for economies located in developing countries (see Table 5.1).

Regarding technological classes, a strong concentration is observed in topics related to materials development, some of them, for example, for prosthetic purposes; the largest number of classes are subclassifications of class A61L 27/00, with the exception of C12N5/00, including inventions related to fundamental elements for the advancement of tissue engineering, such as biologically active materials and hydrogels. The C12N5/00 classification is relevant because it integrates cell line and tissue development and the means to generate or maintain them (Table 5.2). However, the concentration of technological classes can be seen more clearly within the first positions, thereby helping to identify technological trajectories; it is also possible to

TABLE 5.1
Main Jurisdictions and Applicants

No.	Jurisdiction	Patents	Applicants	Patents
1	United States	666	Tepha INC	47
2	European Patents	234	Centre Nat Rech Scient	18
3	China	114	Massachusetts Inst Technology	17
4	Russia	76	Ethicon INC	15
5	Canada	63	Univ Columbia	14
6	Korea, Republic of	52	Univ California	13
7	Australia	51	Agency Science Tech & Res	12
8	Czech Republic	10	Momentive Performance Mat INC	12
9	Mexico	5	Tufts College	12
10	Taiwan	5	Bezwada Biomedical LLC	11

Source: Elaborated by the authors based on Lens [56].

TABLE 5.2
IPC Technological Classes

No.	IPC Code	Description	Patents
1	A61L27/38	"**A61L 27/00** Materials for prostheses or for coating prostheses (dental prostheses A61C 13/00; shape or structure of prostheses A61F 2/00; use of preparations for artificial teeth A61K 6/80; artificial kidneys A61M 1/14) … A61L 27/38 Animal cells (for use in artificial skin A61L 27/60)".	223
2	A61L27/56	"Porous or cellular materials…".	206
3	A61L27/18	"Obtained otherwise than by reactions only involving carbon-to-carbon unsaturated bonds…".	176
4	A61L27/54	"Biologically active materials, e.g., therapeutic substances…".	169
5	C12N5/00	"Undifferentiated human, animal, or plant cells, e.g., cell lines; tissues; cultivation or maintenance thereof; culture media therefor (plant reproduction by tissue culture techniques A01H 4/00) …".	152
6	A61L27/52	"Hydrogels or hydrocolloids…".	126
7	A61L27/22	"Polypeptides or derivatives thereof…".	124
8	A61L27/58	"Materials at least partially resorbable by the body…".	115
9	A61L27/36	"Containing ingredients of undetermined constitution or reaction products thereof …".	113
10	A61L27/20	"Polysaccharides …".	106

Source: Elaborated by the authors based on Lens [56] and WIPO [57].

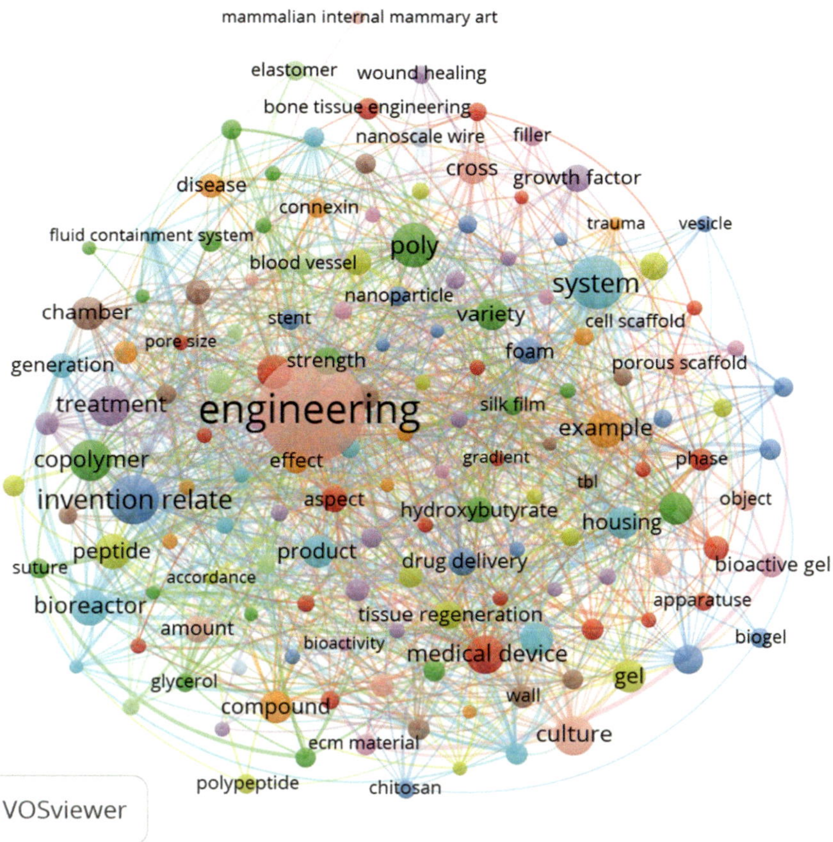

FIGURE 5.1 Network analysis (co-occurrence). (Elaborated by the authors based on Lens [56] and VOSviewer [58].)

evaluate the evolution of other technological sectors that, in some cases, will become more important over time.

On the other hand, when carrying out the network analysis, the results show a significant number of clusters and nodes that allow for a global view of the associated topics (Figure 5.1). A network analysis at the cluster level shows that cluster 1 integrates elements such as medical devices and equipment, as well as items related to artificial tissue and bone tissue and other fundamental topics related to the development of the R&D process, such as angiogenesis or the pore size necessary for scaffold construction. Cluster 2 is clearly associated with multiple materials used in tissue engineering. When the network is analyzed at the node level, the "engineering" node stands out due to its size (occurrence), which is located in cluster 10 with 563 occurrences, 141 links, and a total link strength of 2,782; independently from the size of the nodes to which it is linked, we can find topics as relevant as compounds, materials, medical devices, bioreactors, and many other elements necessary to consolidate this emerging area (for more detail on the network's characteristics, see Annex I).

Regarding the quality of the portfolios, measured as a function of validity, international and technological scope, and citation frequency, the best results are found in the Q1 and Q2 quartiles. These groups of actors are mainly made up of companies (seven), followed closely by academic institutions (five), one medical institution, and one individual. For their part, in the Q3 and Q4 quartiles, the conformation of these groups is characterized by a clear dominance of academia (including one science, technology, and innovation agency) (eleven) and, to a lesser extent, by industry (five). In fact, the first four places in this ranking are occupied by companies that outperform the rest of the other portfolios (Table 5.3). Although an academic entity and even an individual can be positioned within the first quartiles, a possible

TABLE 5.3
Quality of Patent Portfolios by Applicant

Applicants	Patent Quality	Quartile
Polarityte Inc	54.13	1
Acell Inc	48.25	1
Octane Biotech Inc	41.40	1
Bio2 Technologies Inc	40.71	1
University of Missouri	26.80	1
Childrens Medical Center	25.43	1
Contipro Biotech sro	24.86	1
Adhikari Raju	23.71	2
Polynovo Biomaterials Pty Ltd	22.64	2
Ethicon Inc	22.07	2
Tufts College	21.25	2
Institut National de la Sante et de la Recherche Medicale	21.00	2
Massachusetts Institute of Technology	20.88	2
Drexel University	20.00	2
Contipro As	19.75	2
Momentive Performance Materials Inc	19.67	3
Centre National de la Recherche Scientifique	19.11	3
Agency for Science, Technology and Research	16.50	3
Columbia University	16.43	3
Tepha Inc	16.26	3
Bezwada Biomedical Llc	15.36	3
Northwestern University	15.13	3
University of California	14.62	3
The University of Texas	14.44	4
Harvard College	14.00	4
Nanyang Technological University	13.25	4
The University of Akron	11.75	4
University of Pittsburgh—of the Commonwealth System of Higher Education	10.30	4
Empire Technology Development Llc	8.88	4
Fourth Military Medical University	4.50	4

Source: Elaborated by the authors based on Lens [56] and R [59].

explanation in the domain of enterprises is their market experience and knowledge of its customers' needs.

Regarding portfolio strength, evidence shows that measuring the quality of a patent portfolio is insufficient since strength must be associated with the patenting activity to measure its innovative potential more accurately. Thus, it is worth mentioning the case of Tepha, a company located in the Q3 quartile of the previous table, which now occupies the first position, and those of Polarityte, Octane Biotech, and Acell, companies that maintained their position very close to what it was in the previous indicator. These results allow us to further strengthen the hypothesis about a company's advantage in the field of innovation, although regardless of positions, the academy is also beginning to produce exciting outcomes (Table 5.4).

TABLE 5.4
Portfolio Strength

Applicants	Portfolio Strength
Tepha Inc	764.00
Polarityte Inc	433.00
Octane Biotech Inc	414.00
Acell Inc	386.00
Massachusetts Institute of Technology	355.00
Centre Nationale de la Recherche Scientifique	344.00
Ethicon Inc	331.00
Bio2 Technologies Inc	285.00
University of Missouri	268.00
Tufts College	255.00
Polynovo Biomaterials Pty Ltd	249.00
Momentive Performance Materials Inc	236.00
University of Columbia	230.00
Institut National de la Sante et de la Recherche Medicale	210.00
Agency for Science, Technology and Research	198.00
University of California	190.00
Childrens Medical Center	178.00
Contipro Biotech sro	174.00
Bezwada Biomedical Llc	169.00
Adhikari Raju	166.00
Drexel University	160.00
Contipro As	158.00
The University of Texas	130.00
Harvard College	126.00
Northwestern University	121.00
Nanyang Technological University	106.00
University of Pittsburgh—of the Commonwealth System of Higher Education	103.00
The University of Akron	94.00
Empire Technology Development Llc	71.00
Fourth Military Medical University	36.00

Source: Elaborated by the authors based on Lens [56].

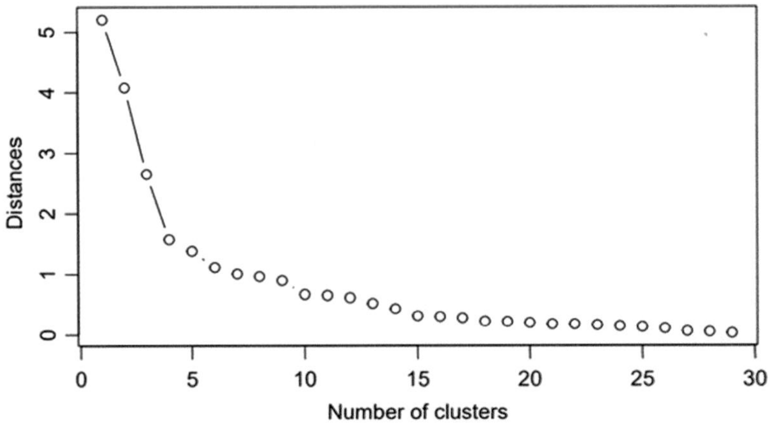

FIGURE 5.2 Sedimentation chart. (Elaborated by the authors using R [59].)

When carrying out the sedimentation analysis, Figure 5.2 shows that, according to the inflection point, the adequate number of clusters to carry out the cluster analysis using the furthest neighbor method was four.

Figure 5.3 shows that, concerning the quality and strength of the portfolios, Thepa formed a cluster of its own. To some extent, this might be because the company showed a value close to twice that of its closest competitor in the strength indicator. Next to this cluster is the one including Acell, Polarityte, Octane Biotech, and Bio2 Technologies, the companies that occupied the first four positions in the portfolio quality indicator, and all are within the first ten positions in the portfolio strength indicator. However, in the largest clusters, the participation of academia and research institutions is a common element.

5.5 CONCLUSIONS

Tissue engineering is an emerging scientific and technological field that has vast potential to provide novel solutions. However, many of these advances are still in experimental phases, and the investment of large sums of capital and considerable research efforts will be needed to bring these inventions to the market. Consequently, the expected clinical applications must comply with robust regulatory frameworks and safety tests that guarantee proper use in medical practice. In addition, the management of these technologies will be essential to take all the necessary steps for this group of inventions to effectively become innovations.

The results obtained in the present study allow for the conclusion that the main technological developments are being generated in the United States and Europe, although China is making significant efforts and is one of the most important players in this technological field. The discipline of tissue engineering is characterized by considerable participation of the academy and some health institutions, although companies will play an important role. Regarding technological classes, this study shows that advances in biomaterials, hydrogels, cell lines, and tissues, among others,

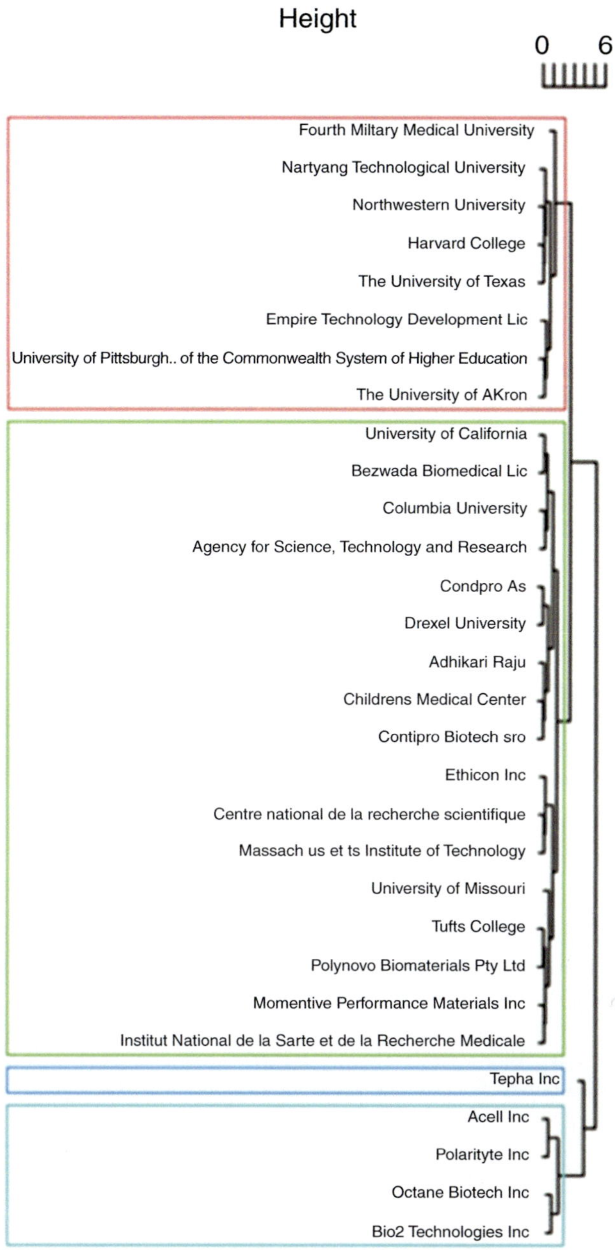

FIGURE 5.3 Furthest neighbor method. (Elaborated by the authors using R [59].)

are helping to consolidate this area of knowledge. Network analysis points out other fundamental technologies for the development of tissue engineering, such as medical devices and equipment, artificial and bone tissues, and others related to scaffold construction.

Concerning the quality of the patent portfolios, since all these technologies have undergone formal and comprehensive examination, their technical compliance, novelty, and industrial application are guaranteed, which helps to obtain their legal protection. However, this study measured the quality of these portfolios based on validity, international and technological scope, and citation frequency to assess the potential of these portfolios to become innovations. Additionally, the study provides a metric to measure the strength of these portfolios. In this regard, we found that the quality and strength of the portfolios are not always equivalent, so it is possible to have a high-quality portfolio with limited strength. However, over academia and medical institutions, the first three positions in these two indicators were firms, which highlights the disadvantages that those other institutions have in order to make their inventions visible and introduce them into the market. Based on the conglomerate analysis, firms coalesced together, forming two groups. The remaining two groups are a mix of firms and academic and medical institutions.

ACKNOWLEDGMENTS

We wish to acknowledge the support provided by the National Polytechnic Institute (Instituto Politécnico Nacional) and the Secretariat for Research and Postgraduate Studies (Secretariat for Research and Postgraduate Studies), grant numbers 20220729, 20230213 and 20240867.

REFERENCES

[1] Goldstein, S. 2002. Tissue engineering. *Annals of the New York Academy of Sciences* 961, no. 1:183–192.
[2] Patrick, C.W. 2001. Tissue engineering strategies for adipose tissue repair. *The Anatomical Record* 263, no. 4:361–366.
[3] Ikada, Y. 2006. Challenges in tissue engineering. *Journal of The Royal Society Interface* 3, no. 10:589–601.
[4] Vunjak-Novakovic, G., Altman, G., Horan, R., and D.L. Kaplan. 2004. Tissue engineering of ligaments. *Annual Review of Biomedical Engineering* 6, no. 1:131–156.
[5] Khademhosseini, A., and R. Langer. 2016. A decade of progress in tissue engineering. *Nature Protocols* 11, no. 10:1775–1781.
[6] Khademhosseini, A., Vacanti, J., and R. Langer. 2009. Progress in tissue engineering. *Scientific American* 300, no. 5:64–71. doi: 10.1038/scientificamerican0509-64.
[7] Rouwkema, J., Rivron, N., and C. Van-Blitterswijk. 2008. Vascularization in tissue engineering. *Trends in Biotechnology* 26, no. 8:434–441. doi: 10.1016/j.tibtech.2008.04.009.
[8] Archer, R., and D. Williams. 2005. Why tissue engineering needs process engineering. *Nature Biotechnology* 23, no. 11:1353–1355. doi: 10.1038/nbt1105-1353.
[9] Mansbridge, J. 2006. Commercial considerations in tissue engineering. *Journal of Anatomy* 209, no. 4:527–532.

[10] Sanzenbacher, R., Dwenger, A., Schuessler-Lenz, M., Cichutek, K., and E. Flory. 2007. European regulation tackles tissue engineering. *Nature Biotechnology* 25, no. 10:1089–1091.

[11] Chapekar, M. 2000. Tissue engineering: Challenges and opportunities. *Journal of Biomedical Materials Research* 53, no. 6:617–620. doi: 10.1002/1097-4636(2000)53:6<617::aid-jbm1>3.0.co;2-c.

[12] Singh, H.B., Jha, A., Keswani, C. 2016. Intellectual property issues in biotechnology. CABI, Wallingford, UK, p. 304.

[13] Black, C.R., Goriainov, V., Gibbs, D., Kanczler, J., Tare, R.S., and R.O. Oreffo. 2015. Bone tissue engineering. *Current Molecular Biology Reports* 1, no. 3:132–140.

[14] Healy, K.E., and R.E. Guldberg. 2007. Bone tissue engineering. *Journal of Musculoskeletal & Neuronal Interactions* 7, no. 4:328–330.

[15] Vunjak-Novakovic, G., Tandon, N., Godier, A., Maidhof, R., Marsano, A., Martens, T.P., and M. Radisic. 2010. Challenges in cardiac tissue engineering. *Tissue Engineering Part B: Reviews* 16, no. 2:169–187.

[16] Nerem, R., and D. Seliktar. 2001. Vascular tissue engineering. *Annual Review of Biomedical Engineering* 3, no. 1:225–243. doi: 10.1146/annurev.bioeng.3.1.225.

[17] Baguneid, M.S., Seifalian, A.M., Salacinski, H.J., Murray, D., Hamilton, G., and M.G. Walker. 2006. Tissue engineering of blood vessels. *British Journal of Surgery* 93, no. 3:282–290.

[18] Böttcher-Haberzeth, S., Biedermann, T., and E. Reichmann. 2010. Tissue engineering of skin. *Burns* 36, no. 4:450–460.

[19] Abou-Neel, E.A., Chrzanowski, W., Salih, V.M., Kim, H.W., and J.C Knowles. 2014. Tissue engineering in dentistry. *Journal of Dentistry* 42, no. 8:915–928.

[20] Kim, B., Baez, C.E., and A. Atala. 2000. Biomaterials for tissue engineering. *World Journal of Urology* 18, no. 1:2–9.

[21] Meijer, G.J., De-Bruijn, J.D., Koole, R., and C.A. Van Blitterswijk. 2007. Cell-based bone tissue engineering. *PLoS Medicine* 4, no. 2:e9.

[22] Chang, C., Lin, F., Kuo, T., and H. Liu. 2005. Cartilage tissue engineering. *Biomedical Engineering: Applications, Basis and Communications* 17, no. 2:61–71. doi: 10.4015/s101623720500010x.

[23] Huang, B.J., Hu, J.C., and K.A. Athanasiou. 2016. Cell-based tissue engineering strategies used in the clinical repair of articular cartilage. *Biomaterials* 98:1–22.

[24] Alsberg, E., Hill, E.E., and D.J. Mooney. 2001. Craniofacial tissue engineering. *Critical Reviews in Oral Biology & Medicine* 12, no. 1:64–75.

[25] Zhang, W.J., Liu, W., Cui, L., and Y. Cao. 2007. Tissue engineering of blood vessel. *Journal of Cellular and Molecular Medicine* 11, no. 5:945–957.

[26] Hosseini, V., Maroufi, N.F., Saghati, S., Asadi, N., Darabi, M., Ahmad, S.N., Hosseinkhani, H., and R. Rahbarghazi. 2019. Current progress in hepatic tissue regeneration by tissue engineering. *Journal of Translational Medicine* 17, no. 1:383.

[27] Bach, A.D., Beier, J.P., Stern-Staeter, J., and R.E. Horch. 2004. Skeletal muscle tissue engineering. *Journal of Cellular and Molecular Medicine* 8, no. 4:413–422.

[28] Stevens, M.M. 2008. Biomaterials for bone tissue engineering. *Materials Today* 11, no. 5:18–25.

[29] Rahaman, M.N., Day, D.E., Sonny-Bal, B., Fu, Q., Jung, S.B., Bonewald, L.F., and A.P. Tomsia. 2011. Bioactive glass in tissue engineering. *Acta Biomaterialia* 7, no. 6:2355–2373.

[30] Khademhosseini, A., and R. Langer. 2007. Microengineered hydrogels for tissue engineering. *Biomaterials* 28, no. 34:5087–5092.

[31] Naahidi, S., Jafari, M., Logan, M., Wang, Y., Yuan, Y., Bae, H., Dixon, B., and P. Chen. 2017. Biocompatibility of hydrogel-based scaffolds for tissue engineering applications. *Biotechnology Advances* 35, no. 5: 530–544.

[32] Chen, H.C., and Y.-C. Hu. 2006. Bioreactors for tissue engineering. *Biotechnology Letters* 28, no. 18:1415–1423.

[33] Haleem, A., Javaid, M., Khan, R.H., and R. Suman. 2020. 3D printing applications in bone tissue engineering. *Journal of Clinical Orthopedics and Trauma* 11:S118–S124.

[34] Li, L.M., Ding, G., Feng, N., Wang, M.H., and Y.S. Ho. 2006. Global stem cell research trend: Bibliometric analysis as a tool for mapping of trends from 1991 to 2006. *Scientometrics* 80, no. 1:39–58.

[35] Huang, X., Liu, X., Shang, Y., Qiao, F., and G. Chen. 2020. Current trends in research on bone regeneration: A Bibliometric analysis. *BioMed Research International* 2020:1–12.

[36] Caló, E., and V.V. Khutoryanskiy. 2015. Biomedical applications of Hydrogels: A review of Patents and commercial products. *European Polymer Journal* 65:252–267.

[37] Li, X., Liu, X., Yu, Y., Qu, X., Feng, Q., Cui, F., and F. Watari. 2009. Recent patents on polymeric scaffolds for tissue engineering. *Recent Patents on Biomedical Engineering* 2, no. 1: 65–72.

[38] Murray, F., 2002. Innovation as coevolution of scientific and technological networks: Exploring tissue engineering. *Research Policy* 31, no. 8–9:1389–1403.

[39] Karimi, F., Green, D., Matous, P., Varvarigos, M., and K.R. Khalilpour. 2021. Network of networks: A bibliometric analysis. *Physica D: Nonlinear Phenomena* 421:132889.

[40] Trappey, A.J., Trappey, C.V., and K.L. Lee. 2016. Tracing the evolution of biomedical 3D printing technology using ontology-based patent concept analysis. *Technology Analysis & Strategic Management* 29, no. 4:339–352.

[41] Kim, T.H., and W.K., Jung. 2019. R&D trends of brown algae as potential candidates in biomedical application. *Journal of Marine Bioscience & Biotechnology* 11, no. 1:1–13.

[42] Morsczeck, C., Frerich, B., and O. Driemel. 2009. Dental stem cell patents. *Recent Patents on DNA & Gene Sequences* 3, no. 1:39–43.

[43] Pabst, P.L. 2003. Tissue engineering: A historical review as seen through the US Patent Office. *Expert Opinion on Therapeutic Patents* 13, no. 3:347–352.

[44] Coccia, M. 2014. Emerging technological trajectories of tissue engineering and the critical directions in cartilage regenerative medicine. *International Journal of Healthcare Technology and Management* 14, no. 3:194.

[45] Charreau, H., Cavallo, E., and M.L. Foresti. 2020. Patents involving nanocellulose: Analysis of their evolution since 2010. *Carbohydrate Polymers* 237:116039.

[46] Li, P.H. 2014. 3D bioprinting technologies: Patents, innovation and access. *Law, Innovation and Technology* 6, no. 2:282–304.

[47] Tseng, Y.H., Lin, C.J., and Y.I. Lin. 2007. Text mining techniques for patent analysis. *Information Processing & Management* 43, no. 5:1216–1247.

[48] Faulkner, A. 2009. Regulatory policy as innovation: Constructing rules of engagement for a technological zone of tissue engineering in the European Union. *Research Policy* 38, no. 4:637–646.

[49] Maine, E., Thomas, V.J., and J. Utterback. 2014. Radical Innovation from the confluence of technologies: Innovation Management Strategies for the emerging nanobiotechnology industry. *Journal of Engineering and Technology Management* 32:1–25.

[50] Fabry, B., Ernst, H., Langholz, J., and M. Köster. 2006. Patent portfolio analysis as a useful tool for identifying R&D and business opportunities – An empirical application in the nutrition and health industry. *World Patent Information* 28, no. 3:15–225.
[51] Higham, K., De-Rassenfosse, G., and A.B. Jaffe. 2021. Patent quality: Toward a systematic framework for analysis and measurement. *Research Policy* 50, no. 4:1–14.
[52] Ernst, H. 2003. Patent information for strategic technology management. *World Patent Information* 25, no. 3:233–242.
[53] Wu, J.L., Chang, P.C., Tsao, C.C., and C.Y. Fand. 2016. A patent quality analysis and classification system using self-organizing maps with support vector machine. *Applied Soft Computing* 41:305–316, 2016.
[54] Jiang, R., Jefferson, G.H., Zucker, S., and L. Li. 2019. The role of research and ownership collaboration in generating patent quality: China-U.S comparisons. *China Economic Review* 58:1–14, 2019.
[55] Schettino, F., Sterlacchini, A., and F. Venturini. 2019. Inventive productivity and patent quality: Evidence from Italian inventors. *Journal of Policy Modeling* 35, no. 6:1043–1056.
[56] Lens 2023. Search, analyze and manage patent and scholarly data. Available in: www.lens.org/ (accessed February 20, 2023).
[57] WIPO 2023. International Patent Classification (IPC). Available in: www.wipo.int/classifications/ipc/en/ (accessed February 20, 2023).
[58] VOSviewer 2022. Vosviewer (Software). Version 1.6.18.
[59] R Core Team (2023). _R Core Team: A Language and Environment for Statistical Computing_. Foundation for Statistical Computing, Vienna, Austria. www.R-project.org/

Annex I

Cluster 1				Cluster 2			
Ítems				Ítems			
	Links	Total link strength	Occurrences		Links	Total link strength	Occurrences
angiogenesis	18	63	13	active agent	33	144	13
apparatuse	20	89	17	biocompatible material	19	67	11
artificial tissue	18	108	14	composite scaffold	13	103	24
aspect	40	431	34	copolymer	40	446	89
bio block	6	98	14	dimensional structure	17	56	10
body	36	209	32	film	54	432	54
bone repair	23	118	19	glycerol	17	271	17
bone tissue engineering	25	75	23	hydroxybutyrate	15	267	34
channel	37	223	25	medical application	31	173	22
computer	14	93	14	p4hb	11	131	12
differentiation	61	525	60	poly	42	581	101
gradient	25	111	14	self	16	202	20
medical device	47	501	75	sheet	17	128	24
metal	14	75	12	silk fibroin	19	100	12
microsphere	14	84	19	silk film	11	294	18
organoid	6	100	11	strength	58	299	41
phage	6	112	13	suture	21	217	20
pharmaceutical composition	26	90	12	thermoform	7	124	10
phase	23	285	29	variety	58	448	53
polymeric region	7	219	15	w w	19	105	11
pore size	14	64	14				
soft tissue	13	105	14				

Cluster 3				Cluster 4			
Ítems				Ítems			
	Links	Total link strength	Occurrences		Links	Total link strength	Occurrences
absorbable polymer	10	203	22	accordance	15	112	11
biogel	9	96	20	amphiphilic peptide	18	202	19
biomedical application	22	72	19	biocompatible polymer	18	103	16
chain extender	6	57	14	drug	52	269	40
chitosan	19	160	19	fibril	6	105	25
composite material	20	205	24	gel	45	428	53
cosmetic	16	54	10	peptide	38	341	65
derivative	37	299	45	peptoid	11	153	13
discovery	24	138	17	polypeptide	12	77	22
drug delivery	67	397	41	portion	30	293	35
epidermal growth factor	8	78	10	ribbon	4	120	27
foam	30	333	34	self assembling	20	129	11
invention relate	101	910	126	tissue regeneration	52	323	38
nanoparticle	31	123	20	tissue replacement	11	95	18
stent	30	184	25	wound dressing	33	141	14
stent coating	12	94	10				
tannin	9	166	14				
vesicle	8	102	15				

Cluster 5				Cluster 6			
Ítems				Ítems			
	Links	Total link strength	Occurrences		Links	Total link strength	Occurrences
binding	13	71	12	bioreactor	39	1302	71
data	19	145	12	fluid communication	18	249	11
domain	15	153	18	fluid containment system	16	468	16
fragment	21	91	20	generation	32	530	32
growth factor	32	213	39	housing	24	1006	55
nanoparticle probe	3	36	12	microprocessor	18	318	12
patient	20	80	16	physiological cellular function	16	468	16
pemf device	6	150	15	product	69	656	56
proliferation	49	392	28	production	61	491	56
protein complex	7	150	15	sensor	22	615	25
replacement	43	187	27	system	92	1136	143
tissue engineering device	14	100	13	tissue construct	29	516	21
treatment	62	511	83	tissue engineering system	23	265	20
wound healing	21	95	19				

Cluster 7				Cluster 8			
Ítems				Ítems			
	Links	Total link strength	Occurrences		Links	Total link strength	Occurrences
activity	39	204	29	chamber	35	607	61
biomolecule	11	35	14	dwg	53	295	30
cell adhesion	21	86	14	fixation	39	237	20
compound	46	345	56	hour	16	87	11
connexin	12	205	21	medicine substance	40	169	20
disease	34	201	33	point	36	349	24
disorder	16	141	18	sample	32	207	22
dual functional nonfouling surface	5	60	12	tbl	37	144	11
effect	54	303	33	tissue engineering structure	34	430	30
example	60	455	71	viability	29	265	19
regenerative medicine application	9	20	11	volume	32	178	17
severity	10	116	11	wall	32	283	28
trauma	14	116	11				

Cluster 9				Cluster 10			
Ítems				Ítems			
	Links	Total link strength	Occurrences		Links	Total link strength	Occurrences
advantage	16	137	13	amount	40	382	39
bioactive gel	7	288	32	aqueous solution	40	298	26
bioactivity	16	110	11	cell delivery	9	162	10
ecm material	7	240	24	cross	31	374	55
fibroin	4	48	12	culture	70	722	79
filler	30	149	20	engineering	141	2782	563
hyaluronate fiber	4	32	10	mammalian internal mammary artery	1	10	10
manufacturing method	25	134	16	object	20	155	25
pressure	39	155	19	polysaccharide	24	271	29
tissue reconstruction	10	117	11	porous scaffold	13	259	23

Source: elaborated by the authors based on Lens [56].

Cluster 11				Cluster 12			
Ítems				Ítems			
	Links	Total link strength	Occurrences		Links	Total link strength	Occurrences
blood vessel	25	119	24	cell scaffold	6	240	20
dimension	31	235	19	electronic circuit	6	144	10
dimensional body	5	156	12	nanoscale wire	6	240	20
elastomer	7	92	21	presence	24	104	13
location	25	154	15				
shape	41	233	29				

Applicants	Number of patents granted	relation of active/grant patents	average citation frequency	average technological scope	average family size	Patent quality	Portfolio strength
Tepha INC	47	1.000	2.170	2.383	10.702	16.255	764
Centre Nat Rech Scient	18	0.944	0.222	5.111	12.833	19.111	344
Massachusetts Inst Technology	17	0.882	7.471	4.294	8.235	20.882	355
Ethicon INC	15	0.933	2.333	4.000	14.800	22.067	331
Univ Columbia	14	0.571	5.071	2.214	8.571	16.429	230
Univ California	13	0.923	1.462	7.154	5.077	14.615	190
Agency Science Tech & Res	12	0.750	2.167	3.750	9.833	16.500	198
Momentive Performance Mat INC	12	1.000	2.667	6.000	10.000	19.667	236
Tufts College	12	1.000	3.167	4.167	12.917	21.250	255
Bezwada Biomedical LLC	11	1.000	2.727	3.636	8.000	15.364	169
Polynovo Biomaterials Pty LTD	11	1.000	0.545	3.182	17.909	22.636	249
Inst Nat Sante Rech Med	10	0.900	0.400	5.100	14.600	21.000	210
Octane Biotech INC	10	0.900	3.600	3.100	33.800	41.400	414
Univ Missouri	10	0.700	3.500	2.000	20.600	26.800	268
Univ of Pittsburgh—of the Commonwealth System of Higher Education	10	0.900	1.100	2.600	5.700	10.300	103
Harvard College	9	1.000	0.778	4.889	7.333	14.000	126
Univ Texas	9	1.000	2.556	3.667	7.222	14.444	130
Acell INC	8	1.000	5.500	1.750	40.000	48.250	386
Contipro As	8	1.000	0.000	3.500	15.250	19.750	158
Empire Technology Dev LLC	8	0.125	0.375	4.125	4.250	8.875	71
Fourth Military Medical Univ	8	0.375	0.125	1.875	2.125	4.500	36
Polarityte INC	8	1.000	0.000	1.125	52.000	54.125	433
Univ Akron	8	0.875	0.375	4.250	6.250	11.750	94
Univ Drexel	8	0.625	10.000	4.000	5.375	20.000	160

Univ Nanyang Tech	8	0.750	2.500	3.875	6.125	13.250	106
Univ Northwestern	8	1.000	2.375	6.375	5.375	15.125	121
Adhikari Raju	7	1.000	1.429	4.000	17.286	23.714	166
Bio2 Technologies INC	7	1.000	1.429	1.857	36.429	40.714	285
Childrens Medical Center	7	1.000	0.714	3.857	19.857	25.429	178
Contipro Biotech Sro	7	0.714	4.571	2.714	16.857	24.857	174

Source: Elaborated by the authors based on Lens [56].

6 Vaccine Preparedness for Highly Pathogenic Avian Influenza

Innovation and Intellectual Property Issues

Cristina Possas, Ernesto TA Marques, Alessandra Oliveira, Suzanne Schumacher, Flavia Maria Lins Mendes, Marilda M Siqueira, Adelaide Antunes, and Akira Homma

6.1 INTRODUCTION

The potential of the highly pathogenic avian influenza (HPAI) pandemic is a critical issue. Outbreaks of the H5N1 avian flu, causing mass die-offs in poultry and other birds, are now affecting mammals from otters to dolphins worldwide. H5N1 and other HPAI viruses have thus far caused rare human infections. However, if they evolve to spread between people, they could trigger major outbreaks. The main question is how to overcome bottlenecks in HPAI vaccine development to accelerate scale-up and implement an effective global vaccine preparedness system. From a patent landscape approach, we identified breakthroughs and gaps in HPAI vaccine development for the search period (2010–2022), discussing technological strategies in patent filings, including universal vaccine patent documents. Our results indicated that patents are a reliable indicator of vaccine innovation: despite recent breakthroughs in HPAI vaccine development, progress has been very slow. From the retrieved patent documents for human HPAI vaccines (683 patent filings), 71 were chosen from the main applicants in the search period. Of them, 31 were in active status (43.7%), with few breakthroughs (15 documents – 21.1%) and in early stages of development. These results indicate a technological lag in HPAI vaccine development and highlight the constraints for scaling up these vaccines when they will be needed. Current egg-based processes are laborious and too difficult to scale-up in the case of a pandemic. Concerning mRNA technology there are still issues regarding the duration and level of protection that mRNA vaccines might induce. Vaccinating farm animals is not sufficient to protect the population: if not properly executed, the virus could continue to circulate at a low level, increasing the chance of mutations

DOI: 10.1201/9781003405665-6

and adaptation to humans. Universal influenza vaccines would be key, but we identified only a few (28 in the search period) patents for these vaccines, still in the early stages of development. In addition, it is very difficult to predict the impact immunological memory would have on the development of protective immunity to a broad spectrum of viruses due to the original antigenic sin mechanism. This study identified critical barriers to overcome in HPAI vaccine development, improving efficacy and reducing costs. It highlights the urgent need for an effective global vaccine preparedness system supported by genomic, antigenic, and epidemiological surveillance, an innovation fund and public–private partnerships. We strongly support the WHO in coordinating a Global Initiative for Influenza Pandemic Preparedness, encompassing an effective global response.

6.2 CONCEPTUAL FRAMEWORK AND OVERVIEW

Influenza viruses are complex and constantly changing, instigating a broad range of scientific issues due to their highly changeable nature and dynamic profile. The four types of influenza virus are A, B, C, and D. Different from endemic seasonal An influenza, highly pathogenic An influenza (HPAI) can spread easily among animals and potentially cause a future pandemic due to the virus' capacity to infect people and to sustain human-to-human transmission [1, 2].

This influenza A virus is considered one of the most important to public health due to its very high pandemic potential. Influenza A is often classified based on host species (Figure 6.1), such as swine influenza and avian influenza. Influenza B virus can also cause seasonal epidemics, circulates among humans and can infect seals. Influenza C virus is normally mild and can infect humans and pigs. Finally, influenza D virus cannot infect humans and primarily infects cattle [1]. These flu viruses change in two different ways ("antigenic drift" – small changes or mutations in viral genes and "antigenic shift" – an abrupt, major viral change, which can happen if an influenza virus from an animal population evolves to infect humans ("spill-over"), as indicated in Figure 6.1.

Human infections with highly pathogenic bird flu viruses have occurred in most cases after unprotected contact with infected birds (wild birds/domestic poultry) or contaminated surfaces. Nevertheless, some infections have emerged where direct contact with infected birds or their environments was not identified. Human-to-human transmission has been very rare, and when it occurs, it has only spread to a few people. However, because of the possibility that bird influenza viruses can mutate and spread easily between people, in outbreaks and even in a human pandemic, genetic and immune type monitoring for human infection and person-to-person spread is extremely important for public health [3, 1]. Concerning potentially pandemic strains, Asian lineage H7N9 and highly pathogenic avian influenza Asian lineage H5N1 viruses have been responsible for most human illnesses from bird influenza viruses worldwide to date, including the most serious illnesses and illness with the highest mortality. With regard to the H5N1 strain producing human disease, to date, more than 50% of the diagnosed infections have been fatal [3]. Although it should be recognized that there is a variety of genotypes and immunotypes of H7N9 and H5N1 strains with relatively low severity and lethality, the high pathogenic and pandemic

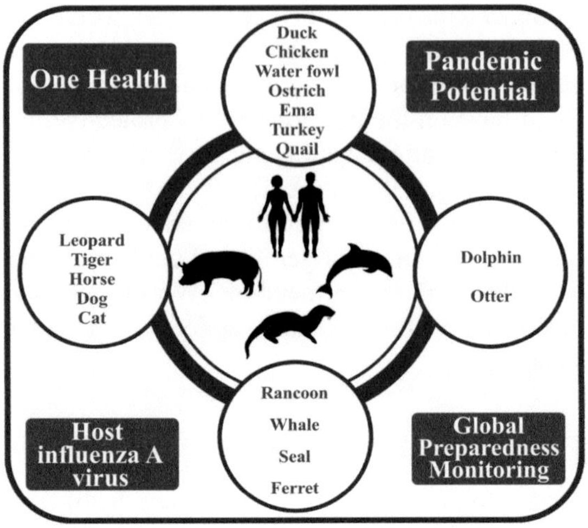

FIGURE 6.1 Host influenza A virus and pandemic potential. (Elaborated by the authors based on data from Kuchipudi and Nissly [2].)

potential of these two lineages has been a matter of global concern by public health authorities.

An influenza pandemic is thus a global outbreak of a new influenza A virus and occurs when novel influenza A viruses emerge and gain the ability to infect humans easily and spread from person to person in an efficient and sustained way [4]. Although influenza usually emerges as seasonal outbreaks, with pandemics being relatively rare, with five influenza pandemics since the beginning of the last century, the possibility of emergence of a new influenza A pandemic has been a matter of great global concern due to their potential for enormous social and economic impacts. This concern has increased in the current global scenario of environmental changes, mass production of animal food and high-density populated cities that facilitate the emergence and transmission of severe respiratory syndromes, as demonstrated by the COVID-19 pandemic.

"An influenza pandemic is a global epidemic caused by a new influenza virus to which there is a little or no preexisting immunity in the human population" [5]. Highly pathogenic influenza viruses are associated with rapid global spread/severe disease and have been responsible for many deaths. Five influenza pandemics have occurred since 1918: H1N1 (1918–1920), when one-third of the world population became infected and at least 50 million people died; H2N2 (1957–1958); H3N2 (1968–1969); a much milder H1N1 (1977–1978); and a recent H1N1 pandemic (2009), which caused between 100,000 and 400,000 deaths globally in the first year [6, 7]. Moreover, H5N1, a highly pathogenic An influenza, emerged in 1997, circulating worldwide in avian populations with reports of human cases and deaths. This strain was observed in 16 countries (649 patients, almost 60% of whom have died), and it was first identified in 1997 in Hong Kong. In Figure 6.2, we provide a global overview of confirmed human cases of avian influenza H5N1 reported to the WHO from 1997 to 2022.

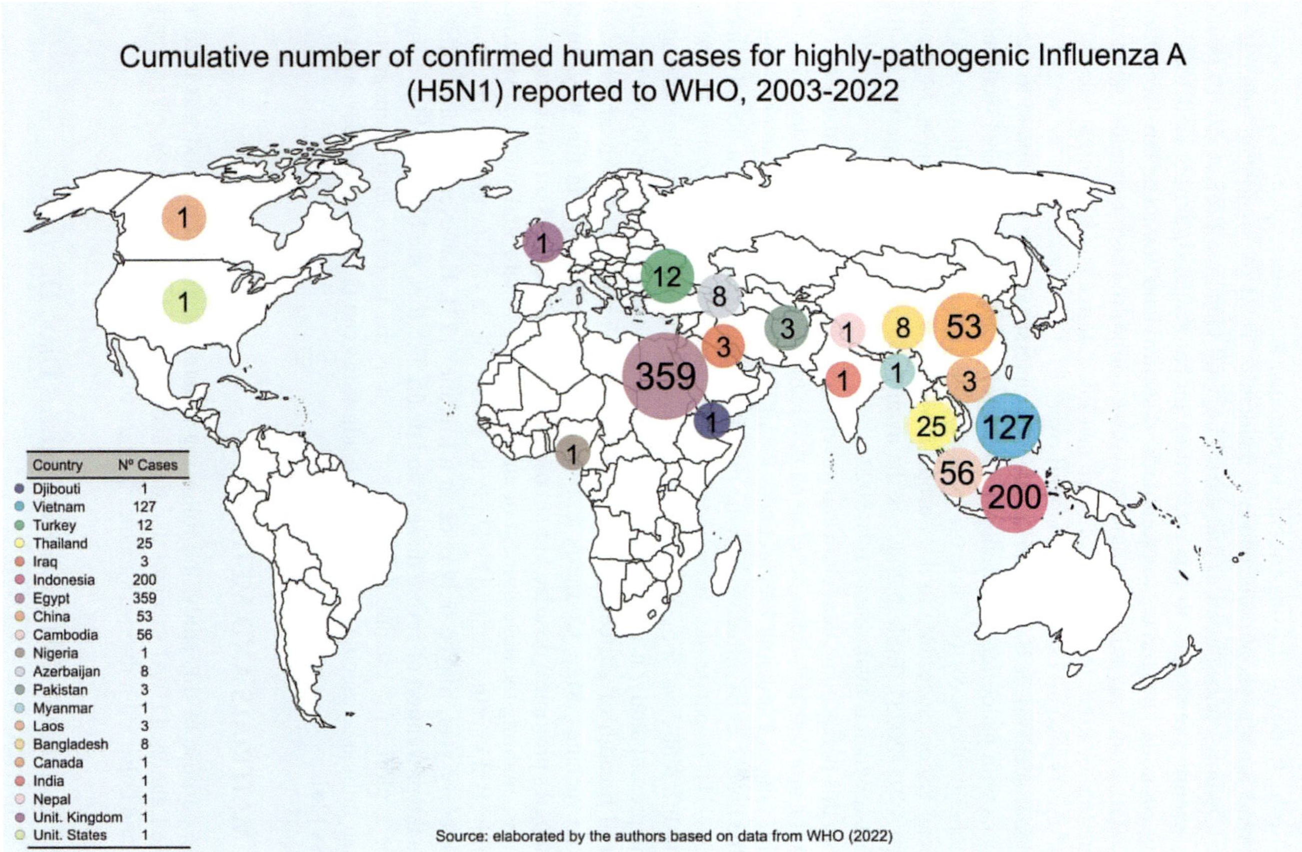

FIGURE 6.2 Reported global human cases with highly pathogenic avian influenza A [H5N1 (HPAI H5N1)] by country, 1997–2022. (Elaborated by the authors based on the WHO [5].)

At the dawn of the COVID-19 pandemic, there has been an increased recognition that more attention needs to be paid to global preparedness and human vaccines against highly pathogenic respiratory viruses. Flu viruses can mutate and gain the ability to be transmitted from animals to humans: a zoonotic spillover process that can spread among people and cause a pandemic [7, 8]. A significant amount of research is now concentrated in this area. The Viral Genomic Project (VGP) is one of the scientific responses to this concerning global scenario, aiming to detect and identify zoonotic viral threats to promote food security and human health. According to the VGP, from 1,67 million unknown viruses from birds and mammals, estimates indicate that 631,000 to 827,000 of them can infect humans with a high risk (99%) of viral emergence and spread from "spill-over" [9–11]. Considering the severity of this global scenario, it is imperative to better understand how to respond to future outbreaks with pandemic potential, which severely compromise peoples' well-being and global health security. Table 6.1 presents four HPAI vaccines authorized in the European Union, according to the European Medicine Agency (European Medicine Agency 1995–2023). Table 6.2 shows the avian influenza vaccine approved by the U.S. FDA, which is available in July 2023.

From this perspective, academia and governments must urgently collaborate in the search for innovative vaccine development, integrating vaccine preparedness into an effective Global Pandemic Preparedness Plan [21]. According to Krammer & Palese [22], the processes of influenza vaccine production have increased significantly since the 2009 H1N1 pandemic. In fact, the authors presented virus vaccine technologies in different development stages, i.e., licensed, clinical, and preclinical stages. For the present study, it is important to note the technologies used in broad/mainly universal breadth of protection vaccines: Modified Vaccinia Virus Ankara (MVA)-vectored vaccines, M2e, Chimeric and Headless Hemagglutinin. In this way, novel technologies promote a detailed human immune response analysis on influenza A and B viruses, enabling the development of an immunogenic, safe, and universal vaccine. The Universal Vaccine, defined by Dr. Peter Palese[1], is "a vaccine that you don't have to take every year, maybe only once in a lifetime" [23]. However, a more realistic goal is the prevention of severe disease and death. For this, the biological basis must also be explored for vaccine development, such as the increase in immunogenicity of avian influenza vaccines. For this goal, adjuvants can be used in the development process [24].

6.3 MATERIALS AND METHODS

For the purpose of this study, the following schematic applied method was used to retrieve Launched HPAI vaccines and data about technological focus on patent landscape and universal influenza vaccines (Figure 6.3).

6.3.1 Launched HPAI Vaccines (Cortellis Drug Discovery Intelligence – CDDI)

Data on launched HPAI vaccines between 2010 and 2023 were extracted from Cortellis Drug Discovery Intelligence (CDDI) (from Clarivate™). This methodology

TABLE 6.1
List of Pandemic Preparedness Vaccines Authorized in the European Union

Pandemic Preparedness Vaccine	Organization: Marketing-Authorization Holder (Country)	Flu Strains (H5N1)	Vaccine Description	Therapeutic Indications	Observations
Foclivia	Seqirus Srl (Italy)	A/Vietnam/ 1194/2004	*"Pandemic influenza vaccine (H5N1) (surface antigen, inactivated, adjuvanted)"*	*"Prophylaxis of influenza in an officially declared pandemic situation"(use in adults against pandemic flu)*	"Foclivia is a '**pandemic preparedness' vaccine**: *"This is a special type of vaccine that can be developed to help with the management of a future pandemic"* (Authorized under exceptional circumstances)
Adjupanrix	GlaxoSmithKline Biologicals (UK)	A/Vietnam/ 1194/2004 NIBRG 14'	*"Previously Pandemic influenza vaccine (H5N1) (split virion, inactivated, adjuvanted)"*	*"It should only be used once a flu pandemic has been officially declared by the World Health Organization (WHO) or European Union (EU)"*	*"Adjupanrix is a '**mock-up' vaccine**'*: "*A mock-up vaccine needs to bring about protective levels of antibodies in at least 70% of adults for it to be considered suitable"*, according to Committee for Medicinal Products for Human Use. (Authorized under exceptional circumstances)

(continued)

TABLE 6.1 (Continued)
List of Pandemic Preparedness Vaccines Authorized in the European Union

Pandemic Preparedness Vaccine	Organization: Marketing-Authorization Holder (Country)	Flu Strains (H5N1)	Vaccine Description	Therapeutic Indications	Observations
Pandemic Influenza Vaccine H5N1 Baxter AG	Resilience Biomanufacturing Ireland Limited (Ireland)	A/Vietnam/1203/2004	*"Pandemic influenza vaccine (H5N1) (whole virion, inactivated, prepared in cell culture)"*	It is *"the same as the Celvapan H5N1 mock-up vaccine, which was previously authorized in the European Union"*	*It is a '**mock-up' vaccine**'* (Authorized under exceptional circumstances)
Pandemic influenza vaccine H5N1 AstraZeneca (previously Pandemic influenza vaccine H5N1 Medimmune)	AstraZeneca AB (Sweden)	A/Vietnam/1203/2004 strain	*"Pandemic influenza vaccine (H5N1) (live attenuated, nasal)"*	It is used *"to protect children aged between 12 months and 18 years against influenza (flu) during a flu pandemic"*	It is **'pandemic preparedness' vaccine"** (Authorized under exceptional circumstances)

Sources: Elaborated by the authors from European Medicines Agency [12–18].

TABLE 6.2
Avian Influenza Vaccines Approved by the U.S. FDA (*)

Approved Influenza Vaccines	Company (Country)	Description	Strain	Use
Audenz™	CSL Seqirus Inc (UK)	*"monovalent, adjuvanted, cell-based inactivated subunit vaccine"*	aH5N1c	To protect people (six months age and older) against influenza A (H5N1 subtype) in the event of bird flu pandemics

Sources: Elaborated by the authors from Carlson and Lutmer [19] and Precision Vaccinations [20].

Note: The vaccines were available in July 2023.

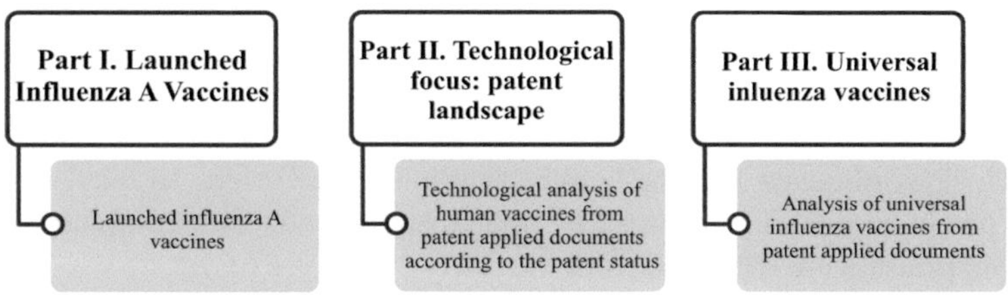

FIGURE 6.3 Schematic applied method (2010–2022/2023). (Elaboration by the authors.)

of data collection was based on strategy 3, which was designed to retrieve HPAI vaccines launched in this period [25] (Figure 6.3). Concerning the inclusion of H1N1 vaccines in our search, one key point should be noted: although the previously pandemic H1N1 virus attenuated and became seasonal due to viral evolution (WHO Declaration in August 10, 2010, of the end of the 2009 pandemic [26]), many technological strategies deposited for the pandemic H1N1 vaccine in the first semester of 2010 were retrieved in our search and persisted in our search period, with significant interfaces with other HPAI vaccines.

6.3.2 Technological Analysis of Human Vaccines from Patent Applied Documents (Derwent Innovation)

According to the methodology, the search was performed in the Derwent Innovation database between 2010 and 2022. The strategy searched for descriptors in the title/abstract/claims in applied patent documents: (Pandemic Potential or H1N1 OR swine flu pandemic OR swine flu OR H2N2 OR Asian flu OR H3N2 OR Hong Kong flu OR H5N1) matching the Derwent Manual Code[2] [B14-S11 – Description: (Vaccine general) AND (vaccine* or vaccination)]. Based on the results, the applied documents from the top applicants were discussed with a focused technological review in a chronological way from the priority year. All the applied documents were retrieved regardless of their patent status, considering the active status only[3]. To identify the legal status of the retrieved applied patent documents, two databases were used: Lens and Indian Patent Advanced Search System (inPASS). Lens is a free database that contains information on more than 144.3 million patent records, in addition to 252 million academic papers [27]. To identify the status of each application, the publication number was used. The identification of the status of Indian documents was carried out using the inPASS database launched in 2015 [28] by the Controller General of Patents Designs and Trademark, which is subordinated to the Government of India. The search was performed using the original title of the patent document and the filing date.

6.3.3 Universal Influenza Vaccines

The method has been conceived to retrieve human universal influenza vaccines following a two-step process. First, the search was performed on the Derwent Innovation Index database[4] according to the strategy developed from Part II ("Technological analysis of human vaccines from patent applied documents"). Second: This simple and practical method allowed us to retrieve documents related to "Universal Vaccines", which are effective against all influenza strains. For this purpose, the terms[5] "universal vaccine", "universal influenza vaccine", "universal flu vaccines", "paninfluenza vaccine", and "panflu vaccine" were adopted in different fields of patent applied documents, i.e., title, abstract, and claims. Influenza vaccines for veterinary use were excluded from the analysis (Figure 6.3).

6.4 RESULTS

6.4.1 Launched HPAI Vaccines

This search methodology retrieved launched HPAI vaccines in the 2010–2023 period. Nineteen (19) drugs and biologics (launched influenza A vaccines)[6] were recorded. A key finding emerged in the drug type classification: most of them were biotechnologies (16), while few (three) were classified as peptides. The United States is the country with the leading number of vaccines (nine vaccines), followed by India, the Republic of Korea, and the United Kingdom (each with three vaccines). The top organizations were SANOFI (from France, with four products), GLAXOSMITHKLINE (from the United Kingdom, three), SEQIRUS (from the United Kingdom, two) and SK BIOSCIENCE (from South Korea, with two products).

In addition, it is important to highlight the low number of launched HPAI vaccines: 19 products in 13 years (2010–2023). Table 6.3 provides an overview of the most recent vaccines, i.e., the vaccines launched in 2017 and 2018.

6.4.1.1 Flublok® Quadrivalent (Launched in 2017: USA)

The influenza vaccine Flubok® Quadrivalent was developed by PROTEIN SCIENCES CORPORATIONS (US), a SANOFI company (France), which is an

TABLE 6.3
Description of the Most Recent Launched HPAI Vaccines

Vaccine Description	Launched Year	Organizations (Origin Country)	Brand Name (Vaccines)	Recommendations
"Quadrivalent influenza vaccine consisting of four recombinant hemagglutinin (rHA0) proteins against two influenza A strains (H2N1 and H3N2) and two influenza virus B strains"	2017	Protein Sciences (The United States) (from Sanofi)	**Flublok Quadrivalent** (Influenza quadrivalent vaccine). **Note**: The vaccine was launched in The United States by Protein Sciences (Sanofi) [25, 34]	The vaccine is used to prevent against influenzas A and B in adults 18 and older [25, 18]
"Quadrivalent influenza vaccine based on two influenza A strains, and two influenza B strains"	2017	ADImmune (*) (Taiwan) [35]	**AdimFlu-S(QIS) Quadrivalent influenza vaccine** [36]	Protection of children (older than three years old) and adults. The vaccine is recommended by WHO [36]
"Quadrivalent inactivated influenza vaccine against two influenza A strains (A/H1N1 and A/H3N2) and two influenza B strains (B/Victoria and B/Yamagata). The vaccine contains the influenza strain composition as recommended annually by the World Health Organization"	2017	Sanofi (France)	**VaxigripTetra®** (quadrivalent inactivated influenza vaccine) [37]	To prevent influenzaA and B in 6 months children or more and adults [38]
"Quadrivalent inactivated subunit influenza vaccine containing antigens of two influenza type A virus strains (H1N1 and H3N2) and two type B virus strains; adjuvanted with polyoxidonium"	2018	NPO Petrovax Pharm (Russia) [39]	**Grippol® Quadrivalent** (Quadrivalent inactivated subunit adjuvanted influenza vaccine) [39]	The vaccine is recommended by WHO, which antigens' doses are about three times less (compared to the standard recommended dose by WHO) [40, 41]

Source: Elaborated by the authors from CDDI [25].

Note: (*) organization (originator): *"The body that invented or created the drug"* [25].

important developer of recombinant vaccines [34, 42]. The vaccine is approved for persons 18 years or older and is used to immunize against influenza A and B [43]. The quadrivalent vaccine is available for the 2021–2022 influenza season and is produced using recombinant technology. In fact, in the manufacturing process, recombinant flu vaccines do not use chicken eggs or flu virus (egg-free flu vaccine). This has significant benefits in terms of the speed of the influenza manufacturing process. Since the product is not dependent on egg supply, it can be produced faster when compared to other egg-based vaccines, which is crucial in a pandemic scenario. Another advantage arising from how recombinant technology can affect how the final vaccine works is that the production process avoids virus mutations that can occur when eggs are needed for virus growth [44].

6.4.1.2 AdimFlu-S® (QIS) – Quadrivalent Influenza Vaccine (Launched in 2017 in Taiwan)

The quadrivalent vaccine, without any live virus particles (H1N1, H3N2, Yamagata, and Victoria type B lineages), protects children (older than 3 years old) and adults for 6–12 months. To prepare the vaccine, the influenza virus is cultured in embryonated eggs and then purified, split, and inactivated [36].

6.4.1.3 VaxigripTetra®: Quadrivalent Inactivated Influenza Vaccine (Launched in 2017 in Europe)

The vaccine against influenza from Sanofi Pasteur (France) contains four-strain influenza vaccine (influenza A strains: H1N1 AND H3N2 and influenza B strains: Victoria and Yamagata). It is indicated for 6-month-old children or more and adults. Available in 20 European markets, it will be planned to be launched in the future worldwide [38, 45]. In fact, the vaccine is a quadrivalent split-virion vaccine as a well-established record of Vaxigrip® (a trivalent split-virion flu vaccine) [37].

6.4.1.4 Grippol® Quadrivalent: Quadrivalent Inactivated Subunit Adjuvanted Influenza Vaccine (Launched in 2018 in Russian Federation)

Grippol® Quadrivalent was developed by Petrovax Pharma, a Russian full-cycle biopharmaceutical company. The product was the first Russian quadrivalent inactivated subunit adjuvanted influenza and prevents against influenza A (H1N1 + H3N2) and influenza B (Yamagata and Victoria) for adults aged 18–60 and for children aged 6 or more [39, 46]. According to the organization, the formulation has low reactogenicity and high efficacy. In fact, the use of a biodegradable and water-soluble adjuvant name, polyoxidonium (azoximer bromide), is responsible for cutting in threefold the antigen load compared to traditional technology, and it (the adjuvant) enhances the immune response [47]. More recent work by Talayev and collaborators [40] concluded that this quadrivalent formulation with polyoxidonium induces the production of antibodies to all influenza hemagglutinins, enhances natural killer cell activity, induces a Th2-polarized T-cell response and partial maturation of dendritic cells, which are stimulated to enhance inducible costimulator-ligant (ICOSL) expression. This is an important event in humoral immune response induction.

6.4.2 Patent Landscape – Human Vaccines

Our strategy search retrieved the global patent applications for vaccines against influenza A in humans for the 2010–2022[7] period, resulting in 683 patent deposits from a broad range of assignees/applicants in different countries, as illustrated in Figure 6.4. The developed strategy also demonstrated that the leading priority country in the number of applications was the United States (US) (with 225 documents), followed by China (191), the Russian Federation (68) and South Korea (60). Based on the available data, the greatest number of documents was applied in 2020 (88 documents). In fact, the average number of documents in each year was 52,5 applications.

These results were obtained using the described methods to retrieve the applied documents from the leading applicants of vaccines against influenza A strains, i.e., H1N1, H2N2, H3N2, and H5N1. Based on the results (683 retrieved documents), the following observations are necessary: first, the leading applicant was from the Russian Federation – The Institute of Experimental Medicine (with 39 applications), SANOFI (12 documents), followed by the American Institution JANSSEN (11 documents), and the British Institution SEQIRUS UK LIMITED with 10 documents (Figure 6.5 and Table 6.4).

New technologies for therapeutic and vaccine development to prevent and address future influenza pandemics are actively being investigated not only to address and control subsequent influenza pandemic strains but also to assure the management of the ongoing COVID-19 pandemic in humans [121]. The method presented in this research was sufficiently general to describe how the patent documents were related to influenza A subtypes, i.e., H1N1, H2N2, H3N2, H5N1, and H7N9. In a second

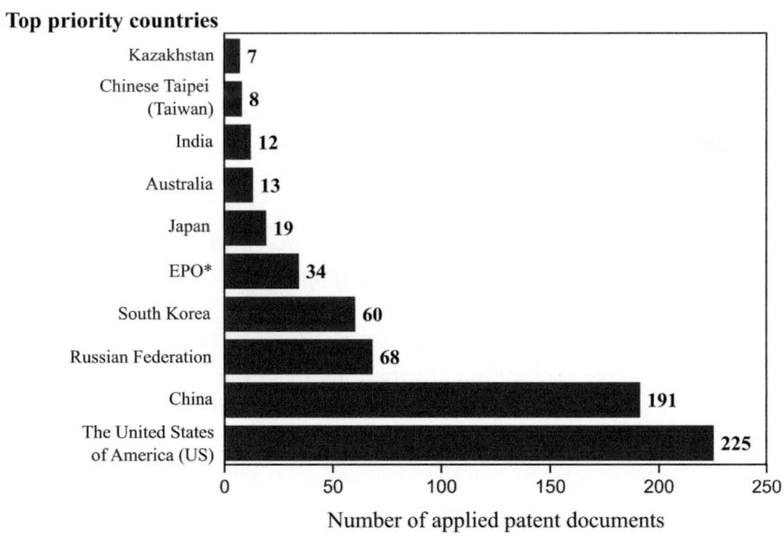

FIGURE 6.4 Top priority countries for vaccines against influenza A according to the number of patent documents applied for during the 2010–2022 period (all patents with different statuses were considered). (Elaborated by the authors from Derwent Innovation Index Database [49].)

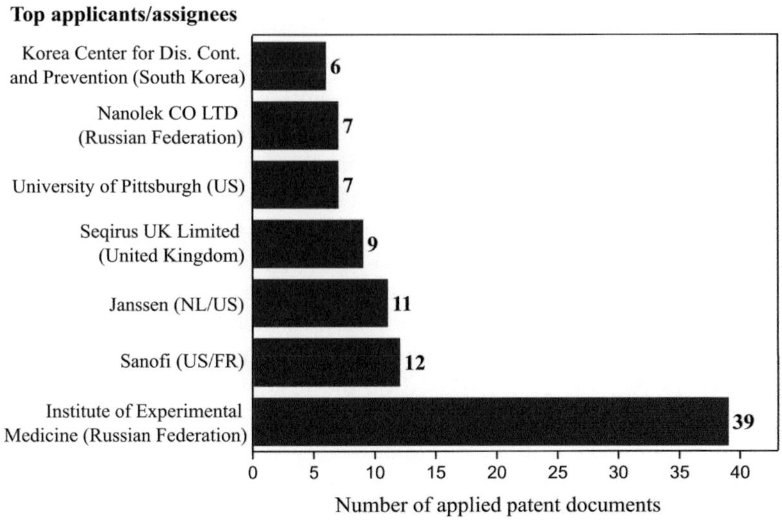

FIGURE 6.5 Top applicants/assignees for influenza A vaccine according to the number of applied patent documents during the 2010–2022 period (all patents with different statuses were considered). (Elaborated by the authors from Derwent Innovation Index [49].)

stage and after rigorous examination of these documents, we identified that most of them were related to vaccines to prevent/control/treat H1N1 influenza in humans. Moreover, we identified, as expected, a considerable number of documents related to more than one influenza A strain (Figure 6.6).

No further increase in the number of applications was observed for the other applicants/assignees. The results of our search have important global implications for influenza pandemic preparedness and technological development of vaccines: there are a low number of documents for vaccines against influenza A strains by the leading applicants. This patent deposit scenario allows us to shed light on understanding the technological evolution of the three top companies in influenza vaccine development, i.e., Institute of Experimental Medicine (from Russian Federation), SANOFI (France), JANSSEN (from US), and SEQIRUS UK LIMITED (United Kingdom), indicating the strength of our methodological approach.

6.4.2.1 Institute of Experimental Medicine (Russian Federation)

The main achievements, including contributions from the Institute of Experimental Medicine, may be summarized as follows. From the 39 retrieved applied documents, it was observed that most of the claimed vaccines were live intranasal, including live quadrivalent influenza vaccines from 2011 to 2022, i.e., almost the entire study period (Figure 6.7).

However, a closer examination of the results reveals different patent statuses that must be highlighted here. Most of the applied Russian documents were inactive (17 documents – 58.6%) or discontinued (two documents – 6.9%). Therefore, the results here deviate significantly from the expected: a few numbers of active documents: 19

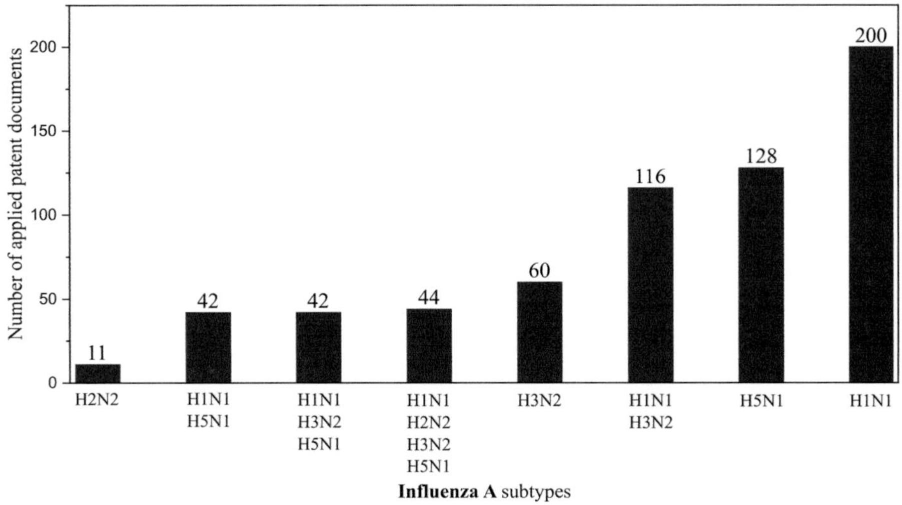

FIGURE 6.6 Influenza A subtypes cited in influenza vaccine documents. Source: elaborated by the authors from the Derwent Innovation Index [49]. (*) For three documents, influenza strains were not described. The authors used the terms "influenza A" (two documents) and "pandemic potential" (one document), as described in the original applied patent documents. N: 683 (all patents with different statuses were considered). Note: The viral strains with less than 10 patent filings were not computed.

(only 48,7%) (Table 6.4). On the other hand, the quality of the results is satisfactory because the documents offer information about the potential benefits of the use of biotechnology for practical health services to prevent seasonal influenza. In 2011, the company developed from the *Orthomyxoviridae* family obtained a new attenuated cold-adapted H1N1 strain and a reassortant H2N2 influenza (vaccine strain) to be used in an influenza intranasal vaccine for children and adults. The vaccine strains (H2N2) were prepared by crossing epidemic viruses H1N1 (cold-adapted) and H2N2 strains. A key strength of the document lies within the fact that the attenuated H1N1 strain (cold-adapted) is new. This one is a donor of attenuation to obtain vaccine strains. In fact, from this process, it is possible to obtain strains of H2N2 (subtype A). Moreover, the H1N1 influenza strain has sensitivity to preparations of adamantane rows and temperature strips (PN: RU2556833C2) [50]. In the same year (2011), another document is also related to a live intranasal influenza vaccine, where the vaccine strains were also obtained as attenuation donors to prepare an H2N2 vaccine strain (PN: RU2556834C2) [51].

Once again, in 2018, the developed vaccine strain H3N2 was a candidate to produce a live intranasal influenza vaccine (PN: RU2683500C1) [52]. In 2019, the close agreement of the results with the last applied documents suggested that the Russian Institute worked in the same line of research: the development of influenza vaccine strains, basically to live intranasal influenza vaccine for adults and children: H3N2 (PN: RU2711101C1) [53]; RU2715674C1 [54]; RU2732610C1 [55]; RU2734897C1 [56] or H1N1 (PN: RU2724706C1 [57]; RU2735291C1 [58]). It was also clear the

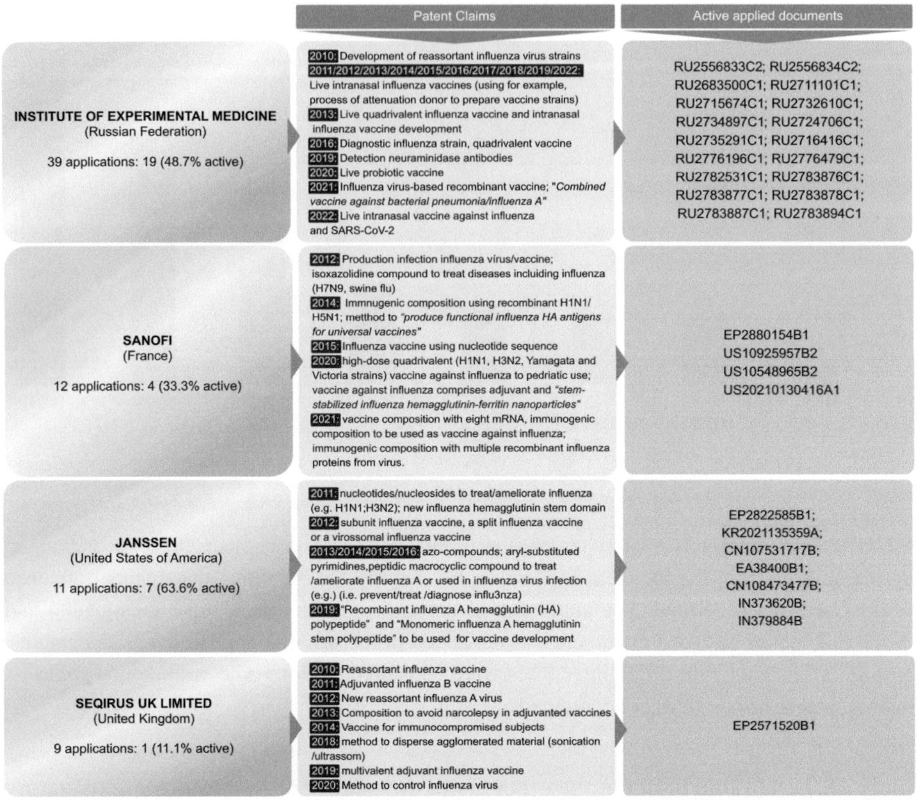

FIGURE 6.7 Main applicants' chronological patent filings related to influenza A vaccines for human use. Source: elaborated by the authors from Derwent Innovation Index Patents. Note: all patents with different statuses were considered (Table 6.4).

potential of this biotechnology to detect antibodies to neuraminidase N2 in blood serum through a solid-phase inhibition reaction of neuraminidase activity using the reassortant strain H6N2 to detect neuraminidase antibodies (PN: RU2716416C1 [59]). In 2020, a live probiotic vaccine was applied using conservative influenza A virus epitopes based on the active strain *Enterococcus faecium* (PN: RU2776479C1 [61]). In 2021, a combined vaccine with high stability was developed to be used against Pneumococcus and influenza virus (PN: RU2776196C1 [60]). Finally, in 2022, applications of live intranasal vaccines against influenza for children and adults used the H3N2 strain (PN: RU2783876C1 [63]; RU2783877C1 [64]; RU2783894C1 [66]; RU2783887C1 [67]) or H1N1 strain (PN: RU2783878C1 [65]; RU2793387C1 [68]). Meanwhile, the company focused on other live intranasal vaccines against influenza and SARS-CoV-2 infection (PN: RU2782531C1 [62]) (Figure 6.7).

6.4.2.2 Sanofi (from France)

From 12 retrieved documents, only 4 (33.3%) were in active status, and most of them (58.3%) were pending. Further information about the active documents applied by the French Company follows: in 2012, an applied document was about methods to produce influenza vaccine and influenza virus using reassortant RNA viruses, including influenza A, B, and C genomes (PN: EP2880154B1) [89]. In 2014, an immunogenic composition comprised recombinant H1N1 (PN: US10925957B2) [90] or H5N1 (PN: US10548965B2) [91] influenza hemagglutin polypeptides or to prevent influenza disease. Moreover, in 2015, an influenza vaccine was obtained using an optimized nucleotide sequence obtained in a simple manner (PN: US20210130416A1) [92].

6.4.2.3 Janssen (from US)

Based on the methodology used, once again, the number of documents retrieved from the second main applicant was low: 11 documents. From these, key findings emerged. In general, and first, most of the documents claimed new compounds used to treat/ameliorate or to be used in influenza virus infection. To demonstrate this, examples of molecules can be presented: azo compounds, aryl-substituted pyrimidine peptidic macrocycle compounds or nucleotides/nucleosides. Second, from the retrieved documents, most of them were active, seven (63.6%) were active, two applications were discontinued (18.2%), and two documents (18.2%) were pending (Table 6.4). From this scenario, we approach the low number of active documents with a broader perspective that the active applied documents are in fact the object of analysis to the present work due to the interest in maintaining these documents by their applicants to protect the claimed products and methods. The company JANSSEN BIOPHARMA INC. (US) claimed in 2011 that new substituted nucleotide and nucleoside compounds can inhibit viral replication and can be used to treat/ameliorate paramyxovirus viral infection (e.g., pneumovirus infection) or orthomyxovirus viral infection (H1N1, H3N2, influenza A, B, C) or treat/ameliorate infections caused by parainfluenza, mumps, and measles (e.g., PN: IN373620B [106]). JANSSEN VACCINES AND PREVENTION B.V., from the Netherlands, is part of JANSSEN PHARMACEUTICALS and has some applications in the development of vaccines against influenza. The company claimed (priority year 2011) for a new influenza hemagglutinin stem domain used to treat influenza. The invention comprises hemagglutinin HA1/HA2 domains and "influenza hemagglutinin stem domain polypeptide is capable of broadly neutralizing antibody response and resistant to protease cleavage" (PN: IN379884B [107]).

The company JANSSEN VACCINES AND PREVENTION B.V. (Netherlands) claimed in 2012 (priority year) for a subunit influenza vaccine, split influenza vaccine or virosomal influenza vaccine that comprises the proteins neuraminidase (NA) and influenza hemagglutinin (HA) from at least three influenza strains. In the interest of clarity, it may be mentioned that the vaccine is able to induce cross-protection against the H5N1 strain (despite there being no NA/HA from the H5N1 strain) due to the proteins from H3N2 and H1N1 and the B strain of influenza (PN: EP2822585B1 [101]). In 2013 and 2015, the company claimed for aza-pyridone compounds to be used in pharmaceutical compositions to inhibit influenza virus replication, such as H1N1, H3N2, H5N1, H7N9, and Influenzas A and B (PN: KR2021135359A [102] and CN107531717B [103]), respectively/JANSSEN BIOPHARMA INC). In 2014,

TABLE 6.4
Human Vaccines: Applied Documents and Their Patent Status from the Main Applicants (*)

Applicants (Country/Total of Retrieved Documents)	Patents' Status				
	According to The Lens/Indian Advanced Search System (*)				
	Active *n* (%)	Discontinued *n* (%)	Pending *n* (%)	Inactive *n* (%)	Indetermined *n* (%)
THE INSTITUTE OF EXPERIMENTAL MEDICINE RU (Russia/39)	RU2556833C2 [50] RU2556834C2 [51] RU2683500C1 [52] RU2711101C1 [53] RU2715674C1 [54] RU2732610C1 [55] RU2734897C1 [56] RU2724706C1 [57] RU2735291C1 [58] RU2716416C1 [59] RU2776196C1 [60] RU2776479C1 [61] RU2782531C1 [62] RU2783876C1 [63] RU2783877C1 [64] RU2783878C1 [65] RU2783887C1 [66] RU2783894C1 [67] RU2793387C1 [68]	RU2013149103A [69] RU2016146943A [70]	-	RU2416640C1 [71] RU2416641C1 [72] RU2422517C1 [73] RU2422518C1 [74] RU2422519C1 [75] RU2464310C1 [76] RU2464311C1 [77] RU2532844C2 [78] RU2563352C2 [79] RU2587629C1 [80] RU2013149106A [81] RU2606026C1 [82] RU2605314C1 [83] RU2606019C1 [84] RU2627188C1 [85] RU2625024C1 [86] RU2653388C1 [87]	RU2021121139A [88]
	19 (48.7)	**2 (5.1)**	**0 (0.0)**	**17 (43.6)**	**1 (2.6)**

SANOFI (FR/12)	EP2880154B1 [89] US10925957B2 [90] US10548965B2 [91] US20210130416A1 [92]	US20170121373A1 [93]	US20230043128A1 [94] WO2023059857A1 [95] WO2023079113A1 [96] WO2023081798A1 [97] WO2021245070A1 [98] WO2021173907A1 [99] WO2021231729A1 [100]	-	-
	4 (33.3)	**1 (8.3)**	**7 (58.3%)**	**0 (0.0)**	**0 (0.0)**
JANSSEN (USA/11)	EP2822585B1 [101] KR2021135359A [102] CN107531717B [103] EA38400B1 [104] CN108473477B [105] IN373620B [106] (*) IN379884B [107] (*)	IN201727012541 [108] (*) US20200048308A1 [109]	CN113597428A [110] WO2021043869A1 [111]	-	-
	7 (63.6)	**2 (18.2)**	**2 (18.2)**	**0 (0.0)**	
SEQIRUS UK LIMITED (United Kingdom/9)	EP2571520B1 [112]	US20190247489A1 [113] US20200000909A1 [114] US20190184007A1 [115] AU2021202331A1 [116]	JP2021531339A [117] WO2022005893A1 [118] AU2020228151A1 [119] AU2019302562A1 [120]	-	-
	1 (11.1)	**4 (44.4)**	**4 (44.4)**	**0 (0.0)**	**0 (0.0)**

Source: Elaborated by the authors from The Lens and Indian Patent Advanced Search System. The search was carried out on July 24, 2023 [27, 28].

Legends:
Active: "Granted patent in force. If an application has been Granted, all published documents associated with the application have a status Active."
Discontinued: "Application discontinued, withdrawn or rejected, i.e., *discontinuation before Grant"*
Pending: "Application is pending (Pending: IP Right has not been granted yet and application is in either filing, examination, pregrant stage)"
Inactive: "Granted patent not in force because of lapse, nonfee payment, etc. The patent has not reached the term date and can be revived".
(*): From the 71 patent filings, 15 documents (21.1%) were related to breakthroughs, and 56 documents (78.9) were related to incremental innovations.

the company claimed that a new influenza hemagglutinin stem with an HA1 domain would induce an immune response against current and future influenza virus strains. Moreover, it provides an effective and safe vaccine to stimulate a robust neutralizing antibody response (PN: EA38400B1/JANSSEN VACCINES & PREVENTION B.V., Netherlands) [104]. In 2016, the company claimed aryl-substituted pyrimidines. The compounds are used in influenza virus infection. The compound was claimed to be used as a medicament (therapeutic agent: influenza vaccine, antiviral agent or both). The compound was tested against H1N1 through a cell-based antiviral assay (PN: CN108473477B/JANSSEN SCIENCES IRELAND UC, Ireland [105] (Figure 6.7).

6.4.2.4 Seqirus UK Limited (from United Kingdom)

SEQIRUS is a CLS company leader provider of pharmaceuticals and vaccines. Seqirus Pty Ltd. is an affiliate from Australia and has applied some documents. In March 2022, the "Cell-based Flucelvax® Quad Influenza Vaccine" was announced. It is approved for children 2 years of age and older, and it comprises a surface antigen prepared in cell cultures and inactivated. In terms of the quality of results, this approach delivers well. In fact, it is the only well-based/influenza vaccine available in Australia [122]. As seen in Figure 6.7, the third main applicant (nine applications) claimed different products/methods, such as reassorted influenza vaccines/virus, multivalent adjuvanted vaccines or methods to disperse agglomerated material or to control the influenza virus. Our findings are somewhat surprising since only one document from 2010 (PN: EP2571520B1 [112]) is active (11.1%), while 44.4% (four documents) are discontinued and 44.4% (four documents) are pending. Moreover, the active application consists of a contribution to existing knowledge, and research is a set of techniques for vaccine production. The vaccine developed by SEQIRUS UK LIMITED contains less than 10 ng of residual host cell DNA per dose. It is important to highlight that the production of reassortant virus was greatly increased due to (during virus production) the reduction of translation/transcription of the backbone strain's neuraminidase/hemagglutinin genes. Furthermore, the claimed vaccine was a virosomal vaccine, surface antigen vaccine, split virion vaccine or whole virion vaccine. The virus strains used were H1N1, H5N1, H5N3, H9N2, H2N2, H7N1, and H7N7.

6.4.3 Universal Influenza Vaccines

The results illustrated in Table 6.2 show that universal influenza vaccines are the object of protection with individual advantages and technologies. In summary, the compositions cover a large subset of influenza virus strains. The adopted method attempted to retrieve 20 documents according to their status (Table 6.5).

6.5 DISCUSSION

The results of our search have important global implications for influenza pandemic preparedness and the technological development of influenza vaccines: the low

TABLE 6.5
Universal Influenza Vaccines: Applied Documents and Their Patent Status

	Patents' status			
	According to The Lens/Indian Patent Advanced Search System (*)			
Applicants (Country/Total of Retrieved Documents)	**Active** *n* (%)	**Discontinued** *n* (%)	**Pending** *n* (%)	**Inactive** *n* (%)
Ertl Hildegund C. J., and Zhou Dongming (US)		US20130209512 A1 [123]		
Temasek Life Sciences Laboratory Ltd. (Singapore)		IN201203735P2 [124] (*)		
The Secretary of State for Health (United Kingdom)		IN201401597P1 [125] (*)		
Streck INC. (US)		US20180243405A1 [126]		
The Wistar Institute of Anatomy and Biology (US)		US20140377295A1 [127]		
Mogam Biotechnology Research Institute (Republic of Korea)	KR1382244B1 [128]			
Henan Agricultural University (China)				CN102600466B [129]
CSG Microbial Immune Preparation Engineering Center Nantong CO. LTD (China)	CN103505725B [130]			
Sanofi Pasteur INC (US)		US20170121373A1 [93]		
Qingdao Agricultural University (China)	CN103948942B [131]			
Nitto Denko Corporation (Japan)	US10456464B2 [132]			
Gamaleya Epidemiology Microbiology FED	RU2618918C2 [133]			

(continued)

TABLE 6.5 (Continued)
Universal Influenza Vaccines: Applied Documents and Their Patent Status

Applicants (Country/Total of Retrieved Documents)	Patents' status			
	According to The Lens/Indian Patent Advanced Search System (*)			
	Active	Discontinued	Pending	Inactive
	n (%)	*n* (%)	*n* (%)	*n* (%)
Sun Yat-Sen University (China)		CN105950646A [134]	CN115894636A [135]	
The Board of Regents of The University of Texas System (US)			US20190351046A1 [136]	
Industry-Academic Cooperation Foundation Yonsei University (Republic of Korea)	US10953086B2 [137]			
Xu Jian Qing (China), Zhang Xiao Yan (China), Wang Jing (China) Zhu Ling Yan (China),Lubit Beverly W (US)			WO2021141758A1 [138]	
Korea Research Institute of Bioscience and Biotechnology (Republic of Korea)	KR2211378B1 [139]			
University of Manitoba (Canada)			WO2021077215A1 [140]	
The Trustees of The University of Pennsylvania; Icahn School of Medicine at Mount Sinai (US)			WO2021202734A3 [141] US20230181715A1 [142]	
Massachusetts Institute of Technology (US)	US20210379179A1 [143]			
Henan University; Military Veterinary Research Institute Academy of Military Medical Sciences (China)			CN111514287A [144]	

Institute for Viral Disease Control and Prevention Chinese Centre for Disease Control and Prevention (China)			CN115947801A [145]	
Icahn School Of Medicine at Mount Sinai (US)			WO2022061216A1 [146]	
Animal and Plant Quarantine Agency; Komipharm International Co. Ltd. (Republic of Korea)			KR2022108440A [147]	
Najit Technologies INC.(US)			WO2023015145A1 [148]	
Ascione Richard (US); QI Shengmei,US			WO2022032274A1 [149]	
Total 28 (100%)	**8 (28.6)**	**7 (25.0)**	**12 (42.8)**	**1 (3.6)**

Source: Elaborated by the author from Derwent Innovation Index [49].

(*) Human universal influenza vaccine.

Note: all patent status.

TABLE 6.6
Summarizes the Documents (in Active Status) About Universal Influenza Vaccines

Human Universal Influenza Vaccines					
Patent Document Number (Earliest Priority Year)	**Applicant/ Assignee (Country)**	**Vaccine/ Antibodies**	**Influenza Strains**	**Considerations**	**Advantages (e.g.)**
KR1382244B1 (2012) [128]	MOGAM BIOTECHNOLOGY RESEARCH INSTITUTE (Republic of Korea)	T-cell-based universal vaccine	*"broad-spectrum defense against various heterosubtypic variants of influenza viruses"*	The universal flu vaccine contains nucleoprotein gene/ polymerase acidic gene	- The vaccine expresses a broad-spectrum against influenza variants (many heterosub typic variants).
CN103505725B (2013) [130]	CSG MICROBIAL IMMUNE PREPARATION ENGINEERING CENTER NANTONG CO. LTD (China)	New cross-protective effect influenza vaccine	*"H1N1 influenza virus, human influenza virus, swine influenza virus or avian influenza virus"*	The vaccine contains *"a common heat shock glyco protein 96"*	- The universal influenza vaccine exhibits cross protection performance.
CN103948942B (2014) [131]	QINGDAO AGRICULTURAL UNIVERSITY (China)	Influenza virus universal gene vaccine	Prevention against *"the occurrence of various influenza virus diseases"*	The vaccine *"is formed by cloning hemagglutinin 2 gene of H1N1 virus strain of influenza with pCAGGS eukaryotic expression vector"*	- The vaccine induces fine cell immune and humoral responses. - The vaccine exhibits less adverse reactions.

					- The vaccine is able to prevent the disease against various influenza disease
US10456464B2 (2014) [132]	NITTO DENKO CORPORATION (Japan)	The vaccine the antigen and antithrombotic drug to induce humoral immunity	H1N1, H3N2, Influenza B virus, Influenza B virus and other antigens, such as *"live attenuated mumps virus, live attenuated measles virus, live attenuated rubella virus, tetanus toxoid-conjugated haemophilus influenzae type b polysaccharide"*, etc.	The vaccine can be used universally due to the induction of humoral immunity against many antigens	The composition is not invasive for the patients
RU2618918C2 (2015) [133]	GAMALEYA EPIDEMIOLOGY MICROBIOLOGY FED	Universal anti-infectious vaccine	H1N1, H2N3, H3N2, H5N2, H9N2	The vaccine is comprised by recombinant pseudoadenovirus particles	The composition has a broad-spectrum of antiviral activity
US10953086B2 (2018) [137]	INDUSTRY-ACADEMIC COOPERATION FOUNDATION YONSEI UNIVERSITY (Republic of Korea)	Universal Influenza vaccine	It produces immune response "*against at least one HA protein (HA group 1) such as H1, H2, H5, H6, H8, H9, H11, H12, H13, and H16; and at least one HA protein (HA group 2) selected from the group consisting of H3, H4, H7, H10, H14, and H15*" and against "*heterologous influenza viruses*"	The vaccine uses cold-adapted live-attenuated virus	The vaccine ensure protection and can exhibit cross-protective effect against influenza virus strains

(*continued*)

TABLE 6.6 (Continued)
Summarizes the Documents (in Active Status) About Universal Influenza Vaccines

Human Universal Influenza Vaccines					
Patent Document Number (Earliest Priority Year)	**Applicant/ Assignee (Country)**	**Vaccine/ Antibodies**	**Influenza Strains**	**Considerations**	**Advantages (e.g.)**
KR2211378B1 (2019) [139]	KOREA RESEARCH INSTITUTE OF BIOSCIENCE AND BIOTECHNOLOGY (Republic of Korea)	Influenza vaccine	H1N1, H1N2, H2N2, H3N1, H3N2, H5N1 and H5N8	The vaccine induces immune response against new mutant influenza virus and influenza virus subtypes	The fusion protein can produce neutralizing antibodies against influenza subtypes and swine mutant influenza virus
US20210379179A1 (2020) [143]	Massachusetts Institute of Technology (US)	Universal prophylactic compositions	H1N1, H3N2, Victoria and Yamagata (influenza B) and or H5N1	The composition induces immune response: antibodies to bind to the B-cell epitope	There is a promotion of immune response against the one B-cell epitope (at least)

Source: Elaborated by the author from Derwent Innovation Index [49].

(*) Human universal influenza vaccine.
Note: patent status – active.

number of documents for innovative vaccines against potentially pandemic influenza A strains by the leading applicants. From 49 deposits in the search period (2010–2022), only 18 products in active status (37%) were retrieved. In addition, most of them referred only to incremental technologies. This scenario is concerning and alerts to the urgent need to design effective strategies to prevent HPAI, considering its pandemic potential and high lethality.

6.5.1 Vaccine Scale-Up: Bottlenecks

These results indicate that it will be very difficult to scale-up these vaccines in the case the world needs them. As indicated previously, current egg-based processes for vaccine production are too laborious and difficult to scale-up in the case of a pandemic. Concerning the use of mRNA technology, there are still many open questions regarding the duration and level of protection mRNA vaccines might induce. Vaccinating just birds, on the other hand, is not easy or sufficient: if not carried out properly, the virus could continue to circulate at a low level, actually increasing the chance of mutations and spread. Universal influenza vaccines will thus be key for HPAI pandemic preparedness. However, our results identified few documents for these vaccines (20 in the search period) and indicate that they are still in early stages of development, facing many gaps and challenges to be introduced into the market. In addition, it is very difficult to predict the impact immunological memory would have on the development of a broad-spectrum vaccine due to the original antigenic sin mechanism.

6.5.2 Applicants' Strategies

Considering the main applicants for human flu vaccines, the Russian Institute (Institute of Experimental Medicine RU) focused its research (among other strategies) on the development of reassortant influenza virus strains and quadrivalent vaccines and vaccines for intranasal use, as observed in the following applied documents: PN: RU2556834C2 [51], RU2683500C1 [52], RU2711101C1 [53]; RU2715674C1 [54]; RU2732610C1 [55]; RU2734897C1 [56] and RU2735291C1 [58].

From the French Company Sanofi, in patent filings, vaccines using recombinant H1N1/H5N1 (PN: US10925957B2 [90]; PN: US10548965B2) [91], optimized nucleotide sequences (PN: US20210130416A1) [92] or reassorted RNA virus genomes (influenza A, B and C) (PN: EP2880154B1) [89] were claimed.

The American company (Janssen), with a more diversified portfolio, acted in different strategies in the development of vaccines by using new compounds (azo-compounds – PN: KR2021135359A [102] and PN: CN107531717B [103]) and aryl-substituted pyrimidines – PN: CN108473477B [105].

Concerning the documents applied by the Institute of Experimental Medicine RU, the documents that claimed intranasal vaccines (PN: RU2556834C2 [51]) were selected. From the American company Janssen, these innovations were observed in a new influenza hemagglutinin stain domain polypeptide (PN: IN379884B [107]), which was also a retrieved innovation from Janssen.

This patent deposit scenario allows us to shed light on the urgent need to accelerate the global effort toward more effective and long-duration influenza vaccines, supporting vaccine innovation and the scientific and technological development of a universal paninfluenza vaccine.

6.5.3 Technological Issues: mRNA and Universal Vaccines

Our results also indicate that the number of patent documents for highly pathogenic influenza A human vaccines, in addition to being very low, are mostly referred to incremental innovations, despite significant recent breakthroughs in cell-based and recombinant vaccines. In fact, an effective influenza prepandemic vaccine is difficult to obtain due to the impossibility of developing a vaccine against the correct virus subtype for the next pandemic. New technologies are available to provide protection against influenza strains, such as the "nucleoside-modified messenger RNA (mRNA)–lipid nanoparticle vaccine", which is able to act in a multivalent way, i.e., the formulation can encode hemagglutinin antigens for influenza A subtypes (20 known) and B lineages [150].

Nevertheless, despite recent studies using mRNA technology for these new influenza vaccines, they have not yet advanced much. The main issue constraining mRNA vaccine development for pathogens that mutate quickly, such as the potentially pandemic influenza H7N9 and H5N1 strains, is producing the correct antigen. The problem with influenza vaccines and other vaccines for potentially pandemic or pandemic diseases, such as HIV, is that thus far, researchers have not been able to engineer the antigen to present the correct pan-neutralizing epitopes.

Consequent to these innovation constraints, the current influenza vaccines examined in our study have important limitations: in addition to their low efficacy and effectiveness, due to the high mutation and viral antigenic variation, the protection conferred is of short duration. It is also important to underline the need for new adjuvant technologies for longer influenza vaccine protection and engineering the antigens to expose to the immune system the pan-neutralizing epitopes in their correct conformation. Finally, there is also a need to accelerate the development of more influenza vaccines with nasal application, in addition to Flumist*. Finally, a novel strategy of combined influenza-COVID-19 vaccines should be developed, and some efforts have already started in this direction.

Considering this scenario, the development of innovative vaccine technologies for large-scale production is urgent for improving vaccine efficacy for seasonal influenza and for global vaccine pandemic preparedness. The complexity of influenza viral evolution, requiring annual shots due to the emergence of new variants, evidence the need for urgently accelerating funding and incentives toward innovative vaccines and a universal paninfluenza vaccine against all influenza strains. A universal vaccine would be very important if it can combine protection for different antigens and viral mutations and confer a long protection, which seems very difficult. The COVID-19 pandemic has highlighted these limitations and provides a critical drive for vaccine development and regulatory innovation.

6.5.4 Human and Veterinary Vaccines: One Health

Accelerating vaccine development for HPAI, reviewing current time frames, has thus proven to be a critical issue. The development process for a human vaccine can take between 10 and 20 years. In contrast, veterinary vaccines usually take between 3 and 6 years, according to the country where the product is registered [151]. Veterinary vaccines must also be urgently improved and the object of accelerated Research, Development, and Innovation (RDI). Numbers from the World Organization for Animal Health (WOAH) indicate that the enormous impacts of HPAI and other zoonotic diseases on wildlife could lead to a devastating effect on the biodiversity of global ecosystems. This scenario alerts to the need for effective and immediate response efforts to zoonotic diseases and for coordinated multisectoral prevention as important tools to the control of HPAI and reinforces the One Health concept: 1. Eighty percent of pathogens of bioterrorism concern originate in animals; 2. Seventy-five percent of the emerging pathogens in humans are from animals, and finally, 3. Sixty percent of pathogens that can cause diseases in humans are from wild or domestic animals. The world faces the COVID-19 pandemic, which resulted from a virus of potential animal origin [152], with 762.791.152 confirmed cases and 6.897.025 deaths globally. Fortunately, a total of 13.340.275.493 COVID-19 vaccine doses have been administered [153], the best and current example of innovation, technological development for pandemic preparedness for future generations. Vaccine preparedness for the influenza pandemic should thus be inspired in these COVID-19 pandemic lessons and success stories.

6.6 CONCLUSION: VACCINE PREPAREDNESS

In contrast to the rapid global vaccine response to the COVID-19 pandemic, in less than six months, due to previous developments and strong funding mechanisms from the US government (Warp Speed Operation) and international agencies, our results indicate that in the case of a future HPAI outbreak or pandemic, this global vaccine response might be much slower and weaker. Even if funding mechanisms similar to the COVID-19 response are immediately put in place for HPAI vaccine development, our results indicate that there are still many technological and production platform gaps and bottlenecks to overcome. These constraints in vaccine development contrast with reports of new HPAI human cases, now also recently identified in the American region. In March 2023, the second H5N1 human case in South America was reported by the Government of Chile. The first case in this region was reported in Ecuador in December 2022 in the US, and one case was already reported in April 2022 and was the first case in the country [154]. Recently, on March 27, 2023, the National Health Commission of the People's Republic of China notified the death from avian influenza (H3N8) of a 56-year-old female from Guangdong Province hospitalized for severe pneumonia on March 3 and subsequently died on March 16, 2023. No close contact of the case developed an infection, symptoms, or illness at the time of reporting. The results of testing in environmental areas from the residence and the wet market where the patient spent time before the onset of illness showed samples positive for influenza An in the wet market (H3) [154].

This concerning scenario highlights the urgent need for an integrated global influenza vaccine preparedness system, supported by extensive genomic, antigenic and epidemiological surveillance, an innovation fund and public–private partnerships,

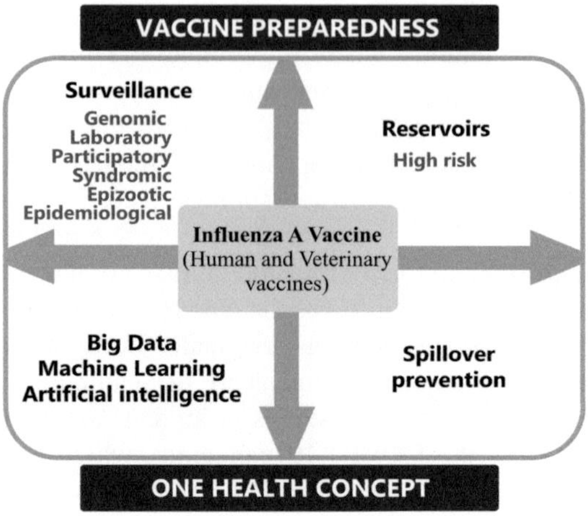

FIGURE 6.8 Toward a new global eco-social influenza pandemic preparedness system. (Elaborated by the authors.)

encompassing an effective global response (Figure 6.8). This system should be based on more sensitive surveillance, more relevant preclinical models and a prototype vaccine bank ready for production and clinical trials. Incentives to accelerate innovation and technology transfer for multi-patented vaccines, such as "patent pools", should also be provided [155]. We strongly support the WHO in coordinating a Global Initiative for Pandemic Preparedness in partnership with the Coalition for Epidemic Preparedness Innovations (CEPI) and other partners: governments, NGOs, and stakeholders. In addition, collaboration with another important initiative should be strengthened and supported: the Influenza Vaccines Research and Development (R&D) Roadmap (IVR), created through an extensive international stakeholder engagement process, to promote influenza vaccine R&D [156].

From this perspective, a combined strategy should be conceived, integrating into a comprehensive pandemic preparedness system, the development of diagnostic kits and access to antiviral drugs, in addition to vaccines. These diagnostic kits should be urgently developed as rapid, sensitive, specific, and easy to use tests. Access to antiviral drugs should also be a concern in the case of a pandemic: even when they are available, their production by local manufacturers in developing countries might be very limited due to patent protection. New incentives such as "patent pools" and other alternatives should be urgently put in place.

In conclusion, our results provide evidence on the main gaps and bottlenecks for HPAI vaccine development and scale-up in the case of a pandemic, stressing the urgent need to accelerate vaccine development and production in a global vaccine preparedness system based on a comprehensive One Health conceptual framework. This system should integrate a global strategic plan to accelerate innovation, technological development, and the production of influenza vaccines. These strategies will require clear priorities, innovative funding, and new business models, integrating academia, pharmaceutical enterprises, and governments.

ACKNOWLEDGMENTS

The authors thank Bio-Manguinhos of the Oswaldo Cruz Foundation, the School of Chemistry of the Federal University of Rio de Janeiro and the National Institute of Industrial Property for support. The authors gratefully acknowledge the financial support of the University of Pittsburgh. We also thank Rafael Cavalcante dos Santos, M. Sc., School of Chemistry, Federal University of Rio de Janeiro for the figures and final art.

NOTES

1 Dr. Peter Palese – Department of Microbiology at Icahn School of Medicine (New York City) [15].

2 Derwent World Patent Index (from Clarivate) Manual is a hierarchical indexing system to be used as an analysis tool and a patent retrieval in patent searching. The strains H1N1, H2N2, H3N2, and H5N1 were chosen according to their pandemic potential [29].

3 Active: "Granted patent in force. If an application has been Granted, all published documents associated with the application have a status Active".

Discontinued: "Application discontinued, withdrawn, or rejected, i.e., discontinuation before Grant".

Pending: "Application is pending (Pending: IP Right has not been granted yet and application is in either filing, examination, pregrant stage)".

Inactive: "Granted patent not in force because of lapse, nonfee payment, etc. The patent hasn't reached the term date and can be revived" (The Lens [27], Indian Patent Advanced Search System [30]).

4 Derwent Innovations Index (from Clarivate) covers over 14.3 million basic inventions. The documents are from almost 60 authorities in worldwide. The search was carried out on February 2, 2022, and it will retrieve dead/alive applied patent documents [22].

5 The terms were chosen from the following references: World Health Organization [31], Medical Subjects Headings (Mesh) from National Library of Medicine [32] and Darricarrère N et al. [33].

6 Here, influenza B vaccines were also retrieved (e.g., quadrivalent vaccines).

7 Deposits referring to 2023 were not considered, due to the secrecy's period (18 months). In most countries, the patent application is published 18 months after the filing date (or the priority date) [48].

REFERENCES

[1] World Health Organization (WHO). 2022a. "Influenza (Avian and Other Zoonotic)". Accessed May 21, 2022. https://apps.who.int/mediacenter/factsheets/avian_influenza/en/index.html.

[2] Kuchipudi, S. V. and Nissly, R. H. 2018. Novel flu viruses in bats and cattle: "Pushing the envelope" of influenza infection. *Veterinary Sciences* 5 (3): 71. https://doi.org/10.3390/vetsci5030071.

[3] Centers for Disease Control and Prevention (CDC). 2022. "Information on Bird Flu". Accessed June 14, 2022. www.cdc.gov/flu/avianflu/index.htm.

[4] Centers for Disease Control and Prevention (CDC). 2021a. "How Flu Viruses Can Change: 'Drift' and 'Shift'". Accessed September 21, 2021. www.cdc.gov/flu/about/viruses/change.htm.

[5] World Health Organization (WHO). 2022b. "Cumulative Number of Confirmed Human Cases for Avian Influenza A(H5N1) Reported to WHO, 2003–2022, 30 August 2022". Accessed 18 September, 2022. www.who.int/publications/m/item/cumulative-number-of-confirmed-human-cases-for-avian-influenza-a(h5n1)-reported-to-who-2003-2022-30-aug-2022.

[6] World Health Organization (WHO). 2022c. "Pandemic Influenza". Accessed April 03, 2023. www.euro.who.int/en/health-topics/communicable-diseases/influenza/pandemic-influenza.

[7] Centers for Disease Control and Prevention (CDC). 2018. "Highly Pathogenic Asian Avian Influenza A(H5N1) Virus Avian Influenza (Flu)". Accessed November 20, 2019. www.cdc.gov/flu/avianflu/h5n1-virus.htm.

[8] Katsnelson, A. 2020. How do viruses leap from animals to people and spark pandemics?. *Chem. Eng. News* 98:33. https://cen.acs.org/biological-chemistry/infectious-disease/How-do-viruses-leap-from-animals-to-people-and-spark-pandemics/98/i33.

[9] Carroll, D., Daszak, P., Wolfe, N. D., Gao, G. F., Morel, C. M., Morzaria, S., et al. 2018. The global virome project. *Science* 359(6378):872–4. doi: 10.1126/science.aap7463.

[10] Global Virome Project. 2021. Accessed September 21, 2022. www.globalviromeproject.org.

[11] Jonas, O. and Seifman, R. 2019. Do we need a global virome project?. *The Lancet Global Health* 7(10):e1314–e6. doi: 10.1016/S2214-109X(19)30335-3.

[12] European Medicine Agency. 2022a. "Pandemic Influenza Vaccine H5N1 AstraZeneca (Previously Pandemic Influenza Vaccine H5N1 Medimmune)". Accessed July 7, 2023. www.ema.europa.eu/en/medicines/human/EPAR/pandemic-influenza-vaccine-h5n1-astrazeneca-previously pandemic-influenza-vaccine-h5n1-medimmune.

[13] European Medicine Agency. 2022b. "Annex I Summary of Product Characteristics". Accessed July 4, 2023. www.ema.europa.eu/en/documents/product-information/foclivia-epar-product-information_en.pdf.

[14] European Medicine Agency. 2022c. "Adjupanrix (Previously Pandemic Influenza Vaccine (H5N1) (Split Virion, Inactivated, Adjuvanted) GlaxoSmithKline Biologicals)". Accessed July 7, 2023. www.ema.europa.eu/en/medicines/human/EPAR/adjupanrix-previously pandemic-influenza-vaccine-h5n1-split-virion-inactivated-adjuvanted#product-information-section.

[15] European Medicine Agency. 2022d. "Foclivia". Accessed July 7, 2023. www.ema.europa.eu/en/medicines/human/EPAR/foclivia#authorization-details-section.

[16] European Medicine Agency. 2023a. "Vaccines for Pandemic Influenza". Accessed July 7, 2023. www.ema.europa.eu/en/human-regulatory/overview/public-health-threats/pandemic-influenza/vaccines-pandemic-influenza.

[17] European Medicine Agency. 2023b. "Pandemic Influenza Vaccine H5N1 Baxter AG". Accessed July 7, 2023. www.ema.europa.eu/en/medicines/human/EPAR/pandemic-influenza-vaccine-h5n1-baxter-ag.

[18] European Medicines Agency. 2022e. "Supemte". Accessed April 03, 2023. www.ema.europa.eu/en/medicines/human/EPAR/supemtek.

[19] Carlson, R. and Lutmer, H. 2023. "Bird Flu Vaccines 2023". Accessed July 11, 2023. www.precisionvaccinations.com/bird-flu-vaccines-2023.

[20] Precision Vaccinations. 2023. "Audenz Avian Influenza Vaccine". Accessed July 11th, 2023. www.precisionvaccinations.com/vaccines/audenz-avian-influenza-vaccine.

[21] Jordan, D., Tumpey, T., and Jester, B. 2019. "The Deadliest Flu: The Complete Story of the Discovery and Reconstruction of the 1918 Pandemic Virus". Accessed June 16, 2022. www.cdc.gov/flu/pandemic-resources/reconstruction-1918-virus.html.

[22] Krammer, F. and Palese, P. 2015. Advances in the development of influenza virus vaccines. *Nature Reviews Drug Discovery* 14(3): 167–82. doi: 10.1038/nrd4529.

[23] Berman, R. 2022. "Is a Universal Flu Vaccine Within Reach?". Accessed March 13, 2023 www.medicalnewstoday.com/articles/is-a-universal-flu-vaccine-within-reach.

[24] Subbarao, K. and Joseph, T. 2007. Scientific barriers to developing vaccines against avian influenza viruses. *Nature Reviews Immunology* 7(4): 267–78. doi: 10.1038/nri2054.

[25] Cortellis Drug Discovery Intelligence. 2022. "CDDI – Home". Accessed March 13, 2023. www.cortellis.com/drugdiscovery/home.

[26] World Health Organization (WHO). 2010. "WHO Declares H1N1 Pandemic Over, Urges Continued Influenza Vigilance". Accessed April 03, 2023. www.paho.org/en/news/10-8-2010-who-declares-h1n1-pandemic-over-urges-continued-influenza-vigilance.

[27] Lens. 2023. "Lens – Home". Accessed March 17, 2023. www.lens.org/lens/.

[28] Intellectual Property in India (inPASS). 2022b. "Patent Search". Accessed November 21, 2022. https://ipindiaservices.gov.in/publicsearch.

[29] Clarivate. 2022a. "Derwent World Patents Index Manual Codes". Accessed March 13, 2023. https://clarivate.com/derwent/dwpi-reference-center/dwpi-manual-code/.

[30] Intellectual Property in India (inPASS). 20221. "Advanced Search System". Accessed March, 13, 2023. https://ipindiaservices.gov.in/publicsearch.

[31] World Health Organization (WHO). 2017. "WHO Preferred Product Characteristics for Next Generation Influenza Vaccines". Accessed March 13, 2023. https://apps.who.int/iris/bitstream/handle/10665/258767/9789241512466-eng.pdf.

[32] National Library of Medicine. 2023. "Pubmed". Accessed March 13, 2023. www.ncbi.nlm.nih.gov/pubmed/.

[33] Darricarrère, N., Pougatcheva, S., Duan, X., Rudicell, R. S., Chou, T. H., DiNapoli, J., Ross, T. M., Alefantis, T., Vogel, T. U., Kleanthous, H., Wei, C. J., and Nabel, G. J. 2018. Development of a Pan-H1 influenza vaccine. *Journal of Virology* 92(22):e01349–18. doi: 10.1128/JVI.01349-18.

[34] U.S. Food and Drug Administration. 2022. "Flublok Quadrivalent". Accessed March 13, 2023. www.fda.gov/vaccines-blood-biologics/vaccines/flublok-quadrivalent.

[35] Adimmune Corporation. 2023a. "Company Profile". Accessed April 05, 2023. www.adimmune.com.tw/en/about_intro.php

[36] Adimmune Corporation. 2023b. "Products. AdimFlu-S (QIS). Quadrivalent Influenza Vaccine (2023)". Accessed March 13, 2023. www.adimmune.com.tw/en/product/detail/adimflu_s__qis_.

[37] Gresset-Bourgeois, V., Leventhal, P. S., Pepin, S., Hollingsworth, R., Kazek-Duret, M. P., Bruijn, I. D., and Samson, S. I. 2018. Quadrivalent inactivated influenza vaccine (VaxigripTetra™). *Expert Review of Vaccines* 17(1): 1–11. doi: 10.1080/14760584.2018.1407650.

[38] Thiem, V. D., Chabanon, A. L., Fournier, M., Lavis, N., Quang, N. D., Ha, V. H., and Sanicas, M. 2021. Safety of a quadrivalent influenza vaccine in Vietnamese healthy subjects aged 6 months and older. *Human Vaccines & Immunotherapeutics* 17(3): 690–3. doi: 10.1080/21645515.2020.1795477.

[39] Petrovax. 2023a. "Grippol® Quadrivalent". Accessed March 13, 2023. https://petrovax.com/medication/catalog/grippol_quadrivalent/.

[40] Talayev, V., Zaichenko, I., Svetlova, M., Matveichev, A., Babaykina, O., Voronina, E., and Mironov, A. 2020. Low-dose influenza vaccine Grippol Quadrivalent with adjuvant Polyoxidonium induces a T helper-2 mediated humoral immune response and increases NK cell activity. *Vaccine* 38(42): 6645–6655. doi: 10.1016/j.vaccine.2020.07.053.

[41] World Health Organization (WHO). 2022d. "Recommendations Announced for Influenza Vaccine Composition for the 2022–2023 Northern Hemisphere Influenza Season". Accessed April 03, 2023. www.who.int/news/item/25-02-2022-recommendations-announced-for-influenza-vaccine-composition-for-the-2022-2023-northern-hemisphere-influenza-season.

[42] Sanofi Pasteur. 2018. “Former Shareholder Information”. Accessed March 13, 2023. www.proteinsciences.com/.

[43] Sanofi Pasteur. 2021. “Flublok Quadrivalent”. Accessed March 13, 2023. www.sanofiflu.com/flublok-quadrivalent-influenza-vaccine.html.

[44] Centers for Disease Control and Prevention (CDC). 2021b. “Recombinant Influenza (Flu) Vaccine”. Accessed October 21, 2022. www.cdc.gov/flu/prevent/qa_flublok-vaccine.htm.

[45] Sanofi Pasteur. 2017. “Sanofi Invests €170 Million in New Vaccine Production Facility in France”. Accessed March 13, 2023. www.sanofi.com/en/media-room/press-releases/2017/2017-10-12-05-00-00-1144755.

[46] Petrovax. 2023b. “Petrovax Today”. Accessed March 13, 2023. https://petrovax.com/about/today/.

[47] Petrovax. 2018. “Grippol® Quadrivalent: The First 4-Valent Influenza Prevention Vaccine Authorized in Russia”. Accessed March 13, 2023. https://petrovax.com/press_center/news/2018/1344/.

[48] World Intellectual Property Organization (WIPO). 2023. “How to Protect Inventions through Patents?”. Accessed April 03, 2023. www.wipo.int/patents/en/protection.html

[49] Clarivate. 2022b. “Derwent Innovations Index on Web of Science”. Accessed May 21, 2021. https://clarivate.com/webofsciencegroup/solutions/webofscience-derwent-innovation-index/ml.

[50] Isakova-Sivak, I. N., and Rudenko, L. G. 2015a. New attenuated cold-adapted strain of H1N1 and reassortant vaccine strain of H2N2 influenza virus belonging to Ortomyxoviridae family useful in vaccine. RU Patent 2556833C2, filed Jul. 12, 2011, and issued Jul. 20, 2015a.

[51] Isakova-Sivak, I. N., and L. G. Rudenko. 2015b. Use of influenza virus A/17/new Caledonia/99/145 hemagglutinin 1 neuraminidase 1 as donor of attenuation for preparing e.g. influenza virus A/17/Tokyo/67/912 H2N2 vaccine strain, used for preparing intranasal vaccine for e.g. infant. RU Patent 2556834C2, filed Jul. 12, 2011, and issued Jul. 20, 2015b.

[52] Larionova, N. V., E. A. Bazhenova, E. V. Krutikova, V. A. Matyushenko, I. V. Kiseleva, and L. G. Rudenko. 2019. A/17/Singapore/2016/3571 (H3N2). 2019. Vaccine influenza strain for producing live influenza intranazal vaccine for adults and children. RU Patent 2683500C1, filed May, 29, 2018, and issued Mar. 28, 2019

[53] Bazhenova, E. A., I. V. Kiseleva, E. V. Krutikova, E. A. Stepanova, N. V. Larionova, and L. G. Rudenko. 2020. Vaccine strain of influenza virus A/17/Brisbane/2017/7178 (H3N2) for production of live influenza intranasal vaccine for adults and children. RU Patent 2711101C1, filed Sep. 2, 2019, and issued Jan. 15, 2020.

[54] Stepanova, E. A., I. V. Kiseleva, T. S. Kotomina, I. N. Isakova-Sivak, E. V. Krutikova, N. V. Larionova, and L. G. Rudenko. 2020b. Vaccine strain of influenza virus A/17/Switzerland/2017/51 (H3N2) for production of live influenza intranasal vaccine for adults and children. RU Patent 2715674C1, filed Sep. 2, 2019, and issued Mar. 2, 2020-b.

[55] Stepanova, E. A., E. V. Krutikova, E. A. Bazhenova, N. V. Larionova, I. V. Kiseleva, T. S. Kotomina, I. N. Isakova-Sivak, and L. G. Rudenko. 2020c. Vaccine Strain of Influenza A/17/Kansas/2017/631 (H3N2) Virus for Production of Live Influenza Intranasal Vaccine for Adults and Children. RU Patent 2732610C1, filed Dec. 17, 2019, and issued Sep. 21, 2020-c.

[56] Stepanova, E. A., E. A. Bazhenova, E. V. Krutikova, N. V. Larionova, I. V. Kiseleva, T. S. Kotomina, I. N. Isakova-Sivak, and L. G. Rudenko. 2020a. Vaccine strain of influenza virus A/17/Brisbane/2018/3610 (H3N2) for production of live influenza

intranasal vaccine for adults and children. RU Patent 2734897C1, filed Dec. 17, 2019, and issued Oct. 26, 2020-a.

[57] Isakova-Sivak, I. N., V. A. Matyushenko, I. V. Kiseleva, E. A. Bazhenova, and L. G. Rudenko. 2020. Vaccine strain of influenza virus A/17/Slovenia/2015/1121 (H1N1) PDM09 for production of live influenza intranasal vaccine for adults and children. RU Patent 2724706C1, filed Sep. 2, 2019, and issued Jun. 25, 2020.

[58] Stepanova, E. A., E. V. Krutikova, E. A. Bazhenova, N. V. Larionova, I. V. Kiseleva, V. A. Matyushenko, I. N. Isakova-Sivak, and L. G. Rudenko. 2020d. Vaccine strain of influenza virus A/17/Brisbane/2018/8175 (H1N1) PDM09 for production of live influenza intranasal vaccine for adults and children. RU Patent 2735291C1, filed Dec. 17, 2019, and issued Oct. 29, 2020-d.

[59] Sychev, I. A., Y. A. Desheva, E. A. Bazhenova, S. A. Donina, and L. G. Rudenko. 2020. Reassortant strain of influenza virus Rn2/14-Human A (H6N2) for detecting neuraminidase antibodies in influenza infection and vaccination. RU Patent 2716416C1, filed Sep. 24, 2019, and issued Mar. 11, 2020.

[60] Bazhenova, E. A., S. I. N. Isakova, T. A. Kramskaya, G. F. Leonteva, A. S. Matushkina, V. A. Matyushenko, D. A. Mezhenskaya, P. I. Prokopenko, Rudenko Larisa Georgievna, Smolonogina T. A., and Suvorov A. N. 2022. Influenza virus strain influbact-h7/pspa for the production of a combined vaccine against influenza a virus and bacterial pneumonia caused by streptococcus pneumoniae. RU Patent 2776196C1, filed Aug 24, and issued Jul. 14, 2022.

[61] Bormotova, E. A., T. V. Gupalova, S. I. N. Isakova, T. A. Kramskaya, E. V. Kuleshevich, G. F. Leonteva, D. A. Mezhenskaya, L. G. Rudenko, and Suvorov A. N. 2022. Live probiotic vaccine containing conservative influenza a virus epitopes (LAH+4M2E) for the prevention of influenza infection. RU Patent 2776479C1, filed Sep. 18, 2022, and issued Jul. 21, 2022.

[62] Isakova-Sivak, I. N., V. A. Matyushenko, D. A. Mezhenskaia, L. G. Rudenko, and E. A. Stepanova. 2022. Recombinant vaccine strain for live intranasal vaccine providing combined prevention of influenza and coronavirus infections. RU Patent 2782531C1, filed Apr. 15, 2022, and issued Oct. 28, 2022.

[63] Bazhenova, E. A., I. V. Kiseleva, E. V. Krutikova, N. V. Larionova, L. G. Rudenko, and E. A. Stepanova. 2022b. Reassortant strain of influenza virus A/17/SYDNEY/2019/156 (H3N2) for the production of live intranasal influenza vaccine for adults and children. RU Patent 2783876C1, filed Apr. 15, 2022, and issued Nov. 21, 2022.

[64] Bazhenova, E. A., I. V. Kiseleva, E. V. Krutikova, N. V. Larionova, L. G. Rudenko, and E. A. Stepanova. 2022. Reassortant strain of influenza virus A/17/South Australia/2019/122 (H3N2) for production of live intranasal influenza vaccine for adults and children. RU Patent 2783877C1, filed Apr. 15, 2022, and issued Nov. 21, 2022.

[65] Bazhenova, E. A., S. I. N. Isakova, I. V. Kiseleva, E. V. Krutikova, N. V. Larionova, V. A. Matyushenko, L. G. Rudenko, and E. A. Stepanova. 2022. Reassortant strain of influenza virus A/17/VICTORIA/2019/276 (H1N1) PDM09 for the production of live intranasal influenza vaccine for adults and children. RU Patent 2783878C1, filed Apr. 15, 2022, and issued Nov. 21, 2022.

[66] Bazhenova, E. A., S. I. N. Isakova, I. V. Kiseleva, E. V. Krutikova, N. V. Larionova, L. G. Rudenko, and E. A. Stepanova. Reassortant strain of influenza virus A/17/NEWCASTLE/2018/22 (H3N2) for the production of live intranasal influenza vaccine for adults and children. RU Patent 2783887C1, filed Apr. 15, 2022, and issued Nov. 21, 2022.

[67] Bazhenova, E. A., S. I. N. Isakova, I. V. Kiseleva, T. S. Kotomina, E. V. Krutikova, N. V. Larionova, L. G. Rudenko, and E. A. Stepanova. 2022. Reassortant strain of

influenza virus A/17/HONG KONG/2019/2573 (H3N2) for the production of live intranasal influenza vaccine for adults and children. RU Patent 2783894C1, filed Apr. 15, 2022, and issued Nov. 21, 2022.

[68] Bazhenova, E. A., S. I. N. Isakova, I. V. Kiseleva, E. V. Krutikova, N. V. Larionova, L. G. Rudenko, and E. A. Stepanova. 2023. Reassortant strain of influenza virus A/17/GUANGDONG-MAONAN/2019/211 (H1N1) PDM09 for production of live intranasal influenza vaccine for adults and children. RU Patent 2793387C1, filed Apr.27, 2022, and issued Apr. 4, 2023.

[69] Rudenko, L. G., Y. A. Desheva, and T. A. Smolonogina. 2015. Live quadrivalent influenza vaccine useful for preventing influenza vaccine virus type AA hemagglutinin type 1 and neuraminidase type 1 (H1N1), A H3N2 and type BB/Victoria and B/Yamagata. RU Patent 2013149103A, filed Nov. 5, 2013, and issued May 10, 2015.

[70] Rudenko, L. G., Y. A. Desheva, and T. A. Smolonogina. 2018. Quadrivalent live influenza intranasal vaccine used for prophylaxis of seasonal influenza, comprises two vaccine strains of type B-V/Victoria and B/Yamagata, and are characterized as a donor attenuation and it uses strain. RU Patent 2016146943A, filed Nov. 29, 2016, and issued May 29, 2018.

[71] Kiseleva, I. V., N. V. Larionova, L. G. Rudenko, and G. I. Aleksandrova. 2011a. Vaccine strain of a/17/brisbane/07/1 (H3N2) influenza virus for preparing live influenza intranasal vaccine for adults and children. RU Patent 2416640C1, filed Jan. 11, 2010, and issued Apr. 20, 2011a.

[72] Kiseleva, I. V., N. V. Larionova, L. G. Rudenko, and G. I. Aleksandrova. 2011b. Vaccine strain of a/17/brisbane/07/28 (H1N1) influenza virus for preparing live influenza intranasal vaccine for adults and children. RU Patent 2416641C1, filed Jan. 11, 2010, and issued Apr. 20, 2011b.

[73] Larionova, N. V., L. G. Rudenko, and G. I. Aleksandrova. 2011a. Vaccine strain of b/60/brisbane/08/83 influenza virus for preparing live influenza intranasal vaccine for adults and children. RU Patent 2422517C1, filed Jan. 28, 2010, and issued Jun. 27, 2011a.

[74] Larionova, N. V., I. V. Kiseleva, L. G. Rudenko, and G. I. Aleksandrova. 2011. Vaccine strain of a/17/solomon islands/06/9 (H1N1) influenza virus for preparing live influenza intranasal vaccine for adults and children. RU Patent 2422518C1, filed Jan. 11, 2010, and issued Jun. 27, 2011.

[75] Larionova, N. V., L. G. Rudenko, and G. I. Aleksandrova. 2011b. Vaccine strain of b/60/florida/04/181 influenza virus for preparing live influenza intranasal vaccine for adults and children. Russian RU Patent 2422519C1, filed Jan. 11, 2010, and issued Jun. 27, 2011b.

[76] Desheva, J. A., L. G. Rudenko, and G. I. Aleksandrova. 2012b. Reassortant a/17/new caledonia/99/79(H1N1) influenza virus strain for producing live influenza intranasal vaccine. RU Patent 2464310C1, filed Jun. 16, 2011, and issued Oct. 20, 2012b.

[77] Desheva, J. A., L. G. Rudenko, and G. I. Aleksandrova. 2012a. Influenza virus strain for producing live influenza intranasal vaccine for adults and children. RU Patent 2464311C1, filed May 24, 2011, and issued Oct. 20, 2012a.

[78] Larionova, N. V., I. A. Dubrovina, I. V. Kiseleva, E. A. Bazhenova, G. I. Aleksandrova, and L. G. Rudenko. 2014. Use of specific strain of influenza virus for production of influenza intranasal epidemic vaccine for adult and children. RU Patent 2532844C2, filed Jul. 17, 2012, and issued Nov. 10, 2014.

[79] Dubrovina, I. A., N. V. Larionova, I. V. Kiseleva, E. A. Bazhenova, L. G. Rudenko, and G. I. Aleksandrova. 2015. New influenza A virus useful for preparing nasal vaccine

composition for treating influenza A viral infection in adult and children. RU Patent 2563352C2, filed Nov. 5, 2013, and issued Sep. 20, 2015.

[80] Smolonogina, T. A., Y. A. Desheva, A. R. Rekstin, and L. G. Rudenko. Reassortant influenza virus strain rn9/13-human a(H6N9) for neuraminidase antibody assay in influenza infection and vaccination. RU Patent 2587629C1, filed Dec. 29, 2014, and issued Jun. 20, 2016.

[81] Rudenko, L. G., G. I. Aleksandrova, N. V. Larionova, I. A. Dubrovina, I. V. Kiseleva, E. A. Bazhenova, and V. A. Kuznetsova. 2015. Strain of influenza virus a/17/indiana/11/72 (H3N2v) for production of liquid influenza intranasal vaccine for adults and for children. RU Patent 2013149106A, filed Nov. 5, 2013, and issued May 10, 2015.

[82] Desheva, Y. A., T. A. Smolonogina, and L. G. Rudenko. 2017. Reassortant influenza virus strain a/17/silver gull/sarma/06/887 (H6N1) for preparing live influenza vaccine. RU Patent 2606026C1, filed Jun. 9, 2015, and issued Jan. 10, 2017.

[83] Bazhenova, E. A., I. V. Kiseleva, V. A. Kuznetsova, E. A. Fedorova, and L. G. Rudenko. 2016. Vaccine influenza virus strain a/17/bolivia/2013/6585 (H1N1)PDM09 for preparing live influenza intranasal vaccine for adults and children. RU Patent 2605314C1, filed Oct 23, 2015, and issued Dec. 20, 2016.

[84] Bazhenova, E. A., I. V. Kiseleva, V. A. Matyushenko, E. A. Fedorova, and L. G. Rudenko. 2017. Vaccine influenza virus strain a/17/hongkong/2014/8296 (H3N2) for preparing live influenza intranasal vaccine for adults and children. RU Patent 2606019C1, filed Mar 10, 2016, and issued Jan. 10, 2017.

[85] Rudenko, L. G., T. A. Smolonogina, I. N. Isakova-Sivak, T. S. Tretyak, and I. V. Kiseleva. 2017. Vaccine influenza virus a/17/south africa/2013/01 (H1N1)PDM09 for production of live intranasal influenza vaccine for adults and children. RU Patent 2627188C1, filed Jun. 28, 2016, and issued Aug. 3, 2017.

[86] Smolonogina, T. A., Y. A. DesSheva, and L. G. Rudenko. 2017. Reassortant strain of rn2/66-human a (H7N2) influenza virus for neuraminidase antibodies identification in case of influenza infection and vaccination. RUS Patent 2625024C1, filed Apr. 7, 2016, and issued Jul. 11, 2017.

[87] Dubrovina, I. A., I. V. Kiseleva, and L. G. Rudenko. 2018. Vaccine strain of influenza a/17/new york/15/5364 (H1N1) pdm09 for production of live influenza intranasal vaccine for adults and children. RU Patent 2653388C1, filed Feb. 13, and issued May 8, 2018.

[88] Aleksandrov, A. A., A. N. Doronin, P. An Iakovlev, S. I. N. Isakova, I. N. Isakova-Sivak, V A. Matyushenko, V. A. Matyushenko, D. V. Morozov, B. O. Neterebskii, S. A. Niskanen, T. I. Ostroukhova, T. U. Ostroukhova, L. G. Rudenko, T. M. Sheudzhen, E. A. Stepanova, I. I. Ustyugov, I. U. Ustyugov, Ya Yu Ustyugov, A. K. Vladimirova, and P. A. Yakovlev. 2023. New isolated nucleic acid encoding recombinant polypeptide for increasing titer of antibodies to influenza virus useful for inducing specific immunity, i.e., specific T-cell immunity to influenza virus and coronavirus-related diseases. RU Patent 2021121139A, filed Jul. 16, 2021, and issued Jan. 16, 2023.

[89] Legastelois, I., J. Medina, and C. Moste. 2019. Producing infectious influenza viruses comprises transfecting cells with expression vectors, and infecting Chinese hamster ovary cells with the produced seed of infectious influenza virus. EP Patent 2880154B1, filed Jul. 29, 2013, and issued May, 15, 2019.

[90] Ross, T. M., T. Alefantis, D. M. Carter Jr., C. A. Darby, and H. Kleanthous. 2021. Immunogenic composition or prophylaxis of influenza infection or disease comprises a combination of recombinant H1N1 influenza hemagglutinin polypeptides. US Patent 10925957B2, filed Jan. 25, 2019, issued Feb. 23, 2021.

[91] Ross, T. M., T. Alefantis, D. M. Carter, C. A. Darby, and H. Kleanthous. 2020. Immunogenic composition used for prophylaxis of influenza infection or disease comprises recombinant H5N1 influenza hemagglutinin polypeptides. US Patent 10548965B2, filed Dec. 18, 2015, and issued Feb. 4, 2020.

[92] Strugnell, T. D., G. Cortes-Garcia, and T. Alefantis. 2021. Generating optimized nucleotide sequence for preparing influenza vaccine composition, involves providing an amino acid sequence of engineered influenza structural protein and reverse-translating amino acid sequence. US Patent 20210130416A, filed Jan. 12, 2021, and issued May 6, 2021.

[93] Cortes-Garcia, G., T. Alefantis, and M. Parrington. 2017. Analyzing expression and conformation of engineered hemagglutinin (HA) antigens, by immunostaining cells comprising nucleic acid sequence encoding engineered HA antigen with neutralizing antibody, and determining expression of antigen. US Patent 20170121373A1, filed May 29, 2015, and issued May 4, 2017.

[94] Alefantis, T. and S. Chivukula. 2023. Vaccine composition used for preparing medicament for eliciting immune response against seasonal and/or pandemic influenza strains, preventing influenza infections or reducing symptoms of influenza infections, comprises eight mRNA comprising open reading frame encoding different influenza antigen. US Patent 20230043128A1, filed Jun. 17, 2022, and issued February 9, 2023.

[95] Alefantis, T., M. Barro, A. Byers, P. Davidson, M. Giel-Moloney, P. Gilbert, H. Kleanthous, A. Naik, K. Pugachev, S Sridhar, W. Warren, Zeldovich and Konstantin. 2023. Immunogenic composition useful as vaccine composition for immunizing subject against influenza virus, and preventing influenza virus infection in subject, comprises influenza virus hemagglutinin from standard of care influenza virus strain. WO Patent 2023059857A1, filed Oct. 7, 2022, and issued Apr. 13, 2023.

[96] Alefantis, T., M. Barro, S. Sridhar, T. Vogel, and W. Warren. 2023. Immunogenic composition used to e.g. prevent influenza virus infection comprises influenza virus protein comprising e.g. influenza virus hemagglutinin (HA) protein, and RNA molecule, encoding e.g. the influenza virus HA proteins. WO Patent 2023079113A1, filed Nov. 4, 2022, and issued May 11, 2023.

[97] Alefantis, T., M. Barro, A. Byers, G. Cortes-Garcia, P. Gilbert, H. Kleanthous, A. Naik, K. Pugachev, S. Sridhar, T. Vogel, and W. Warren. 2023. Immunogenic composition used for e.g. immunizing subject against influenza virus comprises multiple recombinant influenza virus proteins comprising e.g. first recombinant influenza virus hemagglutinin antigens. WO Patent 2023081798A1, filed Nov. 4, 2022, and issued May 11, 2023.

[98] Defossa, E., H. Glombik, U. Heinelt, H. Matter, M. Mendez-Perez, N. Rackelmann, K. Ritter, L. Schwink, H. Szillat, and G. Zech. 2021. New isoxazolidine compound used as medicine or in pharmaceutical composition for treating e.g. necrotizing enterocolitis. WO Patent 2021245070A1, filed Jun. 1, 2021, and issued Dec. 9, 2021.

[99] Chang, L.-J, V. Landolfi, and T. Quinn. 2021. Immunizing pediatric subject against influenza virus, preventing influenza virus infection in pediatric subject, or raising protective immune response preferably antibody response in pediatric subject, involves administering high-dose quadrivalent influenza vaccine. WO Patent 2021173907A1, filed Feb. 26, 2021, and issued Sep. 2, 2021.

[100] Nabel, G. J., C. J. Wei, N. Darricarrere, M. Kanekiyo, J. R. Mascola, and B. S. Graham. 2021. vaccine composition for vaccinating against influenza virus comprises stem-stabilized influenza hemagglutinin-ferritin nanoparticles and adjuvant, which elicits broadly neutralizing influenza antibodies when administered to human subject. WO Patent 2021231729A1, filed May, 13, 2021, and issued Nov. 18, 2021.

[101] Roozendaal, R., A. Roos, and K. Radosevic. 2017. Vaccinating against influenza comprises administering a seasonal influenza vaccine that comprises influenza hemagglutinin and neuraminidase proteins from at least three influenza strains. EP Patent 2822585B1, filed Mar. 5, 2013, and issued May 17, 2017.

[102] Beigelman, L., D. B. Smith, A. D. Stoycheva, and R. T. Hendricks. 2021. New aza-pyridone derivative useful in pharmaceutical composition for ameliorating or treating influenza viral infection by inhibiting replication of influenza virus e.g. influenza A, influenza B, or H1N1, H3N2, H5N1 and H7N9 virus. KR Patent 2021135359A, filed Sep. 10, 2014, and issued Nov. 12, 2021.

[103] Hendricks, R T, L Beigelman, D B Smith, and A D Stoycheva. 2021. New aza-pyridone compound used in pharmaceutical composition useful for terracing orthomyxovirus viral infection e.g. influenza. CN Patent 107531717B, filed Mar. 9, 2016, and issued Jul. 27, 2021.

[104] Impagliazzo, A., J. W. Meijberg, K. Radosevic, M. Wagner, and Z. Ding. 2021. New influenza hemagglutinin (HA) stem domain polypeptide comprises an influenza hemagglutinin HA1 domain useful for inducing an immune response against influenza virus, and as a vaccine. EA Patent 38400B1, filed Jul. 9, 2015, and issued Aug. 23, 2021.

[105] Jonckers, T. H. M., D. C. Mc Gowan, J. E. G. Guillemont, W. C. J. Embrechts, G. J. M. Mercey, C. F. R. N. Buyck, W. M. A. Balemans, and J.-M. Raboisson. 2021. New substituted pyrimidine compounds used to treat influenza, and inhibit the replication of influenza virus(es) in a biological sample or patient. CN Patent 108473477B filed Jan. 19, 2017, and issued Oct. 22, 2021.

[106] Beigelman, L., G. Wang, D. B. Smith, J. Deval, and M. Prhavc. 2021. New substituted nucleoside and nucleotide compounds are viral replication inhibitors useful to ameliorate or treat paramyxovirus viral infection e.g. human respiratory syncytial virus infection or orthomyxovirus viral infection. IN Patent 373620B, filed Jul 17, 2014, and issued Aug. 6, 2021.

[107] Meijberg, J. W., A. Impagliazzo, R. Vogels, R. H. E. Friesen, P. Alard, S. Loverix, and K. Radosevic. 2021. New influenza hemagglutinin stem domain polypeptide comprising influenza hemagglutinin HA2 domain and HA1 domain including HA1 N-terminal segment, linked by specified amino acid residues to HA1 C-terminal segment, used to treat influenza. IN Patent 379884B, filed Jun. 24, 2014, and issued Oct. 29, 2021.

[108] Jonckers, T. H. M., D. C. Mc Gowan, P. J.-M. B. Raboisson, W. C. J. Embrechts, and J. E. G. Guillemont. 2017. New pyrrolopyrimidine derivatives useful for treating influenza. Indian IN Patent 201727012541, filed Apr. 7, 2017, and issued Jun. 16, 2017.

[109] Juraszek, J., M. V. Dongen, and B. Brandenburg. 2020. New peptidic macrocyclic peptide compound capable of specifically binding to hemagglutinin (HA) of influenza A virus strain comprising HA of the H1 subtype, for manufacturing medicament for diagnosing, preventing and/or treating influenza. US Patent 20200048308A1, filed Oct. 26, 2017, and issued Feb. 13, 2020.

[110] Brandenburg, B., T. Ritschel, F. J. Milder, M. A. C. Jongeneelen, I. King, Y I-Qiao Song, and J. P M Langedijk. 2021. Recombinant influenza A hemagglutinin (HA) polypeptide for inducing an immune response against an influenza virus, comprises an HA1 and a HA2 domain of an influenza A virus HA, and comprising an amino acid sequence. CN Patent 113597428A, filed Apr. 23, 2020, and issued Nov. 2, 2021.

[111] Jongeneelen, M. A. Catharina, T. Ritschel, F. J. Milder, I. King, Y. Song, J. P. M. Langedijk, and B. Brandenburg. 2021. Monomeric influenza A hemagglutinin stem polypeptide used as vaccine in pharmaceutical composition for inducing immune

response against influenza virus, where polypeptide has amino acid sequence with deletion of head region from HA1 domain. WO Patent 2021043869A1, filed Sep. 3, 2020, and issued Mar. 11, 2021.

[112] Trusheim, H., P. Mason, M. Franti, B. Keiner, M. Sackal, J. Hunziker, D. Morrissey, and F. J. C. Natt. 2018. Preparing reassortant influenza virus used for vaccine involves infecting culture host with two different influenza strains; and contacting it with inhibitory agent which preferentially reduces transcription/translation of one of strain. EP Patent 2571520B1, filed May 20, 2011, and issued Apr. 4, 2018.

[113] Tsai, T. 2019. Immunizing a child comprises administering to the child adjuvanted influenza B virus immunogenic composition. US Patent 20190247489A1, filed Nov. 16, 2018, and issued Aug. 15, 2019.

[114] Suphaphiphat, P., P. Mason, B. Keiner, P. Dormitzer, and H. Trusheim. 2020. New reassortant influenza A virus comprising backbone segments from two or more donor strains, used for preparing a vaccine used for preventing and treating influenza infection. US Patent 20200000909A1, filed Feb. 11, 2019, and issued Jan. 2, 2020.

[115] Ahmed, S. S., L. Steinman, and W. Volkmuth. 2019. Influenza vaccine composition against influenza strains comprises influenza virus A nucleoprotein which has been modified to reduce its binding to major histocompatibility complex class II receptor comprising human leukocyte antigen alleles. US Patent 20190184007A1, filed Mar. 6, 2019, and issued June 20, 2019.

[116] Del Giudice, G., R. Rappuoli, S. Black, and U. Nicolay. Vaccine useful for enhancing immune response against influenza infection in patient on statin therapy, comprises adjuvant, high-dose antigen, or combination of adjuvant and high-dose antigen. AU Patent 2021202331A1, filed Apr. 19, 2021, and issued May 13, 2021.

[117] Bodle, J, N Guzzo-Pernell, and S Rockman. 2021. Dispersing agglomerated material in preparation comprising influenza proteins used for producing influenza vaccine, involves subjecting preparation to sonication. JP Patent 2021531339A, filed Jul. 10, 2019, and issued Nov. 18, 2021.

[118] Bollands, A., M. Scionti, D. Hofmann, and E. Settembre. 2022. Method for partially matched vaccine supplies use in early stages of pandemic, involves correlating manufacturing schedule to expiration date of finished vaccines, and determining stockpile requirement based on received information regarding one or more factors. WO Patent 2022005893A1, filed Jun. 25, 2021, and issued Jan. 6, 2022.

[119] Otten, G., M. Ciarlet, C. Mandl, A. Feller, and B. Leav. 2021. Vaccine composition used to induce immune response, enhance immunogenicity, treat influenza infection or raise immune response against influenza comprises antigens from different influenza virus strains and oil-in-water emulsion adjuvant. AU Patent 2020228151A1, filed Feb. 24, 2020, and issued Oct. 21, 2021.

[120] Rockman, S., J. Bodle, and N. Guzzo-Pernell. 2021. Dispersing agglomerated material in preparation comprising influenza proteins used for producing influenza vaccine, involves subjecting preparation to sonication. AU Patent 2019302562A1, filed Jul. 10, 2019, and issued Jan. 14, 2021.

[121] Harrington, W. N., and Kackos, C. M., Webby, R. J. 2021. The evolution and future of influenza pandemic preparedness. *Experimental & Molecular Medicine* 53(5): 737–49. doi: 10.1038/s12276-021-00603-0.

[122] Commonwealth Serum Laboratories Seqirus Australia. 2022. "Cell-based Flucelvax® Quad influenza vaccine now approved for Australian children aged two years and older". Accessed April 27, 2023. www.seqirus.com.au/news/cell-based-flucelvax-quad-influenza-vaccine-now-approved-for-au-children-aged-2-years-and-older.

[123] Ertl, H. C. J, and D. Zhou. 2013. New fusion polypeptide comprises first matrix protein, a second matrix protein, a third matrix protein, and a nucleoprotein (NP) from a fourth strain of influenza A virus, useful as medicament for immunization against influenza A virus. US Patent 20130209512A1, filed Aug 16, 2011, and issued Aug. 15, 2013.

[124] Mookkan, P., F. He, and H. S. J. Kwang. 2013. Composition useful in medicament and medicine for modulating immune response against avian influenza A H5N1 virus, and for treating disease associated with avian influenza virus, comprises immunogenic agent and carrier. IN Patent 201203735P2, filed Nov. 29, 2012, and issued Jun. 14, 2013.

[125] Hufton, S. New influenza virus antibody, useful for preventing or treating influenza virus infection and used as a reagent in an assay. IN Patent 201401597P1, filed Mar. 3, 2014, and issued Apr. 22, 2016.

[126] Ryan, W. L. and J. A. Grukemeyer. 2018. Composition useful as or in vaccines for inducing immune response in subject, comprises deactivated swine flu virus having infectious component and surface antigens in contact with formaldehyde donor agent having specific molecular weight. US Patent 20180243405A1, filed Apr. 30, 2018, and issued Aug. 30, 2018.

[127] Ertl, H. C. and D. Zhou. New modified adenovirus hexon protein comprising first matrix protein ectodomain from first strain of influenza A virus inserted in hypervariable region of hexon protein, useful e.g. for inducing immune response to influenza A virus. US Patent 20140377295A1, filed May 23, 2012, and issued Dec. 25, 2014.

[128] Youn, J. W., J. S. Kwon, J. S. Yoon, Y. J. Kim, and Y. Yoon. 2014. T-cell-based universal vaccine useful for treating flu caused by influenza virus, comprises nucleoprotein gene and polymerase acidic gene, or nucleoprotein gene, polymerase acidic gene, polymerase B1 and B2 genes and matrix 1 gene. KR Patent 1382244B1, filed Mar. 30, 2012, and issued Apr. 8, 2014.

[129] Xu, J., Q. Ma, Y. J. Zhang, Z. X. Gao, and W. M. Zheng. 2014. Preparing novel general epitope vaccine for inhibiting influenza A virus, comprises adopting bioinformatics method and combining network server and related software to forecast related nucleoprotein, M1 and hemagglutinin function epitope. CN Patent 102600466B, filed Mar. 23, 2012, and issued Jun. 11, 2014.

[130] Meng, S. D., Y. Ju, J. Hu, L.-Z. Chen, Z. F. Li, and B. Zhao. 2015. New cross-protective effect influenza vaccine with a common heat shock glyco protein 96, useful for treating and/or preventing H1N1 influenza virus, human influenza virus, swine influenza virus or avian influenza virus. CN Patent 103505725B, filed Sep. 30, 2013, and issued Jun. 24, 2015.

[131] Shan, H., G.-M. Li, J. Huang, C.-M. Zhang, and H. Liu. 2016. Influenza virus universal gene vaccine for preventing occurrence of various influenza virus diseases, is formed by cloning hemagglutinin 2 gene of H1N1 virus strain of influenza with pCAGGS eukaryotic expression vector. CN Patent 103948942B, filed Apr. 26, 2014, and issued Mar.23, 2016.

[132] Shishido, T., D. Asari, K. Matsushita, and M. Hori. 2019. Humoral immunity inducing composition useful for preventing infectious diseases caused by virus e.g. adenovirus, herpesvirus and picornavirus, bacterial infection and fungal infection in human patient, comprises antithrombotic drug. US Patent 10456464B2, filed Aug. 4, 2015, and issued Oct. 29, 2019.

[133] Esmagambetov, I. B., E. S. Sedova, D. N. Shcherbinin, A. Lysenko, M. M. Shmarov, D. Y. Logunov, B. S. Naroditskij, and A. L. Gintsburg. 2017. Vaccine useful for exhibiting

antiviral activity against influenza virus, comprises recombinant pseudoadenoviral particles having full-length protein M2 gene and NP gene linked with peptide 2A of foot and mouth disease virus. RU Patent 2618918C2, filed Oct. 14, 2015, and issued May 11, 2017.

[134] Li, L. and J. H. Lu. 2016. Host bacterium used in preparing universal avian influenza vaccine for preventing avian influenza diseases is prepared by linking sequentially CotB gene promoter sequence, CotB coding sequence, linker and interleukin-2. CN Patent 105950646A, filed Apr. 25, 2016, and issued Sep. 21, 2016.

[135] Shu, Y. L., X. J. Liu, C. J. Sun, T. Y. Zhao, C. M. Luo, and M. C. Li. 2023. New common mosaic recombinant antigen used in preparing seasonal influenza A universal vaccine and preparing medicines for preventing and/or treating seasonal influenza A, comprises antigen consisting of specific amino acid sequence. CN115894636A, filed Jul. 15, 2022, and issued Apr. 4, 2023.

[136] Veljkovic, V. and S. Paessler. 2019. Eliciting immune response for treating or preventing influenza infection in subject, involves using synthetic peptide having informational spectrum that overlaps with informational spectrum of one hemagglutinin protein or fragment. US Patent 20190351046A1, filed Feb. 2, 2018, and issued Nov. 21, 2019.

[137] Seong, B. L., Y. H. Jang, and Y. H. Byun. 2021. Universal influenza live vaccine composition comprises a first influenza live vaccine that carries six internal genes, influenza virus-derived surface hemagglutinin gene and influenza virus-derived surface neuraminidase gene. US Patent 10953086B2, filed Feb. 1, 2019, and issued Mar. 23, 2021.

[138] Xu, J. Q., X. Y. Zhang, J. Wang, L. Y. Zhu, and B. W Lubit. 2021. Universal vaccine against immunogen of infectious pathogenic organism is selected from virus, bacteria, fungus or protozoan comprises RNA polynucleotide comprising open reading frame encoding polypeptide antigen or immunogenic fragment. WO Patent 2021141758A1, filed Dec. 18, 2020, and issued Jul. 15, 2021.

[139] Jeong, D. G., S. W. Yun, H. K. Kim, and D. S. Song. 2021. Fusion protein trimer useful as expression vector in vaccine and medicine composition for preventing influenza virus infection e.g. sinusitis, comprises short helical domain of hemagglutinin of influenza virus, and ectodomain of M2 protein. KR Patent 2211378B1, filed Feb. 12, 2019, and issued Feb. 4, 2021.

[140] Yao, X. J. 2021. Fusion protein for targeting influenza virus surface protein peptide to dendritic cell, comprises influenza virus surface protein peptide selected from hemagglutinin inserted in mucin-like domain of Filoviridae virus glycoprotein. WO Patent 2021077215A1, filed Oct. 21, 2020, and issued Apr. 29, 2021.

[141] Nachbagauer, R., P. Palese, F. Krammer, A. Garcia-Sastre, A. W Freyn, N. Pardi, and D. Weissman. 2021. Composition useful for inducing immune response against influenza viruses chosen from influenza virus A, influenza virus B, influenza virus C, and influenza virus D in human, comprises at least one isolated nucleoside-modified RNA encoding at least one influenza virus antigen or its fragment. WO Patent 2021202734A3, filed Mar. 31, 2021, and issued Nov. 11, 2021.

[142] Pardi, N., D. Weissman, R. Nachbagauer, S. Hensley, P. Palese, F. Krammer, A. Garcia-Sastre, and A. W. Freyn. 2023. Composition useful for inducing immune response against influenza viruses chosen from influenza virus A, influenza virus B, influenza virus C, and influenza virus D in human, comprises at least one isolated nucleoside-modified RNA encoding at least one influenza virus antigen or its fragment. US Patent 20230181715A1, filed Mar. 31, 2021, and issued Jun. 15, 2023.

[143] Ricke, D. O. 2021. Immunogenic composition used to e.g. induce immune response against influenza virus comprises polypeptide having amino acid sequence of influenza viral protein having hypervariable amino acid residues and conserved amino acid residues. US Patent 20210379179A1, filed Feb. 26, 2021, and issued Dec. 9, 2021.

[144] Ren, Z. G., Y. W. Gao, X. Z. Xia, Y. Dou, Z. Y Yang, T. C. Wang, M. T. Chen, Y. K. Zhao, Y. G Li, N. Feng, and S. T. Yang. 2020. New universal DNA vaccine useful for influenza A, preparing influenza A universal DNA vaccine, and preparing medicine for preventing or treating influenza A, comprises a hemagglutinin fusion protein overexpression plasmid. CN Patent 111514287A, filed Apr. 29, 2020, and issued Aug. 11, 2020.

[145] Gao, R. B. and Z. H Zhang. 2023. New common influenza virus and coronavirus combined polypeptide comprising 6 polypeptide sequences useful in preparing common influenza virus general and coronavirus combined polypeptide vaccine for preparing medicine for preventing influenza viruses and coronaviruses. CN Patent 115947801A, filed Dec. 29, 2022, issued Apr. 11, 2023.

[146] Tenoever, B. R. 2022. Preventing or treating influenza virus infection or influenza virus disease comprises administering nucleic acid molecule comprising polynucleotide encoding L7Ae protein or fragment/variant or L7Ae protein or fragment/variant. WO Patent 2022061216A1, filed Sep. 20, 2021, and issued Mar. 24, 2022.

[147] Kang, Y. M., H. M. Kang, H. K. Cho, M. H. Lee, S. J. Park, L. T. Thanh, S. C. Moon, M. C. Kim, J. H. Kim, and L. Sujin. 2022. New recombinant vector comprising e.g. chimeric hemagglutinin gene containing hemagglutinin-1 region of H5N1 virus and matrix protein-encoding gene, useful for preventing avian influenza virus infection. KR Patent 2022108440A, filed Jan. 27, and issued Aug. 3, 2022.

[148] Amanna, I. J. and A. Ray. 2023. Multivalent paninfluenza vaccine useful eliciting immune response in subject comprises a viral hemagglutinin protein or HA1-containing portion and/or a nucleic acid encoding the HA protein or portion corresponding to a virus strain. WO Patent 2023015145A1, filed Jul. 29, 2022, and issued Feb. 9, 2023.

[149] Ascione, R. and S. QI. 2022. Polypeptide immunogenic composition used in treatment of influenza virus comprises at least one mimetic peptide comprising consecutive amino acid sequence, amino-acid spacer moiety, and immunogenic carrier coupled to mimetic peptide. WO Patent 2022032274A1, filed Jul. 30, and issued Feb. 10, 2022.

[150] Arevalo, C. P., Bolton, M. J., Sage, V. L., Ye, N., Furey, C., Muramatsu, H., et al. 2022. A multivalent nucleoside-modified mRNA vaccine against all known influenza virus subtypes. *Science* 378(6622): 899–904. doi: 10.1126/science.abm0271.

[151] Francis, M. J. 2020. A veterinary vaccine development process map to assist in the development of new vaccines. *Vaccine* 38(29): 4512–4515. doi: 10.1016/j.vaccine.2020.05.007.

[152] World Organization for Animal Health (WOAH). 2022. "One Health". Accessed March 13, 2023. www.woah.org/en/what-we-do/global-initiatives/one-health/.

[153] World Health Organization (WHO). 2022e. "WHO Coronavirus (COVID-19) Dashboard". Accessed March 13, 2023. https://covid19.who.int.

[154] Centers for Disease Control and Prevention (CDC). 2023 "Chile Reports First Case of H5N1". Accessed March 30, 2023. www.cdc.gov/flu/avianflu/spotlights/2022-2023/chile-first-case-h5n1.htm.

[155] Possas, C., Antunes, A. M. S., Mendes, F. M. L., Schumacher, S. O. R., Martins, R. M., and Homma, A. 2015. Access to new technologies in multipatented vaccines: challenges for Brazil. *Nature Biotechnology* 33(6): 599–603. doi: 10.1038/nbt.3244.

[156] Moore, K. A., Ostrowsky, J. T., Kraigsley, A. M., Mehr, A. J., Bresee, J. S., Friede, M. H., Gellin, B. G., Golding, J. P., Hart, P. J., Moen, A., Weller, C. L., and Osterholm, M. T. 2021. A Research and Development (R&D) roadmap for influenza vaccines: Looking toward the future. *Vaccine* 39(45): 6573–6584. doi: 10.1016/j.vaccine.2021.08.010.

7 Intellectual Property Issues in Life Sciences

Contemporary Challenges and Considerations

Rabiatul Basria S. M. N. Mydin and Adam Azlan

7.1 INTRODUCTION

Intellectual property refers to creations such as inventions, designs, and artistic work that can be protected under the law, which enables the creator to earn benefits and be recognized for their creations (World Intellectual Property Organization, WIPO). Intellectual property can be categorized into four items, namely, copyrights, patents, trade secrets, and trademarks. Depending on the laws in specific regions of the world, the definition of these criteria can also influence IP rights issued to a specific body [1,2]. Ownership of these rights, especially in datasets and big data, still serves as a pressing subject, as several aspects need to be taken into account for proper IP awards [3]. Molecular tools or techniques under IP protection should not impede the progress of scientific development; although innovation deserves IP protection, fair implementation should be taken into consideration for proper dissemination of research and development [4]. Although stakeholders' interests should also be taken into consideration as investments were made in the development of the tools, a fair agreement should be emphasized to not limit the use of the tools for researchers.

7.2 LABORATORY-BASED MOLECULAR BIOLOGY TECHNIQUES

Laboratory-based molecular biology techniques refer to several new protocols or tools discovered in molecular biology that allow for the detection of specific patterns that could contribute to deducing information. Tools in life science include polymerase chain reaction (PCR), expressed sequence tags (ESTs), and functional genes, and these tools are also subject to IP rights [5–7]. These tools enable further progress in research and development, allowing for improvements in diagnostics. In the case of PCR, this technique enables life science research to progress as such a significant phase, especially in molecular biology, which enables applications such as gene quantification, isolation, and characterization. A crucial component of this technique, the DNA polymerase enzyme, serves as the pivotal issue in patenting, leading to several arguments that perpetuated court proceedings and limited patent application [6,8,9].

DOI: 10.1201/9781003405665-7

Proceedings issued several claims to the DNA polymerase that would involve other polymerases and several issues of rights given prior to the commercialization of PCR.

Fore *et al.* (2006) discuss the procurement of Cetus Corporation by Roche in which the PCR was invented led to Roche US and Europe patenting of the *Taq* polymerase comes with some issues in IPs [10,11]. A concern was raised in a correspondence in *Nature* stating that the Europen patenting was too broad. The patent claims RNA and DNA polymerase that are between 86 and 90 kDa. The patent claims would not only encompass *Taq* but also involve other competitor polymerases, such as the *Pfu* from Stratagene and Vent from New England Biolabs [8]. Making matters worse, a court ruling was made stating that Roche's claims on the Taq patent were made in fraudulent manners [5]. However, Roche also filed a lawsuit against Promega stating that the Taq used by Promega encourages scientists to use PCR, which infringes on Roche's PCR patent [10]. Promega rights over Taq were obtained earlier from Cetus, in which Promega would be given the rights to manufacture, sell, and use Taq, which has led to Roche's suit [10]. This further adds to the controversies around *Taq* polymerase sparking several litigations [10,12]. The issue with broad patent claims and prior rights could limit the practicality of DNA polymerase.

Additionally, an almost similar case was observed in the patenting of expressed sequence tags (ESTs) [13]. ESTs are naturally occurring sequences flanking the mRNA sequence or coding sequence of a gene that can be used to trace for the presence of the gene in the genome [14], bypassing complex protocols such as genomic sequencing and allowing for rapid detection [15]. An issue that arises from this patenting is the 1999 patent award to Incyte Pharmaceuticals Inc. (IPI) by the United States Patent and Trademark Office (USPTO). This issue seems to arise from the broad claims of the EST, which could also include other ESTs that are yet to be discovered [7]. Moreover, in the dispute between the USPTO and *Monsanto*, the EST patent claimed by Monsanto was rejected, as these ESTs do not convey any substantial function [16]. Patenting the DNA sequence without substantial evidence of its function could place any other sequence discovered adjacent to the ESTs into the same patent [17,18]. The nature of ESTs itself would limit claims as these sequences flank gene regions, patenting this sequence could indirectly subject any functional sequence between the ESTs to the same IPs, which could limit applications.

The issue in IP in life science is that patenting would limit access to crucial information that is pivotal for biological research (Figure 7.1). The argument supporting this notion was stated in the proceedings of the Organization of Economic Co-Operation and Development (OECD), in which patenting biological information would hamper progress, as patenting would exclude others from using the patented information [19]. The OECD also states that the increase in research cost could also be attributed to patenting, which will contribute to the slower pace of research. Furthermore, an abuse of monopoly could also arise, as was seen in the case of Myriad. Myriad patented BRCA1/2 and has exclusive rights in diagnosing this gene, leading to an expensive diagnostic method [20] for patients with BRCA1/2 disorder in breast cancer [19]. Thus, the patenting of functional genes should come with a fair implementation, and considerations should be made, as these genes could be pivotal in diagnosing patients where limitations would impact proper testing.

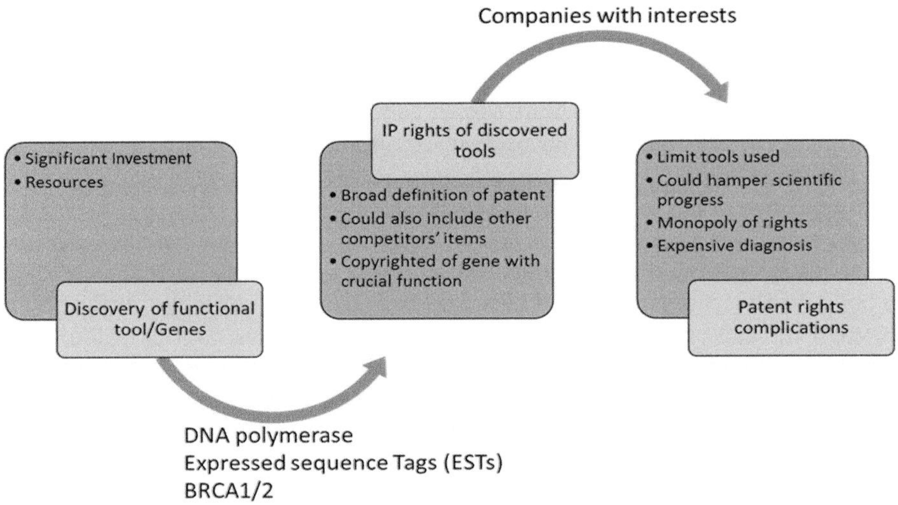

FIGURE 7.1 IP issues in molecular tools. Functional molecular tools begin with significant investments and resources that led to the discovery of tools. This eventually led to the application of patents and copyrights from respective investors, which limited the usage of the technology.

7.3 PUBLICLY AVAILABLE DATASETS

Intellectual property rights over datasets remain elusive and context dependent, as no clear ground rule has been established for the propagation of datasets. Complexity in issuing IP could also be observed in datasets and big data. Datasets in life science refer to a collection or an array of data that can be utilized for obtaining information on a specific subject [21]. These datasets are usually in the form of collections deposited in a database that is by itself a Big Data Hub for life sciences [22]. Depending on the data, different IP rights could be used to segregate them. Patent rights to datasets depend on the context of the "data". If the data could be used to derive a genetically modified organism, it is patentable. However, in the case of inventions arising from the datasets, in this context, the article published does not disclose the invention, but datasets made public do nullify patenting [23]. Therefore, a proper categorization should be implemented in this issue whereby a proper definition or expansion of the categories would facilitate IP of the dataset.

Moreover, datasets could also be classified as trade secrets. This falls within the scope of the commercial sponsor and the depositor. For example, companies may invest in research and generate datasets that are pivotal for the development of a product. In this situation, based on the draft agreement between the depositor and sponsor, the datasets may not be immediately made public until a patentable product is obtained; this would require the researcher to halt the release of such data [23]. This may impair scientific progress, as datasets generated could expand scientific knowledge [24]; however, due to sponsorship, commercial sponsors have rights, as a significant amount of investment was made to acquire such data [25]. This could be observed in the case of Myriad, in which the company withheld its variants of unknown significance (VUS) data from the Breast Cancer Information Core (BIC), an open-access database even after patent rights expire, which could be due to the company concerns with competitors abroad and patenting issues [20]. In this issue, the company made significant investments to procure such data; however, crucial data that could help in proper patient diagnosis should be made available; thus, a fair implementation of the rights could be practiced.

Copyrighting data and datasets, on the other hand, remain elusive and subject to debate. Copyright refers to the rights of exclusivity given to an original work of an author, a creative representation of ideas and facts [23]. Moreover, the elusiveness in copyrighting datasets remains, as the concept of data is a new concept of the digital age since copyrighting originally refers to printed materials. However, databases could be subject to copyright since creative arrangements and compilations could be considered original [23]. Moreover, depending on the law of copyrighting in different regions, a rather different definition was given to the works falling under copyrighting [26]. The difference in these definitions puts copyrighting datasets under unclear categorization and would further complicate this issue (Figure 7.2).

Australia's definition of a database generally states that the database must be generated by a human, not copied and involved intellectually independent effort [23]. The United States, on the other hand, states that a database must be compiled from independent author works and have some degree of creativity, while the European Union involves a more complex description in which it generally states that rights

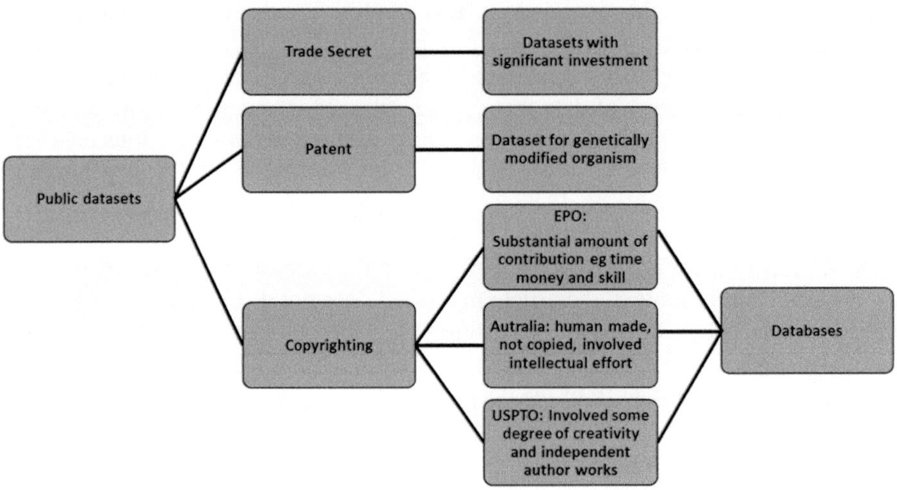

FIGURE 7.2 Categorizing public datasets. Depending on the criteria defining copyrights in several countries, publicly available datasets can fall in all this definition. However, copyrights and patents are usually awarded to the database, which contains a substantial amount of creativity in storing and indexing these datasets. Datasets for genetically modified organisms could fall under patenting, while datasets with investment from sponsors would be considered trade secrets.

will be given to the creators of databases who invest enough skill, time, and money for the generation of their databases [26]. Note that these criteria refer to a database in which datasets are included not directly to the dataset itself. To add to the vagueness, the UK Parliament ruled out that mining of research literature is permissible [23] (Intellectual Property Office, United Kingdom, Exceptions to Copyright: Research, 2014). In this case, whether datasets fall within the scope of the research literature still needs proper elucidation.

Patient privacy issues in datasets could also complicate intellectual property claims. If these datasets contain primary patient information, ethical considerations should be taken since these datasets were generated from primary individual samples [24]. Informed consent should be given during patient declarations [24] for these datasets to be deposited, but whether these datasets protect individual rights to their genomic or transcriptomic data is still ambiguous [27]. To further complicate the issue, the use of the same datasets for multiple publications could also complicate intellectual property claims. This would affect copyright claims, as a derivative from the same datasets could further add to the complication of the layers of copyrights [23,24]. The issue here is whether the copyrights on the original data would also include information derived from the original datasets.

7.4 THE BIG DATA REVOLUTION COMES WITH ISSUES

Intellectual property rights awarding for big data is a pressing concern, and the issue arising from this point is that it is uncertain whether big data, usually composed of datasets collected in a database, could fall within the aspects of IP [28]. Although the database is copyrightable, the copyright of datasets within these databases is still vague. Big data also refers to the collection of data from multiple sources or individuals, for example, the sequence read archives of the NCBI. These datasets are not copyrightable; however, a proposal was made if copyright was to be applied to DNA sequences [29]. In this context, patents are not applicable once the DNA sequence is made available publicly, to overcome this, depositors can convert the DNA sequence into a file, and issue a copyright to that file, it was proposed to convert the DNA sequence into a music file. In this case, a copy of the music file is copyrightable, and the sequence could be available by sharing these files [29]. This, however, remains a proposal, as implementation of this method would add to the complexity of data sharing (Figure 7.3).

Issues also seem to arise during the human genome boom. This was observed in the case of *Celera,* in which the company patented its human genome assembly. In this case, if a company develops a diagnostic method based on a gene patented by *Celera,* a problem would arise as licensing should be discussed before the development of the diagnostic method, as licensing post-development would delay the company's progress in developing the diagnostic method to negotiate the licensing, the company had to share its research ideas, which would also put the company at risk of competition with *Celera* [30]. Some issues also arise during the sharing of the data, as researchers were reluctant to share these data with colleagues to not infringe on contractual agreements [30]. This would also lead to the slowing down of research progress due to the uncertainties in data sharing.

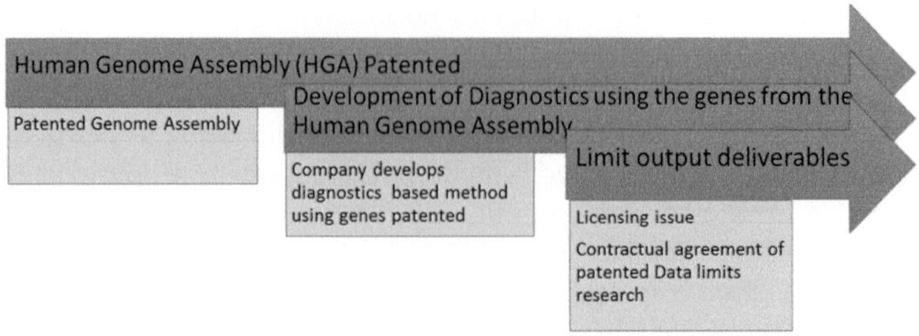

FIGURE 7.3 Common IP issue related to big data. Big data such as the HGA served as a major point for the development of genome-based diagnostics. However, due to the patenting of the HGA, the development of these diagnostics would be delayed by several issues regarding licensing and contractual agreements, which could put a significant delay in the advancement of technology.

7.5 CONCLUSION

IP rights are crucial for the interest of the creators; nevertheless, it is crucial to determine to what extent they can be implemented to not impose on scientific progress. Additionally, IPs should not limit access to information that is crucial in treatment or progress; a rather lenient approach in sharing resources should be made. A clear contractual agreement could address this issue. Moreover, countries could develop a unified article of law regarding IPs that could standardize and allow for a better classification and rights awarded. Despite this, the current understanding of IPs in life science still needs to be debated for fair issuance to be made.

ACKNOWLEDGMENTS

We would like to acknowledge the Ministry of Higher Education for the Fundamental Research Grant Scheme (Project Code: FRGS/1/2021/SKK03/USM/02/3) and for the facilities provided by Universiti Sains Malaysia.

REFERENCES

[1] Council NR. Global dimensions of intellectual property rights in science and technology. National Academies Press, Washington, DC; 1993.

[2] Upreti PN. The role of national and international intellectual property law and policy in reconceptualising the definition of investment. *IIC –Int Rev Intellect Prop Compet Law*. 2021 Feb 1;52(2):103–136.

[3] Mittelstadt BD, Floridi L. The Ethics of Big Data: Current and Foreseeable Issues in Biomedical Contexts. In: Mittelstadt BD, Floridi L, editors. The Ethics of Biomedical Big Data [Internet]. Cham: Springer International Publishing; 2016. p. 445–80. (Law, Governance and Technology Series) [cited 2023 Sep 26]. Available from: https://doi.org/10.1007/978-3-319-33525-4_19

[4] Lee P. Patents, paradigm shifts, and progress in biomedical science. *Yale Law J.* 2004;114(3):659.

[5] Dalton R. Roche's Taq patent 'obtained by deceit', rules US court. *Nature*. 1999;402(6763):709.
[6] Dalton R. Promega and Roche take up battle over PCR patents. *Nature*. 2000 Mar 1;404(6773):7.
[7] Patenting ESTs: is it worth it? *Nat Genet*. 1999 Feb;21(2):145–6.
[8] Crowther SM, Burke JF, Thomas SM. Polymerase patent problems. *Nature*. 1994 Nov 24;372(6504):312.
[9] U.S. Supreme Court Strikes Down Human Gene Patents | Science | AAAS [Internet]. [cited 2023 Sep 26]. Available from: www.science.org/content/article/us-supreme-court-strikes-down-human-gene-patents
[10] Fore J, Wiechers IR, Cook-Deegan R. The effects of business practices, licensing, and intellectual property on development and dissemination of the polymerase chain reaction: case study. *J Biomed Discov Collab*. 2006 Jul 3;1:7.
[11] McDowell D. The polymerase chain reaction patents: going, going,... still going. *J R Soc Med*. 2006 Feb;99(2):62–4.
[12] Dutfield G. Intellectual property rights and the life science industries: a twentieth century history [Internet]. Routledge, London; 2017 [cited 2023 Sep 26]. Available from: www.taylorfrancis.com/books/mono/10.4324/9781315252131/intellectual-property-rights-life-science-industries-graham-dutfield
[13] Calvert J, Joly PB. How did the gene become a chemical compound? The ontology of the gene and the patenting of DNA. *Soc Sci Inf*. 2011 Jun 1;50(2):157–77.
[14] Parkinson J, Blaxter M. Expressed sequence tags: an overview. *Methods Mol Biol Clifton NJ*. 2009;533:1–12.
[15] Murray D, Doran P, MacMathuna P, Moss AC. In silico gene expression analysis – an overview. *Mol Cancer*. 2007 Aug 7;6(1):50.
[16] Brazil M. ESTs must pass specific and substantial utility test. *Nat Rev Drug Discov*. 2005 Dec 1;4(12):950.
[17] Lawrence S. US court case to define EST patentability. *Nat Biotechnol*. 2005 May 1;23(5):513.
[18] Hopkins MM, Mahdi S, Patel P, Thomas SM. DNA patenting: the end of an era? *Nat Biotechnol*. 2007 Feb;25(2):185–7.
[19] OECD. Genetic Inventions, Intellectual Property Rights and Licensing Practices Evidence and Policies: Evidence and Policies. OECD Publishing, France; 2003. 112 p.
[20] Bakshi AM. Gene patents at the supreme court: association for molecular pathology v. myriad genetics. *J Law Biosci*. 2014 Jun 1;1(2):183–9.
[21] Abugessaisa I, Kasukawa T, editors. Practical Guide to Life Science Databases [Internet]. Singapore: Springer Nature; 2021 [cited 2023 Sep 19]. Available from: https://link.springer.com/10.1007/978-981-16-5812-9
[22] Members CN. Database resources of the national genomics data center, China National Center for Bioinformation in 2022. *Nucleic Acids Res*. 2022;50(D1):D27.
[23] Carroll MW. Sharing research data and intellectual property law: a primer. *PLoS Biol*. 2015 Aug 27;13(8):e1002235.
[24] Sielemann K, Hafner A, Pucker B. The reuse of public datasets in the life sciences: potential risks and rewards. *PeerJ*. 2020 Sep 22;8:e9954.
[25] Thessen AE, Patterson DJ. Data issues in the life sciences. *ZooKeys*. 2011; 150:15–51.
[26] Lipton V. Legal Issues Arising in Open Scientific Data. In: Open Scientific Data – Why Choosing and Reusing the RIGHT DATA Matters [Internet]. IntechOpen; 2020 [cited 2023 Sep 13]. Available from: www.intechopen.com/chapters/71064
[27] Oestreich M, Chen D, Schultze JL, Fritz M, Becker M. Privacy considerations for sharing genomics data. *EXCLI J*. 2021 Jul 16;20:1243–60.

[28] Lu M. Intellectual property protection of big data. *J Phys Conf Ser*. 2020 Dec;1693(1):012012.
[29] Stemmer WPC. How to publish DNA sequences with copyright protection. *Nat Biotechnol*. 2002 Mar;20(3):217–217.
[30] Williams HL. Intellectual property rights and innovation: Evidence from the human genome. *J Polit Econ*. 2010 Jul 1;121(1):1–27.

8 Navigating Regulatory Challenges and Intellectual Property Rights for Gene-Edited Crops

Implications for Plant Breeders, Farmers, and Consumers in a Changing Climate

Elisabeth Abergel

8.1 INTRODUCTION

The prospect of genetic editing is expected to usher in a novel Green Revolution [1] while aligning with the objectives of the United Nations' Sustainable Development Goals (SDGs), particularly "SDG 2: Zero Hunger" and "SDG 13: Climate Action" [2]. According to researchers involved in the development of NPBTs, gene-editing technologies promise significant advancements in agriculture. First, they offer the potential for enhanced crop yields by rendering crops more resilient to pests and diseases [3], ultimately fostering heightened agricultural productivity and bolstering global food security [3]. Additionally, gene-editing techniques facilitate nutritional improvement through the development of biofortified crops [4] enriched with augmented vitamin or mineral content as a solution to global malnutrition. Moreover, these technologies contribute to a diminished environmental footprint in agriculture by reducing the dependency on chemical pesticides and fertilizers, thus alleviating the ecological pressures associated with farming practices [5]. It is also claimed that gene editing can increase farmer profitability through enhanced yields and healthier, more resilient crops [6]. This is especially important in economies where agriculture is a major sector. Additionally, the promises of precision and speed include the fact that unlike traditional breeding methods, gene editing offers fast and predictable

DOI: 10.1201/9781003405665-8

outcomes [7]. However, critical ethical, regulatory, and safety outlooks related to gene editing in agriculture must be considered to responsibly develop and use these technologies [8].

The utilization of genetic-editing methods in plant breeding has triggered global legal and ethical discussions, although these debates can vary depending on the specific type of genetic editing. Furthermore, the multitude of regulatory strategies and the intricacy of the intellectual property rights (IPR) framework necessitate a rigorous evaluation of the most suitable mechanisms for governing new plant breeding techniques (NPBTs). These technologies are on the verge of widespread introduction into both commercial markets and natural ecosystems within the coming years, underscoring the need for comprehensive assessment.

NPBTs represent emerging methodologies distinct from conventional breeding and traditional genetically modified (GM) approaches [9]. These new techniques are also known as gene-editing or genome-editing techniques because they enable the intended alteration of genetic materials within an organism's genome either by substituting, deleting, or adding DNA sequences. Genome-editing techniques in plants have revolutionized the field of plant biology and breeding. These tools allow scientists to precisely modify the DNA of plants, leading to targeted alterations of specific genes. The primary focus of this chapter is on the clustered regularly interspaced short palindromic repeats (CRISPR) and CRISPR-associated protein (Cas) system, a widely used genome-editing method in plants [10].

Gene editing encompasses a variety of techniques, with the clustered regularly interspaced short palindromic repeats (CRISPR)-Cas system taking the forefront. CRISPR–Cas technology employs a guide RNA molecule to direct a Cas protein to a specific location in an organism's genome. This enzyme then cleaves the DNA at that location, and the cell's natural repair mechanisms are utilized to introduce precise genetic changes. Alongside CRISPR–Cas, there are other nuclease-based methods, including meganucleases, zinc finger nuclease (ZFN) technology, and transcription activator-like effector nucleases (TALENs), all of which share the capacity to target specific DNA sequences, cleave DNA, and facilitate editing through the cellular repair process [11,12].

Three broad categories of genome-editing technologies warrant in-depth discussion in the context of regulatory considerations. More specifically, SDN-1 and SDN-2 are two different types of genome-editing techniques that differ in the way they introduce changes to the DNA sequence. SDN-1, or site-directed nuclease-1, involves the use of a single nuclease enzyme, such as Cas9, to introduce a double-strand break at a specific location in the genome. The cell's natural DNA repair mechanisms then repair the break, often introducing small insertions or deletions that can disrupt the function of the targeted gene. SDN-1 is relatively simple and efficient, but it can also introduce unintended mutations at off-target sites in the genome [13]. This technique is particularly effective for eliminating undesirable genetic sequences. SDN-2, on the other hand, relies on creating a double-strand break in the DNA at the target site, but instead of precise repair, it relies on the cell's error-prone repair mechanisms. This often leads to insertions or deletions, resulting in gene disruption. This approach is more complex and less efficient than SDN-1, but it can be more precise and reduce off-target mutations (Ibid.). The regulatory treatment of SDN-1 and SDN-2 differs from country to country depending on the presence of foreign DNA in the gene-edited plant. This will be discussed later in the chapter.

In addition to CRISPR–Cas and nuclease-based techniques, there are other advanced methods, such as base editing and prime editing. Base editing allows for the specific replacement of one DNA letter with another, offering high precision. Prime editing, a further refinement of the CRISPR–Cas system, enables the rewriting of a short piece of DNA using a synthetic RNA template [15]. Finally, the SDN-3 technique involves the insertion of additional genes into an organism's genome at predetermined locations. This approach is not found in nature and is a more complex manipulation that can result in the introduction of new functions or traits (Ibid.). This category of gene-edited plants falls under GMO regulations, as they contain foreign DNA and are considered transgenic.

The accelerated pace of research and commercialization of genome-edited crops underscores the strategic importance of regulatory frameworks and social acceptance for both developers and societal movements. Notably, the timeframes for crop breeding programs have drastically shortened, ranging from 7 to 25 years to as few as 2–3 years, depending on plant reproductive systems [14]. Furthermore, research and development costs for introducing a new trait have significantly diminished, with the use of NPBTs reducing expenses to less than a tenth of transgenic methods [15]. Considering the economic dimensions, regulatory frameworks play a pivotal role. It is estimated that genome-edited plants, if regulated as non-GM plants, would incur costs of approximately USD 10.5 million and take 5 years to reach the market, in contrast to the estimated USD 24.5 million and approximately 14 years for regulation as GM organisms [15]. The length and costs associated with regulatory systems constitute significant factors for developers in this evolving landscape.

It is important to highlight some of the debates centering around the safety and environmental concerns associated with gene-editing technologies, highlighting the spectrum of views from all parties involved—scientists, policymakers, and consumer advocacy groups. In analyzing the potential impact of these technologies on both small- and large-scale farming operations, the focus shifts to understanding their potential impact on agriculture and food security issues. While the benefits for large farms are obvious, small-scale farmers may face challenges accessing and utilizing them, but more importantly and for the purposes of this chapter, it is important to underscore that small holders may face problems surrounding the detectability and traceability of genome-edited crops [16].

8.2 NAVIGATING THE GLOBAL REGULATORY LANDSCAPE FOR GENE-EDITED PRODUCTS

Many countries are in the process of revising their regulatory frameworks to govern the use of NPBTs, deliberating on whether genome editing should comply with existing GMO regulations. In instances where they are categorized as exempt from GM biosafety assessments, mechanisms are being devised to oversee their use, implementation, and commercialization. Currently, there is a lack of international regulatory harmonization, and the different regulatory approaches adopted by different countries may present challenges in the areas of trade, research, and development, and the commercialization of gene-edited products [17]. Furthermore, this "regulatory patchwork" also determines the scope of intellectual property rights applicable to NPBTs.

The United Nations Cartagena Protocol on Biosafety (CPB) could play a role in shaping the regulatory landscape for new plant breeding techniques such as CRISPR–Cas. The CPB primarily addresses the transboundary movements of living modified organisms (LMOs) resulting from modern biotechnology, which encompasses genetically modified organisms (GMOs). It provides a global framework for GMOs and safety assessment, which may serve to harmonize approaches for genome-edited organisms. The CPB has an impact on the classification and regulation of new plant breeding techniques within its scope. Nonetheless, the treaty's broad use of terminology, such as "modern biotechnology," does not adequately encompass the nuances linked to methods such as CRISPR–Cas [18]. Furthermore, the diverse approaches taken by individual nations in classifying and regulating GMOs result in varying interpretations among countries about how to govern NPBTs.

The European Union has been the setting for a substantial portion of the debate on the status of NPBT-produced plants, leading to the decisive ruling by the European Court of Justice in July 2018 to regulate gene-edited products as GMOs. The initiation of this legal process was prompted by an investigation launched by the agricultural union Conféderation Paysanne in France, which advocates for the rights of small-scale farmers. This inquiry aimed to determine whether organisms produced through mutagenesis could be classified as genetically modified organisms (GMOs) due to the technique's capacity to modify an organism's genetic material in a manner that is not naturally occurring. As per the Court's decision, the European Union's directive on GMOs is applicable to organisms derived through recent mutagenesis techniques that have arisen since the directive's adoption. Regarding the matter of whether the EU regulations on genetically modified organisms are applicable to organisms acquired through NPBTs, the Court deliberated that the risks associated with the utilization of these methods are akin to those of transgenesis. Transgenesis involves the introduction of a foreign gene into an organism. The Court's rationale is that the direct alteration of an organism's genetic material through mutagenesis allows for the achievement of analogous effects, rendering the risks comparable [19].

Since then, a request made in 2019 by the European Council to the Commission clarified the legal status of the plants obtained via genome editing within the framework of EU law, taking into account the Court's judgment. The study, released in 2021, determined that organisms derived from new genetic technologies (NGTs), particularly targeted mutagenesis, cis-genesis, and intra-genesis, should be classified as genetically modified organisms [17]. Moreover, the aim of the study was to obtain information regarding NPBTs in terms of their status within EU member states, detection methods, risk evaluation, and ethical considerations.

Even though the European Union designated NPBT-derived organisms as GMOs, necessitating labeling and mandatory biosafety assessments, changes in their legal status might be coming. In July 2023, the European Commission introduced a proposal to deregulate gene-edited plants to facilitate research and commercialization

[20], namely, allowing field testing of gene-edited crops. The proposed legislation would exclude gene-edited plants from the existing regulations on genetically modified organisms (GMOs) if they exhibit the same characteristics achievable through traditional plant breeding methods. Without going into the details of the proposed legislation, it establishes two categories of gene-edited plants: those already mentioned that are considered comparable or "equivalent" to conventional plants for which induced genetic mutations occur in nature regardless of human intervention and plants that have undergone more complex genetic modifications [19]. In the first case, these plants would not be considered GMOs and would not be subject to risk assessment, while the second category of plants would fall under the GMO directive. The report also states that NPBTs would not be permitted in organic agriculture, as their use is incompatible with the current definition of organic production as well as consumers' perception of organic products. Generally, it would appear that the EU is moving toward a product-based system rather than the process-based system in place for GMOs. The outcome of these deliberations within the EU bears significant importance for the global landscape of agricultural biotechnology even though it might take some years before the proposed law comes into effect.

In the United Kingdom, a new category of organism called "precision bred organisms" was introduced in a new bill in 2022 [21], which includes gene-edited plants. The bill distinguishes "precision breeding" from genetic modification because its techniques do not insert functional DNA from an unrelated species into a given organism. According to Tachikawa and Matsuo, the framework allows genome-edited organisms "through a simplified procedure in safety assessment while including them in the regulatory framework for GMOs" [17]. China is adopting a similar approach to regulating NPBTs.

South American countries such as Argentina, Brazil, Chile, Paraguay, and Colombia are likely to adopt a regulatory position that encourages the commercialization of gene-edited plants (Ibid.). Argentina was the first country to adopt a "prior consultation procedure" for crops derived from genome edition. Like Canada and the United States, Argentina does not regulate plants produced via genome-editing techniques unless they contain foreign DNA, in which case GMO law applies [22]. Other Latin American countries have followed Argentina's lead and use a case-by-case system to determine the foreign DNA content of gene-edited crops.

In contrast, Japan, Canada, and the United States do not subject NPBT-derived crops to explicit regulatory oversight, as they do not meet the legal definition of GMOs in these jurisdictions. North American regulatory authorities approach GMOs from the perspective that genetic modifications do not introduce novel risks, and therefore, their focus is on the final products [23, 24]. The presence of external DNA in the end-product instigates the novelty criterion within the Canadian regulatory framework, including Seed Regulation. Specifically, the DNA responsible for encoding the gene-editing components, such as CRISPR–Cas proteins and related guide RNAs, is regarded as foreign DNA. If these sequences are eliminated from the final products through conventional plant breeding and selection methods, they qualify for an exemption [25]. In these countries, SDN-1 and SDN-2 are not subject to the same level of regulatory oversight as traditional GMOs, as long as they do not contain foreign DNA. Other countries, such as the European Union, have taken a more precautionary approach, treating genome-edited organisms as equivalent to traditional GMOs and subjecting them to the same level of regulatory oversight. Until the proposed law passes, the EU considers SDN-1 and SDN-2 as GMOs whether or not they contain foreign DNA [17].

Australia regulates gene-edited organisms under the Gene Technology Act of 2000 and the Gene Technology Regulations enacted in 2001. The Office of Gene Technology Regulator (OGTR) oversees the implementation of these regulations. In April 2019, the OGTR revised their regulations to clarify that genome-editing technologies that fall under SDN-1 are exempt from the current regulation, while genome-editing technologies that use artificially created templates outside the cell (SDN-2) are subject to regulation [17].

Finally, another major agricultural country to have also exempted gene-edited plants not containing foreign DNA from GM regulations is India. In 2022, the Department of Biotechnology, Ministry of Science and Technology published detailed guidelines for the safety assessment of genome-edited plants [17].

The four main global approaches to regulating NPBTs can be summarized as follows:

"Approach 1: GMO regulations are applied as it is to genome-edited products. As a result, prior safety assessment and approval by the government are needed.

Approach 2: Simplified GMO regulations will be applied to genome-edited products. As a result, simplified safety review and approval procedures will be applied.

Approach 3: Genome-edited products are exempt from GMO regulations. However, confirmation by the government is needed before being placed on the market.

Approach 4: Genome-edited products are exempt from GMO regulations. Prior confirmation is not needed by the Government" [17].

The authors identify a "limited convergence" in international regulatory policies for genome-edited organisms and GM food products based on a middle ground between Approach 1, which regulates all genome-edited organisms as GMOs, and Approach 4, which exempts all genome-edited organisms from regulations. They contend that the convergence is influenced by factors such as policy harmonization, international organizations, and leading nations such as the E.U. and the U.S.

8.3 DEFINING THE REGULATORY STATUS OF GENE-EDITED CROPS AND GOVERNANCE ISSUES

A pivotal issue in the global discourse surrounding NPBTs revolves around their classification as genetically modified organisms (GMOs) [24, 25], an aspect that is key to developing a comprehensive intellectual property rights (IPR) regime for gene-edited plant varieties. The determination of whether NPBTs are categorized as GMOs bears substantial consequences for crop developers, plant breeders, farmers, and consumer groups. The complexity and difficulty posed by gene-edited plants and varieties in terms of their constructs and their traceability can also have serious impacts.

The classification of gene-edited plants and animals as genetically modified organisms (GMOs) carries several implications from a legal, regulatory, and intellectual property perspective. First, if classified as GMOs, gene-edited organisms are subject to strict regulations that have been established to control the use and distribution of GMOs. These rules vary by country but often involve safety testing, environmental assessments, and authorization procedures prior to market approval. Second, when viewed from an industrial standpoint, it is contended that such categorization would necessitate that enterprises allocate substantial resources toward protracted and frequently expensive regulatory clearance procedures. This serves as a rationale for advocating the deregulation of gene-edited organisms and their exclusion from the ambit of rules governing genetically modified organisms (GMOs). Industrial stakeholders argue that subjecting gene-edited crop varieties to the same regulatory framework as genetically modified organisms (GMOs) has the potential to impede the rapid adoption of gene-editing technologies. In their view, this is particularly burdensome for smaller enterprises and research institutions with finite resources. In essence, it is believed that designating gene-edited crops as GMOs would substantially increase the pace of innovation.

From an intellectual property perspective, classifying gene-edited plants as GMOs means that they would be subject to the same patent protections and licensing agreements as GMOs. This signifies that companies developing and commercializing gene-edited organisms could claim ownership, restrict their use, or charge licensing fees. Hence, the issues affecting plant breeders and small-scale farmers can vary in regard to the patenting of genetically modified organisms (GMOs) versus the patenting of gene-edited plants and plant varieties. While there are similarities in some challenges, there are also distinct differences, primarily related to the regulatory and intellectual property landscape.

A parallel challenge associated with the patenting of gene-edited crops, akin to genetically modified organisms (GMOs), is the potential restriction placed on farmers to save and replant seeds derived from these organisms, a customary practice typically allowed with conventional crops. The GMO classification could further increase market domination by large-scale biomedical and agricultural corporations that hold patents on these technologies, potentially limiting access to genetic materials for plant breeders, smaller-scale farmers, or researchers. Therefore, how gene-edited organisms are classified has far-reaching consequences for how patents might be applied for in various jurisdictions.

Generally, information about whether an organism has been gene-edited tends to come from regulatory agencies and product suppliers. Some countries require the labeling of genetically modified organisms, so theoretically, this information should be available on packaging. However, labeling laws vary significantly around the world, and not all nations require labels for gene-edited organisms in food products [22].

Overall, these factors affect the agricultural sector, influencing farming practices and the autonomy of farmers. They also carry important implications for food safety and security, particularly as an increasing number of these products become commercially available in the near future. Finally, bypassing the regulatory frameworks designed to manage the environmental and health risks posed by GMOs downplays the inherent challenges and potential risks associated with NPBTs. Notably, this includes controversies involving the unintentional "off-target effects" associated with CRISPR–Cas technology. Off-target effects refer to instances where the tools used to induce cell repair mechanisms mistakenly bind to and modify genomically similar yet different sites than the intended target, potentially resulting in harmful genetic alterations [26]. Mosaicism is another issue wherein not all cells of an organism achieve the desired modification, resulting in a mixture of edited and unedited cells within a single organism [27]. Furthermore, gene-editing tools, while being on-target, may still yield unintended insertions or deletions at the edit site or even complex rearrangements such as large deletions and chromosomal translocations [28]. These concerns underscore the need for rigorous precision testing and safety measures when employing gene-editing techniques and deploying gene-edited crops into the environment.

8.4 GENE-EDITED CROPS AND IPR COMPLEXITY

NPBTs, which have the potential to revolutionize plant breeding by enabling complex targets at lower costs, raise various considerations, particularly in the realm of intellectual property (IP). These plants are patentable in the same way as GMOs. Patent applications for gene-edited plants and animals can generally be categorized into two broad types. Patents on the gene-editing techniques themselves that generally aim to obtain the exclusive rights to utilize a specific gene-editing tool (such as CRISPR–Cas9) for various applications, including in agricultural biotechnology. Notable examples of these patents include the CRISPR–Cas9 patent held by the Broad Institute in the U.S. and the Cpf1 patent held by Vilnius University in Lithuania [29]. The second type of patent application involves patents on gene-edited organisms: these patent applications are made by companies or researchers who have used gene-editing tools to develop an organism with a specific trait or feature. This could be, for instance, a plant that has been edited to be more drought-resistant or an animal with enhanced production capabilities (such as a chicken that produces eggs containing a specific human protein). For example, agricultural companies such as Corteva Agriscience and Cibus hold patents on certain types of gene-edited crops. In October 2023, Corteva Agriscience filed a lawsuit against Inari Agriculture Inc. (a gene-editing startup and plant trait developer) for patent infringement, accusing them of stealing seeds deposited at the American Type Culture Collection (ATCC). Corteva's seed sample was placed in compliance with IPR protection. However, after obtaining the patented seed, Irani is alleged to have genetically altered some traits of the Corteva variety, smuggling it into Europe and then applying for a US patent [30]. The lawsuit between Corteva and Inari underscores the "murky world of IP protection for gene-edited plants" [31].

These technologies are promoted as democratizing plant biotechnology and stimulating competition in an industry traditionally dominated by a few multinational corporations. In this case, intellectual property can play a significant role in incentivizing innovation. However, the increasing complexity of patents within the

NBT landscape can become a barrier to industry and agriculture more broadly. This complexity may hinder innovation by increasing transaction costs, causing market entry delays, and reducing the overall incentive for research and development [32]. This can also deter smaller players from entering the field, lead to inefficiencies in resource allocation, and hinder collaboration among stakeholders in regard to exchanging knowledge and germplasm[1]. More significantly, it raises concerns for farmers and plant breeders who may encounter challenges related to what Kock calls "patent thickets" as well as patent stacking [33].

The term "patent thickets" refers to a situation in which a particular technological field or industry becomes densely populated with a multitude of patents, often owned by various parties, covering different aspects or components of a technology or innovation. These patents may overlap, creating a complex web of intellectual property rights that can be challenging to navigate [33]. Patent thickets in the field of plant biotechnology occur when numerous patents are granted for various elements of plant-related innovations, including specific traits, genes, sequences, and methods. These patents encompass a wide range of technologies, such as CRISPR–Cas and other genome-editing techniques, delivery mechanisms, and methods for enhancing the repair of DNA sequences (SDN-2 and SDN-3). The complexity arises from the sheer number of patents, their overlap, and their potential impact on the freedom to operate in the field of plant breeding and biotechnology. Patent thickets require potential developers of new crops to obtain rights from each patent holder to minimize patent infringement [34], leading to extremely complicated legal arrangements.

The presence of patent thickets can have several consequences, including increased transaction costs, delays in market entry, and potential innovation stifling due to difficulties in obtaining licenses or navigating the complex patent landscape [33]. These challenges may discourage smaller entities, such as small and medium-sized enterprises, from engaging in research and innovation in the plant biotechnology sector. Strategies such as cross-licenses, patent pools, and open-source models can be used to mitigate the negative effects of patent thickets on plant-related innovations; however, there is a real need to address patent thickets in a manner that promotes innovation and accessibility to genetic materials for all.

Kock explores the concept of patent stacking, which involves the accumulation of multiple patents covering various traits and technologies within a single crop variety. As technology advances and the pace of development accelerates, this practice can lead to what he terms a "patent minefield." This scenario envisions a single seed being subject to eight to ten patents [35] or a plant variety containing numerous patented traits. Consequently, plant breeders seeking to utilize such plant material to introduce new varieties may find themselves entangled in highly intricate licensing agreements involving multiple patent holders. This complexity in patent arrangements poses a barrier to plant breeders who wish to employ seed material acquired from seed companies or conventional methods to develop new breeds. The fear of inadvertently infringing on existing patents can deter them from utilizing such resources [35].

When farmers purchase seeds or crop varieties covered by numerous patents, it can lead to increased costs, as they may be needed to pay royalties to multiple patent holders. This can potentially reduce the economic viability of farming for small-scale and resource-limited farmers, leading to concerns about the concentration of power and profits in the hands of a few large seed and technology companies. For consumers, patent stacking can indirectly impact them through potential price increases for agricultural products. Farmers passing on the additional costs associated with patent royalties may result in higher prices for food and other goods. Additionally, if patent-related complexities limit the availability of diverse crop varieties, they may affect the overall food supply and biodiversity, potentially limiting consumer choices and food security.

Moreover, plant breeders and farmers may unknowingly be put at risk of patent infringements due to the lack of regulatory coherence concerning the status of NPBTs across the world, and determining whether some plant materials may have been obtained using genome-editing technology may prove to be very complex [33]. Concerns relating to intellectual property rights and the ownership of gene-edited organisms are key for food security and agricultural economies that rely on traditional breeding and nongenetically modified varieties because gene-editing technology is moving at a fast pace and patent systems are outdated [35]. According to Michael Kock, a patent attorney who worked at Syngenta, the accelerated pace of new crop development enabled by the technology means the following:

> If the technology advances and more governments allow genome editing, 50% of new seed varieties on the global market will have at least one genome-edited trait in 20 years' time, [...] all seeds with these traits will be covered by at least one patent.
>
> quoted in [35]

Although very few genome-edited food products have been introduced to the market, genome editing has already resulted in a significant increase in the quantity of patents. Kock's estimation suggests that approximately 50% of all global plant patent applications filed in 2021 pertained to various aspects of genome editing (Ibid.). This will have serious impacts on agrobiodiversity, as the pace of commercial releases of gene-edited crops accelerates at a much faster rate than crops obtained via non-NPBTs.

The lack of traceability and transparency of NPBTs in the field and in the marketplace poses significant challenges for farmers and consumers. Because they do not contain easily identifiable elements and theoretically do not incorporate foreign DNA, NPBTs are difficult to trace and identify. Determining whether an organism has been gene-edited can be a significant challenge for both farmers and consumers. This is mainly because gene editing results in changes within the existing DNA of the organism. These changes often do not introduce any foreign DNA, making the modifications virtually indistinguishable from natural mutations [36]. In cases where gene editing does not leave any foreign DNA behind, typical genetic tests may not detect the changes brought upon by gene editing because tests, such as polymerase chain reaction (PCR) or sequencing, primarily identify foreign DNA sequences. However, advanced testing methods such as whole-genome sequencing (WGS) and detailed bioinformatics analysis can potentially find small insertions or deletions or single nucleotide changes created by gene editing when compared to the original genome [32]. In terms of consumer awareness and public perception about gene-edited foods, even in jurisdictions that regulate them as GMOs that require labeling, exemptions have been granted contrary to strict legislation but in all likelihood because clear detection and identification of gene-edited products are currently impossible without technical details or unique analytical reference data [25].

The practical implementation of coexistence policies between crops that are intentionally genetically altered and nongenetically altered may be jeopardized because of the lack of regulatory oversight of certain categories of gene-edited crops [16]. The challenge is the lack of detection and transparency about the technology, particularly in identifying off-target effects as well as potential admixture between gene-edited material into plants that do not permit the presence of GMOs [37]. This makes the source of adventitious material difficult if not impossible to identify. This may have economic consequences for farmers who may be involved in organic production, as their products must adhere to strict certification guidelines. This raises the question as to how the substantial expenses linked to establishing detection systems for monitoring the utilization of genome-edited plants will be transferred to farmers and consumers. The implementation of systematic detection methods for all commercialized gene-edited products should become a regulatory requirement as a fundamental condition of their release. This emphasizes the need for comprehensive procedures for establishing IPR ownership to mitigate the substantial risks associated with patent infringement for both farmers and plant breeders. While corporations may be reluctant to adopt these measures due to potential technical complexities associated with different NPBTs, the ability to trace the origin of genetic alterations could prove useful in identifying and enforcing potential patent infringements [33] and/or biopiracy claims, especially in the context of international trade where countries may have different regulatory frameworks.

8.5 CONCLUSION

While gene-editing technologies may hold great promise for the development of next-generation agricultural crops, they also give rise to significant concerns. The technology is relatively new and still experimental; few products have been commercially released to date, so knowledge about their safety is lacking. One prominent concern pertains to the potential unintended consequences of gene editing, including off-target effects, which could result in unintended and potentially harmful genetic changes [22]. Additionally, gene editing may have unforeseen ecological impacts, such as the potential loss of agricultural biodiversity and disruptions to ecosystems, particularly due to genetic flow between gene-edited organisms and conventional varieties.

Ethical dilemmas arise, with questions regarding the lack of traceability and detection methods to identify the presence of products created using NPBTs that carry no foreign DNA at all stages of the supply chain, from seed to plate. From a consumer's perspective, this makes labeling of gene-edited products practically impossible. Furthermore, while the lower costs of genome-editing technology are thought to "democratize" research by small- and medium-sized companies and plant breeders [38], it is believed that the power exerted by large corporations such as Bayer and Corteva (the holders of a large number of patents on plant traits, varieties and processes) might actually open the way for dominance by a few key players. The issues of patent thickets and patent stacking act as a deterrent for traditional plant breeders to access genetic materials for the development of new varieties. As intellectual property rights are concentrated in the hands of large corporations or affluent countries, this restricts access for lower-income farmers or countries to access new plant varieties. However, the increased use of gene-edited crops in these countries might threaten traditional varieties, putting food systems at risk. This can disempower farmers, particularly small-sized farmers who may not be able to afford licensing fees or face legal consequences if they save and replant patented seeds. It effectively hampers traditional agricultural practices of reusing seeds, causing an overreliance on companies for each planting cycle [39]. As Kuzma and Rawls point out, the use of these technologies could exacerbate economic disparities in agricultural communities and between developed and developing countries [40]. Smaller farmers who cannot afford gene-edited seeds could find themselves unable to compete with larger operations, thus widening income inequality within farming communities. Ultimately, the search for sustainable solutions to food security and the empowerment of farmers and communities in the context of climate change are severely impacted. Additionally, GMO labeling laws vary significantly from one country to another; not all countries require the declaration of gene-edited organisms on food products. This can disempower consumers who have the right to know what they are eating but are not provided with the information to make informed choices.

Governments around the world have issued patents covering various forms of CRISPR-based technology. Patenting of gene-edited products can be seen as a form of governance, as these legal instruments are designed to limit the use of a given technology. This means that patents control who can use a given technology, on what terms, where and when [41].

The lack of a coordinated framework for regulating NPBTs and the trend toward deregulating gene-editing products among leading agricultural exporters reveals the inefficacy of patents as legal instruments for safeguarding the interests of plant breeders, small-scale farmers, and consumers. Instead, biotechnological interventions in agriculture transform into a *de facto* governance strategy that promotes the interests of powerful actors and major agribiotech corporations. This chapter has attempted to show the interplay between the governance of NPBTs and the governance through NPBTs as illustrated by patents on gene-edited crops.

NOTE

1 It should be noted that if gene-edited organisms are not classified as genetically modified, they may be protected under plant variety rights rather than patents. Plant variety rights, as stipulated by the International Union for the Protection of New Varieties of Plants (UPOV), permit farmers to save, use, exchange, and sell farm-saved seeds, subject to certain restrictions. We will not deal with this aspect of IPR as the trend has been to a patenting regime for gene-edited varieties rather than Plant Breeders Rights.

REFERENCES

[1] C. Bain, S. Lindberg, and T. Selfa, "Emerging sociotechnical imaginaries for gene edited crops for foods in the United States: implications for governance," *Agric Hum Values*, vol. 37, no. 2, pp. 265–279, Jun. 2020, doi: 10.1007/s10460-019-09980-9.

[2] S. H. Idris, N. S. Mat Jalaluddin, and L. W. Chang, "Ethical and legal implications of gene editing in plant breeding: a systematic literature review," *J Zhejiang Univ Sci B*, vol. 24, no. 2, pp. 1093–1105, Jul. 2023, doi: 10.1631/jzus.B2200601.

[3] K. P. Voss-Fels, A. Stahl, and L. T. Hickey, "Q&A: modern crop breeding for future food security," *BMC Biol*, vol. 17, no. 1, p. 18, Dec. 2019, doi: 10.1186/s12915-019-0638-4.

[4] Y.-C. Kim *et al.*, "Applications and major achievements of genome editing in vegetable crops: a review," *Front Plant Sci*, vol. 12, p. 688980, Jun. 2021, doi: 10.3389/fpls.2021.688980.

[5] G. J. Knott and J. A. Doudna, "CRISPR-Cas guides the future of genetic engineering," *Science*, vol. 361, no. 6405, pp. 866–869, Aug. 2018, doi: 10.1126/science.aat5011.

[6] A. Kamthan, M. Kamthan, and A. Datta, "Biotechnology for drug discovery and crop improvement," *Nucleus*, vol. 60, no. 2, pp. 237–242, Aug. 2017, doi: 10.1007/s13237-016-0192-1.

[7] E. Waltz, "With a free pass, CRISPR-edited plants reach market in record time," *Nat Biotechnol*, vol. 36, pp. 6–7, 2018.

[8] N. Kofler *et al.*, "Editing nature: local roots of global governance," *Science*, vol. 362, no. 6414, Art. no. 6414, 2018.

[9] K. M. Esvelt and H. H. Wang, "Genome-scale engineering for systems and synthetic biology," *Mol Syst Biol*, vol. 9, p. 641, 2013.

[10] S. Schiml and H. Puchta, "Revolutionizing plant biology: multiple ways of genome engineering by CRISPR/Cas," *Plant Methods*, vol. 12, no. 8, Art. no. 8, 2016.

[11] M. F. Hamdan, C. K. S. Karlson, E. Y. Teoh, S.-E. Lau, and B. C. Tan, "Genome editing for sustainable crop improvement and mitigation of biotic and abiotic stresses," *Plants*, vol. 11, no. 19, p. 2625, Oct. 2022, doi: 10.3390/plants11192625.

[12] S. Saurabh, "Genome editing: revolutionizing the crop improvement," *Plant Mol Biol Rep*, vol. 39, no. 4, pp. 752–772, Dec. 2021, doi: 10.1007/s11105-021-01286-7.

[13] H. Schinkel and S. Schillberg, "Genome editing: intellectual property and product development in plant biotechnology," *Plant Cell Rep*, vol. 35, no. 7, pp. 1487–1491, Jul. 2016, doi: 10.1007/s00299-016-1988-9.

[14] A. Ricroch, "Global developments of genome editing in agriculture," *Transgenic Res*, vol. 28, no. Suppl 2, Art. no. Suppl 2, 2019.

[15] R. Lassoued, P. W. B. Phillips, S. J. Smyth, and H. Hesseln, "Estimating the cost of regulating genome edited crops: expert judgment and overconfidence," *GM Crops Food*, vol. 10, no. 1, Art. no. 1, 2019.

[16] European Parliament. Directorate General for Parliamentary Research Services. *Genome-edited crops and 21st century food system challenges.* LU: Publications Office, 2022. Accessed: Oct. 22, 2023. [Online]. Available: https://data.europa.eu/doi/10.2861/290440

[17] M. Tachikawa and M. Matsuo, "Divergence and convergence in international regulatory policies regarding genome-edited food: how to find a middle ground," *Front Plant Sci*, vol. 14, p. 1105426, Jan. 2023, doi: 10.3389/fpls.2023.1105426.

[18] D. Eriksson *et al.*, "A comparison of the EU regulatory approach to directed mutagenesis with that of other jurisdictions, consequences for international trade and potential steps forward," *New Phytol*, vol. 222, no. 4, pp. 1673–1684, Jun. 2019, doi: 10.1111/nph.15627.

[19] I. Katsarova, "Plants produced by new genomic techniques," European Parliament, EU Briefing PE754.549, Oct. 2023.
[20] Stokstad, "European Commission proposes loosening rules for gene-edited plants." Accessed: Oct. 26, 2023. [Online]. Available: www.science.org/content/article/european-commission-proposes-loosening-rules-gene-edited-plants
[21] S. Coe, "Genetic technology (Precision Breeding) Act 2023," House of Commons Library, Research Briefing CBP 9557, Mar. 2023. [Online]. Available: https://researchbriefings.files.parliament.uk/documents/CBP-9557/CBP-9557.pdf
[22] D. Zhang *et al.*, "Genome editing with the CRISPR-Cas system: an art, ethics and global regulatory perspective," *Plant Biotechnol J*, vol. 18, no. 8, pp. 1651–1669, Aug. 2020, doi: 10.1111/pbi.13383.
[23] K. Barrett and E. Abergel, "Defining a safe genetically modified organism: boundaries of scientific risk assessment," *Sci Public Policy*, vol. 29, no. 1, Art. no. 1, 2002.
[24] I. Braverman, "Editing the Environment: Emerging Issues in Genetics and the Law," in *Gene Editing, Law, and the Environment: Life Beyond the Human*, Edited by Irus Braverman., New York, NY: Routledge, 2018, pp. 1–11.
[25] T. Sprink, R. Wilhelm, and F. Hartung, "Genome editing around the globe: an update on policies and perceptions," *Plant Physiol*, vol. 190, no. 3, pp. 1579–1587, Oct. 2022, doi: 10.1093/plphys/kiac359.
[26] X.-H. Zhang, L. Y. Tee, X.-G. Wang, Q.-S. Huang, and S.-H. Yang, "Off-target effects in CRISPR/Cas9-mediated genome engineering," *Mol Ther Nucl Acids*, vol. 4, p. e264, 2015, doi: 10.1038/mtna.2015.37.
[27] H. Ma *et al.*, "Correction of a pathogenic gene mutation in human embryos," *Nature*, vol. 548, no. 7668, pp. 413–419, Aug. 2017, doi: 10.1038/nature23305.
[28] M. Kosicki, K. Tomberg, and A. Bradley, "Repair of double-strand breaks induced by CRISPR–Cas9 leads to large deletions and complex rearrangements," *Nat Biotechnol*, vol. 36, no. 8, pp. 765–771, Sep. 2018, doi: 10.1038/nbt.4192.
[29] J. S. Sherkow, "Patent protection for CRISPR: an ELSI review," *J L Biosci*, vol. 4, no. 3, pp. 565–576, Dec. 2017, doi: 10.1093/jlb/lsx036.
[30] S. Zimmerman, "Corteva lawsuit accuses gene-editing startup of stealing seeds," *Agriculture Dive*. Accessed: Oct. 27, 2023. [Online]. Available: www.agriculturedive.com/news/corteva-lawsuit-inari-steal-seeds-gene-editing/695605/
[31] T. L. Tischner and A. P. Zappia, "The Murky World of IP Protection for Gene-Edited Plants," *Troutman Pepper – The Murky World of IP Protection for Gene-Edited Plants*. Accessed: Oct. 27, 2023. [Online]. Available: www.troutman.com/insights/the-murky-world-of-ip-protection-for-gene-edited-plants.html
[32] J. B. Hurlbut, K. Saha, and S. Jasanoff, "CRISPR democracy: gene editing and the need for inclusive deliberation," *Issues Sci Technol*, vol. 32, no. 1, 2015.
[33] M. A. Kock, "Open intellectual property models for plant innovations in the context of new breeding technologies," *Agronomy*, vol. 11, no. 6, p. 1218, Jun. 2021, doi: 10.3390/agronomy11061218.
[34] E. N. Rissberger, "The future of biotechnology: accelerating gene editing advancements through non-exclusive licences and open-source access of CRISPR-Cas9," *Santa Clara High Technol L J*, vol. 38, no. 1, pp. 97–121, 2022, [Online]. Available: https://digitalcommons.law.scu.edu/chtlj/vol38/iss1/3
[35] J. Davis Plüss, "Genome editing's patent problem fuels concern for the future of food," SWI swissinfo.ch. Accessed: Oct. 22, 2023. [Online]. Available: www.swissinfo.ch/eng/business/genome-editing-s-patent-problem-fuels-concern-for-the-future-of-food/47287668

[36] H. Ledford, "CRISPR, the disruptor," *Nature*, vol. 522, no. 7554, pp. 20–24, Jun. 2015, doi: 10.1038/522020a.

[37] European Non-GMO Industry Association ENGA, "Deregulation: ENGA European Non-GMO Industry Association AISBL." Accessed: Oct. 26, 2023. [Online]. Available: www.enga.org/deregulation

[38] M. Montenegro de Wit, "Democratizing CRISPR? Stories, practices, and politics of science and governance on the agricultural gene editing frontier," *Elementa Science of the Anthropocene*, vol. 8, no. 1, Art. no. 1, 2020.

[39] M. S. Clancy and G. Moschini, "Intellectual property rights and the ascent of proprietary innovation in agriculture," *Annu Rev Resour Econ*, vol. 9, no. 1, pp. 53–74, Oct. 2017, doi: 10.1146/annurev-resource-100516-053524.

[40] J. Kuzma and L. Rawls, "Engineering the wild: gene drives and intergenerational equity," *Jurimetrics*, vol. 56, no. 3, pp. 279–296, 2016.

[41] N. Scheinerman and J. S. Sherkow, "governance choices of genome editing patents," *Front Polit Sci*, vol. 3, p. 745898, Sep. 2021, doi: 10.3389/fpos.2021.745898.

9 IP Rights and Food

Michael Blakeney

9.1 INTRODUCTION

At a time when the world's population is expected to increase from 8 billion in 2023[1] to 9.7 billion in 2050[2], the global demand for food is expected to increase between 36 and 56% by 2050 in 2010.[3] This is occurring in a situation of climate change[4], which is causing significant agricultural stress[5] through the depletion of natural resources[6], water shortages[7], unprecedented seasonal heat[8], salinity[9] and an increasing number of crop pests[10] and diseases.[11] Each of these stresses negatively impacts agricultural productivity.[12] Solutions to these agricultural problems involve innovations that may be protected by intellectual property (IP) rights, for example, as patents or as plant varieties. This chapter focuses on patent-based solutions to climate change problems, the associated privatization of those rights and the consequences of patenting for food security.

It should be noted that seed markets are complex value chains involving developers and breeders of improved crop varieties and seed treatments that establish a chemical barrier over the surface of germinating seeds to protect them from chewing insects, such as wireworms. Improved varieties can be obtained from classical breeding or by genetic modification (GM). For example, patented GM traits conferring insect resistance reduced the need for insecticide applications, and GM traits for herbicide tolerance provided a substitute for mechanical tillage, reducing labor inputs, machinery, and fuel previously used for weeding. Firms with large patent portfolios have also entered into cross-licensing agreements with one another to acquire each other's technologies and, through cross-licensing agreements, have been able to reduce or even avoid paying royalties or licensing fees.[13] In a situation of climate change, patented GM seeds that are resistant to the agricultural stresses listed above will be important for farmers around the world. The global patent system will govern the availability of proprietary seeds.

The membership of countries of the World Trade Organization (WTO) obliges them to implement, as a membership obligation, the IP norms prescribed by the WTO Agreement on Trade Related Aspects of Intellectual Property Rights (TRIPS). As most countries are members of the WTO, there is a significant degree of harmonization of IP laws between countries. The category of IP rights in the TRIPS Agreement that will be of the greatest significance in promoting gene-based

 DOI: 10.1201/9781003405665-9

agricultural innovation is patents (Art. 27.1). The TRIPS Agreement in Article 27.1 provides that "patents shall be available and patent rights enjoyable without discrimination as to … the field of technology" *it should be noted, however, that Oxfam reported in 2018 that plants were excluded from patentability in 40% of the 126 countries for which legal information was available.[14] This can be explained by reference to Article 8.1 of TRIPS, which permits WTO members in formulating or amending their laws and regulations to "adopt measures necessary to protect public health and nutrition and to promote the public interest in sectors of vital importance to their socioeconomic and technological development". However, in some of the countries identified in the Oxfam report, the picture is not entirely clear. Brazil is identified as a country that does not permit the patenting of plants, but it allows the patenting of methods to transform plants and expression vectors containing heterologous sequences, although it does not allow the patenting of genetic material isolated from plants or any parts of plants.[15] In contrast, China allows the patenting of isolated and modified DNA sequences and transformed plant cells but excludes the patentability of plants and plant varieties.[16]

The restrictions on the patenting of plants and plant materials in developing countries have been identified as affecting the willingness of private firms to invest in agricultural R&D.[17] This will be of increasing significance as climate change compels the exploration of GM crops as a technological solution.

9.2 GENE PATENTING

The patentability of genetic materials and gene fragments, such as expressed sequence tags (ESTs) and single nucleotide polymorphisms (SNPs), as well as enabling gene-based technologies, led to what has been described as a "genomic gold rush"[18] in the 1990s as vast numbers of gene-based patent applications were filed, particularly in the USA. These patents were filed both on improved plants and on the tools and enabling technologies by which those plants were engineered. Genetic engineering has permitted the expeditious introduction of a wide range of desirable traits into plants. These include:

- pest control traits such as insect, virus, and nematode resistance as well as herbicide tolerance;
- postharvest traits such as delayed ripening of spoilage prone fruits;
- agronomic traits such as nitrogen fixation and utilization, restricted branching, and environmental stress tolerance;
- male and/or seed sterility for hybrid systems; and
- output traits such as plant color and vitamin enrichment.[19]

These developments in genetic engineering in agriculture have been accompanied by the sequencing of the genomes of crops. In 2002, rice (*Oryza sativa)* was the first major crop genome to be fully sequenced and the first food crop genome. In 2006, Cambia, an Australian NGO that promotes transparency in IP, used its Patent Lens project to conduct an in-depth analysis of U.S. patents and patent applications that made claims on the rice genome. Patent Lens revealed that by 2006, approximately

74% of the rice genome was named in the claims of U.S. patent applications and that every segment of the rice genome's 12 chromosomes was recited in patent applications.[20]

Significant misgivings with these developments have been expressed by numerous commentators. Most likely, the most influential among these are Heller and Eisenberg[21], who suggested that genetic research tool patents could create a "tragedy of the anticommons" in which multiple patent owners would tie-up genetic materials in a thicket of IP patent rights. This was perceived by Correa to be a particular problem for the genetic improvement of crops since this is an incremental process and each new patent would constrain the "freedom to operate", particularly of public agricultural research institutes.[22] Significant transaction costs attend navigation through patent thickets, and access to research tools comes with some significant license fees. This is obviously a problem for researchers in developing countries.[23]

Another aspect of this patenting is that it was undertaken by a handful of companies. Jefferson et al, in a 2015 study of patenting in relation to maize, rice, and soybean genomes, identified DuPont and its affiliates as the holders of the largest collection of gene patents in relation to these crops, followed by Monsanto, Syngenta, and BASF.[24] Significantly, from the 1990s to 2014, the study observed a gradually widening gap in the ownership of plant gene patents between the private and public sectors.[25] The strategic use of patent rights has enabled some large-scale private patent holders to exclude or deter others from using critical research tools and materials. This is obviously of significance to developing countries that have relied upon the free availability of research tools.[26] The entry of private sector patenting in the seed industry has pushed patent disputes beyond the resources of the public sector. Bowring gives the example of a soybean patent filed by Agracetus that would grant it a monopoly on all transgenic soybeans, regardless of the genes inserted into the target organism.[27] This patent claim was opposed by Monsanto, which filed a 292-page statement listing its objections, until Monsanto acquired Agracetus for USD 150 million and immediately put its resources into supporting the broad patent application it now owned.

Additionally, these patent owners have enforced their rights against farmers who save harvested proprietary seed for replanting. The Center for Food Safety was reported to estimate that Monsanto collected USD 23.7 million in recorded judgments against farmers by 2013 and received additional payments of USD 85 to USD 160 million from out-of-court settlements by 2006.[28]

9.3 CLIMATE-READY GENES

Plants exhibit stress tolerance through adaptation mechanisms that have evolved through natural selection, for example, drought tolerance through deeper rooting, resistance to salinity through vacuolar ion compartmentalization in shoot cells and resistance to ozone through elevated antioxidant capacity.[29] Researchers have sought to identify plants and crops naturally resistant to one or more agricultural stresses and then to identify the DNA that confers resistance with a view to patenting and exploiting it. Somvanshi, in a 2008 study, identified 30 patents relating to drought-tolerant genes.[30] These included (i) patents related to proline biosynthesis; (ii) patented dehydration responsive element binding factors (DREB) and C-repeat sequence binding factors (CBF); (iii) patents related to protein kinases; (iv) various patents awarded for transcription factors involved in improving drought stress tolerance in plants; and (v) patents related to miscellaneous drought tolerance genes.[31] Two years later, the ETC Group noted "a dramatic upsurge in the number of patents published (both applications and issued patents) related to 'climate-ready' genetically engineered crops" from June 30, 2008, to June 30, 2010, identifying 262 patent families and 1663 patent documents.[32] Equally significant to the growth of the patenting of climate-ready genes was the small number of companies involved. The ETC report noted that the top ten seed companies account for 67% of the global proprietary seed market, with Monsanto accounting for 23% of that market.[33] This report referred to a 2006 study that revealed that approximately 74% of each of the 12 chromosomes in the rice genome was recited in US patent applications by Monsanto.

Examples of these climate-ready patents include the 2004 US patent granted to Pioneer Hi-Bred International in relation to "transcriptional activators involved in abiotic stress tolerance".[34] This patent sought to address reduced yield resulting from drought stress.[35] The international exploitation of this patent was facilitated by an application under the Patent Cooperation Treaty designating more than 100 countries in which the patent would apply.[36]

Rimmer, in a survey of the range of patents obtained on climate-ready genes, referred to the large portfolio of patents held by BASF and Monsanto relating to plants and environmental stress.[37] Rimmer also refers to portfolios of patents on climate-ready genes held by US corporations: Ceres, Inc, Mendel Biotechnology, Inc, Dow Agrosciences, Inc; German company Bayer CropScience, which in 2019 acquired Monsanto; Swiss company Syngenta AG and Israeli company EvoGene, which is collaborating with Bayer CropScience in incorporating climate-ready genes into rice.[38]

In addition to the patents held by these companies over climate-ready genes, they have also obtained patents over the associated enabling technologies. For example, in 2010, Monsanto obtained a patent over "recombinant polypeptides associated with plants".[39] The patent application disclosed that the ability to develop transgenic plants with improved traits depends in part on the identification of polynucleotides that are useful for the production of transformed plants with desirable qualities. It explained that "the recombinant polynucleotides and recombinant polypeptides of the present invention find a number of uses, for example in recombinant DNA constructs, in physical arrays of molecules, for use as plant breeding markers, and for use in computer-based storage and analysis systems."[40] The patent specification claimed that

> As a result of such biotechnological applications, plants, particularly crop plants, with improved properties are obtained. Crop plants of interest in the present invention include but are not limited to soy, cotton, canola, maize, wheat, sunflower, sorghum, alfalfa, barley, millet, rice, tobacco, fruit and vegetable crops, and turf grass.[41]

Given the broad scope of this patent, it is understandable that concerns might be expressed by NGOs about the concentration of market power over food by a small group of companies in industrialized countries.[42] Rimmer refers to responses from executives of these corporations that, for example, Monsanto deserved praise for developing crop varieties that will survive climate change because

> …everyone recognizes that the old traditional ways just aren't able to address these new challenges. The problems in Africa are very severe. We aim to be at once cognizant of our obligations to shareholders who have paid for our research.[43]

In a situation of climate change, food insecure countries can rely on the benevolence of private corporations to address the food security concerns of those countries.

9.4 GENE EDITING

The genomic gold rush described above received a rude check in the USA when the Supreme Court in *Association for Molecular Pathology v. Myriad Genetics Inc.*[44] was called upon to decide whether a naturally occurring segment of deoxyribonucleic acid (DNA) was patent eligible under 35 U.S.C. §101 [US Patents Code] by virtue of its isolation from the rest of the human genome. This case concerned two genes associated with breast and ovarian cancers. The Court held that a naturally occurring DNA segment was a product of nature and not patent eligible merely because it had been isolated. The Court pointed out that the patentee did not create or alter any of the genetic information encoded in the two genes. The location and order of the nucleotides had existed in nature before the patentee found them. Its principal contribution was uncovering the precise location and genetic sequence of the two genes within chromosomes 17 and 13.

Significantly, the Court observed that the patentee did not create or alter the genetic structure of DNA. It went on to say that "Scientific alteration of the genetic code presents a different inquiry, and we express no opinion about the application of §101 to such endeavors."[45] This dictum has generated considerable speculation as to the amount of manipulation that is needed to render modified genes patentable.[46] It also stands behind the explosion of patenting in relation to gene-editing technologies.

Gene-editing technologies allow the alteration, deletion and/or addition of genetic material at specific locations in the gene of a living organism. This enables more accurate site-directed insertions in the DNA than that involved in genetic engineering mentioned above.[47] Examples of existing gene-editing technologies are transcription activator-like effector nucleases (TALENs), zinc-finger nucleases (ZFNs), and clustered regulatory interspaced short palindromic repeats (CRISPR). This section of the chapter will focus on CRISPR, which has been widely adopted.

CRISPR uses pieces of the genetic sequences of invading viruses taken up by bacteria and archaea as a mechanism of their adaptive immune system. This helps them to recognize and target the same or any related future viral infections. In plant cells, the technology can be applied to knock out genes by using CRISPR–Cas nucleases that can alter coding gene regions or regulatory elements, precisely alter a genome by base editing to delete or regulate gene expression, and precisely edit a genome by homology-directed repair mechanisms.[48] Potential agricultural applications of CRISPR technology in a climate change environment[49] include drought resistance,[50] salinity tolerance,[51] and resistance to insects,[52] viruses[53] and fungi.[54]

As with the other gene-based technologies, the early licensees of CRISPR technology are DuPont and Monsanto/Bayer[55] and this carries the same implications for food security as applies in their patenting activities in relation to climate-ready genes.

9.5 INDUSTRY CONSOLIDATION

Concern has been expressed about the concentration of IP rights in new plant breeding technologies in the hands of a few multinational companies. In part, this is because they have the financial capacity to engage in the R&D described above. In the 1970s, 30 separate firms were engaged in crop development. By 2011, this number had reduced to six: Monsanto, DuPont, Syngenta, Bayer, Dow, and BASF ("the Big Six"). They controlled 60% of the global proprietary seed market, with the top three (Monsanto, DuPont, and Syngenta) controlling more than 53%.[56] By 2009, GM traits owned or licensed by Monsanto were used on 95% of soybean acres, 81% of corn acres, and 79% of cotton acres in the United States.[57] Licensing and cross-licensing of GM traits are widely practiced in the industry, and a single variety may have "stacked traits", such as herbicide resistance and drought tolerance, licensed from multiple companies, although information regarding licensing fee practices is not readily available.[58] In addition, the Big Six firms controlled 76% of the global agricultural chemical market.[59] In December 2015, Dow Chemical and DuPont merged. Two months later, the State-owned Chinese company China National Chemical Corporation ("ChemChina") purchased Syngenta. In September 2016, Bayer acquired Monsanto. In 2018, Corteva Agriscience was spun off from the agricultural divisions of Dow and DuPont. Thus, with the exception of BASF, all of the major global seed and chemical firms were parties to the mergers. Each firm had a distinctive profile, focusing on either seeds and traits or chemicals. For example, seeds and traits accounted for approximately 70% of agricultural product sales at DuPont and Monsanto, while Dow Chemical and Bayer derived nearly 80% of their agricultural sales from chemicals.[60]

Each of the merging firms had a global presence, with a wide variety of products sold in multiple countries. Many of the specific seed and chemical markets where these firms operated were already highly concentrated, and the mergers would reduce the number of competitors in some of these markets. As of 2021, ChemChina and Bayer had become the world leaders in agricultural chemicals, while Bayer and Corteva held the leading positions in the crop seeds and traits business.[61]

Market concentration coincided with significant increases in prices of GM seeds.[62] In the USA, between 1990 and 2020, the average price farmers paid for seed rose by 270%, compared with commodity price inflation of 56%. For crops planted predominantly with GM seed (corn, soybeans, and cotton), seed prices rose by an average of 463%.[63] However, US farmers were able to offset the increased prices of GM crop varieties with productivity gains through higher yields and the reduction of inputs such as fertilizers and herbicides and reduced labor costs.[64] Whether these costs could be recouped by farmers outside the USA is questionable, as with the exception of large-scale farmers in countries such as Australia, Brazil, Canada, and China, the smaller scale of farming in the rest of the world is unlikely to attract productivity gains to underwrite expensive proprietary GM seed.

Another problem with market concentration is that it facilitates the implementation of anti-competitive practices. For example, in 2022, the US Federal Trade Commission joined with 10 States in an action against the pesticide manufacturers Corteva and Syngenta claiming that they had sought to exclude potential rivals from pesticide markets through the operation of loyalty programs with pesticide distributors. This allowed the two firms to maintain high prices for herbicides and insecticides after patents on the products had expired by blocking the entry of rival generic products and raising the marketing costs for generic producers.[65]

9.6 COMPULSORY LICENSING

The difficulties confronting countries in securing access to proprietary GM seeds are similar to those confronted by countries in securing access to patented antiretroviral medicines. They had sought to employ the compulsory licensing provisions of the TRIPS Agreement. However, it was inadequate for countries that did not have a domestic pharmaceutical manufacturing capacity, as the TRIPS Agreement only permitted compulsory licensing for domestic use and not for securing generic pharmaceuticals from outside a country. This inadequacy was remedied by the amendment of the compulsory licensing provisions of TRIPS in 2005 by the insertion of Article 31*bis,* which permitted the recruitment of overseas producers to supply patented pharmaceuticals. Article 31*bis* is of no assistance in the current agricultural context, as it only applies to pharmaceuticals. Until that is remedied, the general compulsory licensing provisions of the TRIPS Agreement apply. Article 31 permits the use of the subject matter of a patent without the authorization of the right holder by the Government or third parties authorized by the Government, "in the case of a national emergency or other circumstances of extreme urgency". The food security crisis that is being precipitated by climate change seems to meet this characterization. Article 31(b) provides that such use may only be permitted if, "prior to such use, the proposed user has made efforts to obtain authorization from the right holder on reasonable commercial terms and conditions and that such efforts have not been successful within a reasonable period of time", and Article 31(f) requires that "any such use shall be authorized predominantly for the supply of the domestic market of the member authorizing such use". Thus, the preconditions for the use of compulsory licensing to permit the use of GM technologies for agriculture are first that there is a domestic patent covering the GM technology and that there is a domestic capacity to utilize the technology. This is particularly difficult in the agricultural context where there are patent thickets surrounding the various patented products and the enabling technologies associated with them.

9.7 OPEN-SOURCE BIOTECHNOLOGY

In the absence of an effective compulsory licensing solution, the suggestion has been made that the Open Source philosophy could be applied to biotechnology as a counterbalance to the influence of multinational life sciences companies in the same way that the Open Source approach to software development was proposed as a way of overcoming the domination of the information technology sector by large software companies.[66] Often cited as a successful example of open source biotechnology is the Biological Innovation for Open Society ('BiOS') initiative of CAMBIA ('Center for Application of Molecular Biology in International Agriculture') based in Canberra, Australia.[67] This Centre was established in Canberra by Richard Jefferson, who had patented a gene transfer technology.[68] CAMBIA used patent revenues to create a funding base to underpin an open source system applicable to biotechnologies. To address the transaction costs of negotiating open-source licenses, in 2005, CAMBIA made available to researchers three different strains of bacteria that could be substituted for existing proprietary methods of introducing genetic material into plants. This technology was used as a nucleus to generate a protected commons for researchers in the life sciences, which became known as BiOS. Under the BiOS licenses, products derived from the licensed technology could be patented and commercialized subject to a number of constraints, including a grant back to CAMBIA, which was said to take the BiOS licenses outside the open-source paradigm and in part explained the lack of success of this model.[69]

In September 2021, Wageningen University & Research announced that it will provide nonprofit organizations with free licenses to use its CRISPR–Cas gene-editing technology for noncommercial applications. The University expressed the hope that its action would "inspire a worldwide change in CRISPR–Cas intellectual property policy."[70]

9.8 PUBLIC AGRICULTURAL RESEARCH

The development of GM seeds and securing access to the associated enabling technologies involve substantial financial resources, which is why research in this area has shifted to the private sector. At the time of the Green Revolution in the late 1960s, the key research had been undertaken by public agricultural research institutes, particularly those associated with the Consultative Group on International Agricultural Research ("CGIAR"). It had made improved seeds available gratis to farmers.[71] Some of its biological resources have been used by third parties in their patenting activities.[72] This led to calls for controls to be placed on the germplasm collections of the institutes within the CGIAR.[73] At the same time, it has been suggested that these collections exist as bargaining chips to be deployed in securing access to the patents that they have generated.[74] It has been pointed out that without widespread adoption, by at least 40% of farmers, the genetic gains achieved with improved crop varieties can never be actualized.[75] Thus, the CGIAR, through its credibility with Third World agricultural countries, can leverage trial plantings of improved crops with access to genetic improvements on behalf of those countries.

A number of obstacles exist to this approach, principally because the CGIAR is not a unitary organization but a constellation of 15 separate international research institutes with varying commitments to development outcomes.[76] However, "genetic innovation" has been identified as an "action area" in the CGIAR's 2030 Research and Innovation Strategy[77] published in 2020. Work in this "action area" involves the stewarding of genetic resources and addressing barriers to achieving rapid improvement of crop varieties, thereby increasing genetic gain and varietal turnover in farmers' fields. This is to be accomplished through:

- Conserving genetic resources in genebanks and facilitating and expanding their use by breeders and other stakeholders;
- Prioritizing breeding investments in market segments with the highest potential for impact;
- Development of new crop varieties that deliver higher rates of genetic gain in farmers' fields;
- Accelerating crop improvement through precision genetic technologies;
- Enabling tools, technology, and shared services to enhance the efficiency and effectiveness of CGIAR and partners' breeding programs; and
- Increasing seed sector actors' investments and effectiveness in scaling up access for farmers to new varieties.

Investments in genetic innovations "will provide foundational and multiplier outputs and outcomes for many crop-related innovations such as increased yield, biofortification, pest and disease resistance, and improved environmental tolerances."

NOTES

1 https://worldpopulationreview.com/, accessed 23 July 2023.

2 www.un.org/en/global-issues/population#:~:text=The%20world%20population%20is%20projected,surrounding%20these%20latest%20population%20projections, accessed 23 July 2023.

3 M. van Dijk, T. Morley, M. Rau, et al, 'A meta-analysis of projected global food demand and population at risk of hunger for the period 2010–2050', (2021) 2(7) *Nature Food*, 494.

4 See C. Müller, W. Cramer, W.L. Hare et al 'Climate change risks for African agriculture' (2011) 108(11) Proceedings of the National Academy of Sciences of the United States of America, 4313; T. Hasegawa, G. Sakurai, S. Fujimori, et al, 'Extreme climate events increase risk of global food insecurity and adaptation needs' (2021) 2(8) *Nature Food*, 587

5 See M.L. Parry, *Climate Change and World Agriculture*, London, Routledge, 2019.

6 See A. Steer, 'Resource Depletion, Climate Change, and Economic Growth' in F. Allen, J.R. Behrman, N. Birdsall *et al., Towards a Better Global Economy: Policy Implications for Citizens Worldwide in the 21st Century,* Oxford, OUP, 2014, 381–426.

7 See N. Hughes, D. Galeano and S. Hatfield-Dodds, 'The effects of drought and climate variability on Australian farms' (2019) *Abares Insights* issue 6; G. Leng and J. Hall, 'Crop yield sensitivity of global major agricultural countries to droughts and the projected changes in the future' (2019) 654 *Sci. Tot. Environ.* 811; I.R. Orimoloye, 'Agricultural Drought and Its Potential Impacts: Enabling Decision-Support for Food Security in Vulnerable Regions', (2022) 6 *Front. Sustain. Food Syst.* 838824.

8 See D.S. Battisti and R.L. Naylor, 'Historical warnings of future food insecurity with unprecedented seasonal heat' (2009) 323 *Science* 240; J. Lipiec, C. Doussan, A. Nosalewicz, et al, 'Effect of drought and heat stresses on plant growth and yield: a review' (2013) 27(4) *Int. Agrophys.*, 463.

9 See D.L. Corwin, 'Climate change impacts on soil salinity in agricultural areas', (2021) 72(2) *European Journal of Soil Science* 842.

10 See E. Barford, 'Crop pests advancing with global warming' (2013) *Nature* 13644; M.L.A. Battisti and S. Larson, *Climate Change and Insect Pests*, Wallingford UK, CABI, 2015; M. Skendžić, I.P. Zovko, et al, 'The impact of climate change on agricultural insect pests', (2021) 12 *Insects* Article 440; L. Schneider, M. Rebetez and S.Rasmann, 'The effect of climate change on invasive crop pests across biomes' (2022) 50 *Current Opinion in Insect Science* 100895.

11 S. Chakraborty and A.C. Newton, 'Climate change, plant diseases and food security: an overview' (2011) 60. *Plant. Pathol.*, 2; Y. Elad and I. Pertot, 'Climate change impacts on plant pathogens and plant diseases', (2014) 28 *J Crop Improvement*, 99; B.K. Singh, M. Delgado-Baquerizo, E. Egidi, et al 'Climate change impacts on plant pathogens, food security and paths forward' (2023) *Nat Rev Microbiol* .

12 FAO, *The State of Food Security and Nutrition in the World 2022*, Rome, FAO, 2022.
13 J.M. MacDonald, X. Dong and K.O. Fuglie, Concentration and competition in U.S. agribusiness. U.S. Department of Agriculture, Economic Research Service, EIB-256, 2023 at 8.
14 Oxfam, *Status of Patenting Plants in the Global South. Position Paper and Research Report*, The Hague, OxfamNovib, 2018.
15 L.A.B. de Castro, 'The Patent System in Brazil', *Springer Nature*, Mar 08, 2019.
16 M. Hu and J. Feng, 'Legal Protection of China's Biotechnology Patents' (2014) *Biotechnology Law Report*.103.
17 C. Pray and K. Fuglie, 'Agricultural Research by the Private Sector' (2015) *7 Ann Rev Resource Econ* 399.
18 E.J. Topol, S.S. Murray, K.A. Frazer, 'The genomics gold rush' (2007) 298(2) *JAMA*.:218.
19 See R. Lindner, 'Prospects for Public Plant Breeding in a Small Country' paper at ICABR Conference, The Shape of the Coming Agricultural Biotechnology Transformation: Strategic Investment and Policy Approaches from an Economic Perspective, University of Rome "Tor Vergata", June 17–19, 1999.
20 See www.patentlens.net/daisy/RiceGenome/3662/3108.html, accessed 26 July 2023.
21 M.A. Heller and R.S. Eisenberg, 'Can Patents Deter Innovation? The Anticommons in Biomedical Research'(1998) 280 *Science* 698.
22 Carlos M. Correa, 'Trends in Intellectual Property Rights Relating to Genetic Resources for Food and Agriculture, FAO, Commission on Genetic Resources for Food and Agriculture, Background Study Paper No. 49, Rome FAO, 2009.
23 See Ademola A. Adenle, Sulayman K. Sowe, Govindan Parayil, Obijiofor Aginam, 'Analysis of open source biotechnology in developing countries: An emerging framework for sustainable agriculture' (2012) 34 *Technology in Society* 256.
24 O.A. Jefferson, D. Köllhofer, T.H. Ehrich, et al, 'The ownership question of plant gene and genome intellectual properties' (2015) 33(11) *Nature Biotechnology*, 1138 at 1139–40.
25 Ibid at 1141.
26 See P.H. Howard, 'Intellectual Property and Consolidation in the Seed Industry', (2015) 55(6) *Crop Science*, 2489.
27 F. Bowring, *Science, seeds and cyborgs: Biotechnology and the appropriation of life*, New York, Verso, 2003.
28 D. Barker, B. Freese, G. Kimbrell, *Seed giants vs. U.S. farmers*, Washington, DC, Center for Food Safety, 2013, referred to in P.H. Howard, 'Intellectual Property and Consolidation in the Seed Industry', (2015) 55(6) *Crop Science*, 2489 at.
29 See M.V. Mickelbart, P.M. Hasegawa and J.B. Serres, 'Genetic mechanisms of abiotic stress tolerance that translate to crop yield stability', (2015) 16 *Nature Review. Genetics* 237.
30 V.S. Somvanshi 'Patenting Drought Tolerance in Organisms' (2009) 3 *Recent Patents on DNA & Gene Sequences* 16, Table 2.
31 Detailed in M. Blakeney, 'Climate change and gene patents' (2012) 2 *Queen Mary Journal of Intellectual Property*, 2 at 10.
32 ETC Group, 'Gene Giants Stockpile Patents on "Climate-ready" Crops in Bid to become "Biomassters": Patent Grab Threatens Biodiversity, Food Sovereignty' Issue no 106, October 2010, 52. Ibid., Appendix A.
33 Ibid.
34 US Patent 7,317,141.
35 See: S. Lindemose, C. O'Shea, M.Krogh, et al, 'Structure, Function and Networks of Transcription Factors Involved in Abiotic Stress Responses', (2013) 14 *Int. J. Mol. Sci* 5842; H. Wang, H. Wang, H. Shao et al, 'Recent Advances in Utilizing Transcription Factors to Improve Plant Abiotic Stress Tolerance by Transgenic Technology' (2016) 7 *Front Plant Sci*. 67.
36 PCT/US2005/014145
37 M. Rimmer, Climate-ready crops: intellectual property, agriculture and climate change' in M. Rimmer and A. McLennan, *Intellectual Property and Emerging Technologies: The New Biology*, Cheltenham, Edward Elgar, 2012, 320–360.
38 Ibid. at 335–346.
39 US Patent 7,834,146.

40 Ibid.
41 Ibid.
42 Eg ETC Group, 'Who Owns Nature? Corporate Power and the Final Frontier in the Commodification of Life', (2008) *Communiqué* #100.
43 Quoted Rimmer, n.37 supra at 339.
44 569 U.S. 576 (2013)
45 Ibid. at 596.
46 E.g. see H.J. Coad, "The Degree of Alteration Necessary for Patent Eligibility of Synthetically Printed DNA Post Myriad Genetics," (2019) 14(1) *Liberty University Law Review*. Article 8.
47 See M. Qaim, 'Role of New Plant Breeding Technologies for Food Security and Sustainable Agricultural Development' (2019) 42(2) *Applied Economic Perspectives and Policy*, 129; A. Ricroch, 'Global developments of genome editing in agriculture' (2019) 28 *Transgenic Research*, 45.
48 See Y. Zhang, A.A. Malzahn, S. Sretenovic, et al, 'The emerging and uncultivated potential of CRISPR technology in plant science', (2019) 5 *Nature Plants* 778; G. Song, M. Jia, K. Chen, et al, 'CRISPR/Cas9: a powerful tool for crop genome editing' (2016) 4 *Crop J* 75.
49 See S.S. Zaidi, A, Mahas, H. Vanderschuren, et al, 'Engineering crops of the future: CRISPR approaches to develop climate resilient and disease resistant plants' (2020) 21(1) *Genome Biology*, 289.
50 D. Chilcoat, Z-B, Liu and J Sander, 'Use of CRISPR/Cas9 for Crop Improvement in Maize and Soybean' (2017) 149 *Prog Mol Biol Transl Sci* 27; G. Rai, D.M. Kanday, P. Kumar, et al, 'Enhancing Crop Resilience to Drought Stress Through CRISPR–Cas9 Genome Editing' (2023) 12(12) *Plants* (Basel) 2306.
51 S. Farhat, N. Jain, N. Singh, et al, 'CRISPR–Cas9 directed genome engineering for enhancing salt stress tolerance in rice' (2019) 96 *Seminars in Cell and Developmental Biology* 91; R. Nazir, S. Mandal, M.Sicon et al, 'Clustered regularly interspaced short palindromic repeats (CRISPR)/CRISPR-associated genome-editing toolkit to enhance salt stress tolerance in rice and wheat' (2022) 174(2) *Physiologia plantarum*, e13642-n/a.
52 M.K. Zahoor, A. Ahmad, M.A. Zahoor et al 'CRISPR/Cas based Insect Resistance in Crops' (2021) *CRISPR Crops*, 117; Y. Yan, R.A. Aumann, I. Häcker, et al, 'CRIPR-based genetic control strategies for insect pests' (2023) 22(3) *Journal of Integrative Agriculture*, 651.
53 A. Mahas and M. Magdy, 'Engineering virus resistance vis CRISPR–Cas Systems'(2028) 32 *Current opinion in virology*, 1; M.Z. Khan, I. Amin, A. Hameed, et al, 'CRISPR–Cas13a: Prospects for Plant Virus Resistance' (2018) 36(12) *Trends in Biotechnology*, 1207.
54 H. Boubakri, 'Recent progress in CRISPR/Cas9-based genome editing for enhancing plant disease resistance', 866 *Gene*, 147334.
55 See M. Gupta, M. Gerard, S. Surendran, et al 'Trends of CRISPR technology development and deployment into Agricultural Production-Consumption Systems', (2020) 60 *World Patent Information* 101944; O. A. Jefferson, S. Lang, K. Williams et al, 'Mapping CRISPR–Cas9 public and commercial innovation using The Lens institutional toolkit', (2021) 30 *Transgenic Res* 585.
56 I.M. Moretti, *Tracking the trend toward market concentration: The case of the agricultural input industry*, Geneva, UNCTAD, 2006.
57 G. Moschini, 'Competition Issues in the Seed Industry and the Role of Intellectual Property' (2010) 25(2), *Choices: The Magazine of Food, Farm, and Resource Issues, Agricultural and Applied Economics Association*, 1.
58 O.A. Jefferson, D. Kollhofer, T. Ehrich, et al, 'The ownership question of plant gene and genome intellectual properties' (2015) 33 *Nature Biotechnology,* 1138.
59 P.H. Howard, 'Intellectual Property and Consolidation in the Seed Industry', (2015) 55(6) *Crop Science*, 2489 at 2491.
60 McDonald et al., n.13 supra at 20.
61 Ibid. at Table 5.
62 G. Shi, J. Chavas and K. Stiegert, 'Pricing of herbicide-tolerant soybean seeds: A market-structure approach', (2009) 12 *AgBioForum* 326; G. Shi, J. Chavas and K. Stiegert, 'An analysis of the pricing of traits in the U.S. corn seed market', (2010) 92 *Am. J. Ag. Econ.* 1324.
63 McDonald et al., n13 supra at 15.

64 W. Klümper and M. Qaim, 'A meta-analysis of the impacts of genetically modified crops' (2014) 9 *PLOS One*, e111629; F. Ciliberto, G. Moschini and E. Perry, 'Valuing product innovation: Genetically engineered varieties in U.S. corn and soybeans' (2019) 50 *RAND Journal of Economics*, 615.

65 *Federal Trade Commission* et al *v. Syngenta Crop Protection Ag* et al (1:22-Cv-00828), District Court, M.D.N.C. North Carolina, 2022.

66 See M. Blakeney, 'Linux For Lettuce': Open Source Biotechnology and Agricultural Innovation', (2017) 41(2) *The University of Western Australia Law Review* 139.

67 See Hassan Masum et al, 'Open Source Biotechnology Platforms for Global Health and Development: Two Case Studies' (2011) 7 *Information Technologies & International Developmen*t 61, 62–5.

68 Described in Richard A Jefferson, Tony A Kavanagh, and Michael W Bevan, 'GUS fusions: betaglucuronidase as a sensitive and versatile gene fusion marker in higher plants' (1987) 6 *European Molecular Biology Organization Journal* 3901.

69 See J.E. Hope, Open Source Biotechnology, PhD Thesis, Australian National University, 2005.

70 Correspondence, (2021) 597 Nature, 9 September. 178.

71 See M. Blakeney, 'Agricultural Research: Intellectual Property and the CGIAR System' in P. Drahos and R. Mayne (eds), *Global Intellectual Property Rights: Knowledge, Access and Development*, London, Palgrave Macmillan, 2002, 108–24.

72 See M. Halewood, N. Jamora, I. Noriega, et al 'Germplasm acquisition and distribution by CGIAR genebanks' (2020) 9 *Plants* 1296.

73 M. Blakeney, 'Plant Variety Protection, International Agricultural Research, and Exchange of Germplasm: Legal Aspects of Sui Generis and Patent Regimes' in Anatole Krattiger et al (Eds) *Intellectual Property Management in Health and Agricultural Innovation: A Handbook of Best Practices,* Oxford, MIHR, 2007, 401–418.

74 M. Blakeney, 'Recent developments in intellectual property and power in the private sector related to food and agriculture' (2011) 36 *Food Policy* Supplement 1, 109.

75 M.A. McEwan, C.J. Almekinders, J.J., Andrade-Piedra, et al 'Breaking through the 40% adoption ceiling: Mind the seed system gaps. A perspective on seed systems research for development', (2021) 50(1) One CGIAR. *Outlook on Agriculture*, 5.

76 See D.J. Spielman, 'Pro-poor agricultural biotechnology: can the international research system deliver the goods?', (2007) 32(2) Food Policy 189; D.G. Dalrymple, 'International agricultural research as a global public good: concepts, the CGIAR experience and policy issues' (2008) 20 *J. Int Dev*.347; C. Leeuwis, L. Klerkx, M. Schut, 'Reforming the research policy and impact culture in the CGIAR: Integrating science and systemic capacity development' (2018) 17 *Global Food Security* 16.

77 CGIAR, CGIAR 2030 Research and Innovation Strategy. Transforming food, land, and water systems in a climate crisis, 2020, available at https://cgspace.cgiar.org/bitstream/handle/10568/110918/OneCGIAR-Strategy.pdf

10 Patents in Microbial β-Glucans and Pigments

Potential Commercial Products

Alane Beatriz Vermelho, Felipe R. P. Mansoldo, Jean Vinícius Moreira, Verônica S. Cardoso, Maria Do Socorro Santos Rosa, and Flavia Lima do Carmo

10.1 INTRODUCTION

We live in a new phase on Earth, with climate changes due to greenhouse gases (GHGs) and the accelerated loss of forests, impacting biodiversity and exacerbating this problem. In addition to GGH, other pollution types, such as plastic and other industrial wastes, are in progress. In this context, searching for eco-friendly, nontoxic, sustainable, and biodegradable products has been a target of industrial sectors.

Microbiology arises as a source of biosolutions. Several microbial products, such as polysaccharides, pigments, antimicrobial compounds, enzymes, and other bioactive substances with anti-inflammatory, immunomodulatory, anticancer, and antioxidant actions, have been identified. Microorganisms have been studied and used in industrial bioprocesses based on growth in bioreactors with physical-chemical conditions controlled in a small space with a short production time and high reproducibility. Bacteria, fungi, and microalgae have been used for this purpose. Metabolic engineering and *omic* sciences amplify the possibilities of viable applications. One cause of the increase in the bioreactor market is the microbiology inside the biological products segment. According to Allied Market Research, the market was valued at USD 2,615.4 million in 2020 and is estimated to reach USD 7,328.4 million by 2030, growing at a CAGR of 10.7% from 2021 to 2030 [1].

Due to their properties, two microbial products with potential applications are the β-glucan polysaccharide and pigments. This chapter will focus on them, which are studied for different uses and patented.

10.2 MICROBIAL β-GLUCANS

β-glucans are multifunctional molecules synthesized by cereals and microorganisms such as bacteria, yeasts, microalgae, and fungi [2]. In cereals, extensive production

 DOI: 10.1201/9781003405665-10

and extraction processes are necessary. However, yeast and mushrooms can produce glucan in bioreactors independent of the extraction process. In some fungi, such as basidiomycetes, glucan can be obtained from the fruiting body or the mycelium [3].

Regarding the way it is produced, its properties may differ based on chain conformation, monomeric configuration, sequence of the bonds, frequency, and conformation of chains, branching lengths, and backbone structures. The physiological characteristics of β-glucan may vary according to chemical composition, glycosidic bonds, degree of polymerization, and rheological properties, which relate to the microorganism that synthesizes the biomolecule [4, 5]. Cereal β-glucans are based on linear glucose molecules linked by β-(1,3)- and β-(1,4)-glycosidic bonds with relatively high water-soluble properties, whereas β-glucans from yeast and fungi are constituted by β-(1,3) and β-(1,6) linkages, possessing insoluble properties[6]. Beta-glucan from microalgae is another possible alternative source. They produce ß-glucans in different forms: (i) unbranched 1,3-glucan (paramylon) or (ii) branched 1,3–1,6- β-glucan (chrysolaminarin), with differences in solubility [7]; examples of microalgae producing beta-glucan are *Phaeodactylum tricornutum, Monodopsis subterranea, Cylindrotheca fusiformis* and *Nannochloropsis oculate.* This biopolymer *accumulates* during nitrogen limitation or depletion [8]. Figure 10.1 shows the structure of β-glucans from cereal, yeast, and bacteria.

In bacteria, these molecules were found to have diverse functions. Linear (1→3)-β-glucans and side-chain-branched (1→3,1→2)-β-glucans are significant constituents of capsule materials with roles in bacterial aggregation, virulence, and carbohydrate storage. Predominantly, periplasmic glucans are cyclic (1→3,1→6)-β-glucans with a role in osmotic adaptation. They are found in legume symbionts *Bradyrhizobium japonicum*, *Rhizobium loti*, and *Azorhizobium aulinodans* and the free-living, root-colonizing diazotroph *Azospirillum brasilens* [9]. *Agrobacterium* and marine bacteria isolated from red seaweeds and others have a linear (1→3)-β-glucan named curdlan with unique rheological and thermal gelling properties, with applications in the food industry and other segments [10].

β-glucans structural configurations and properties are primarily influenced by their source of derivation. The simplest form of β-glucan from microalgae is β-(1, 3) glucan. Paramylon, found in microalgal species such as *Euglena* sp. and *Astasia longa*, is an example of this structural type. β-(1,3;1,4)-glucan is located in the cell walls of certain microalgae, such as *Micrasterias* sp. and *Monodus subterraneus*. Last, the branched β-(1,3;1,6)-glucan structure has been identified in specific groups of microalgae, notably in diatoms and chrysophytes [11].

It is important to note that fungi have been used for centuries for biotechnological purposes. In the food and beverage industry, bread, cheese, beer, wine, and many other products use bioproducts from fungi [12]. A great variety of glucans are found in basidiomycete fungi (mushrooms), as shown in Figure 10.2.

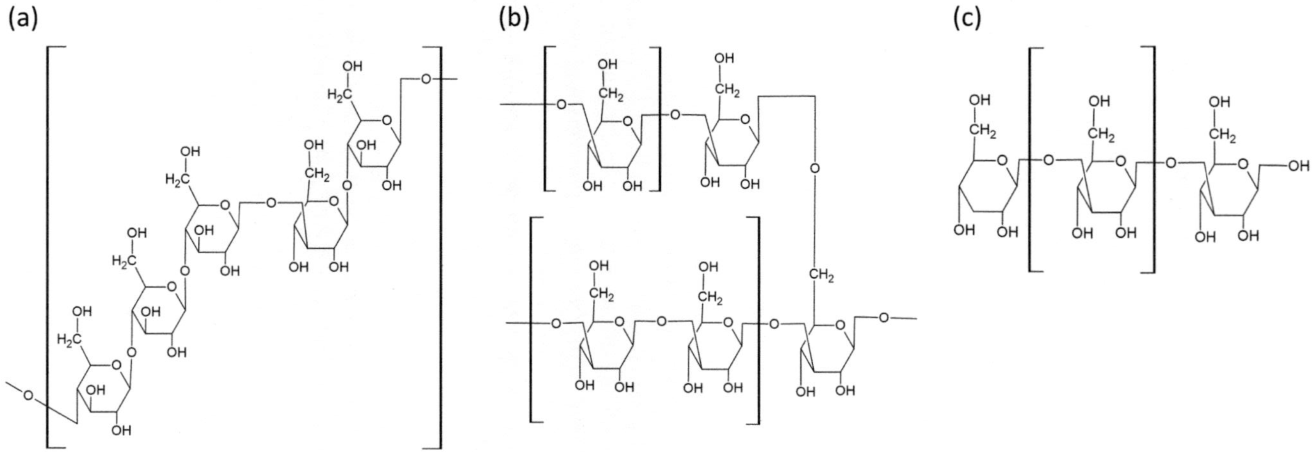

FIGURE 10.1 Structure and branching of β-glucans from (a) cereal β-glucans (linked with β-(1,3)- and β-(1,4)-glycosidic bonds); (b) yeast β-glucans (linked with β-(1,3)- and β-(1,6)-glycosidic bonds); and (c) bacterial β-glucans (linked with β-(1,3)-glycosidic bonds) [2].

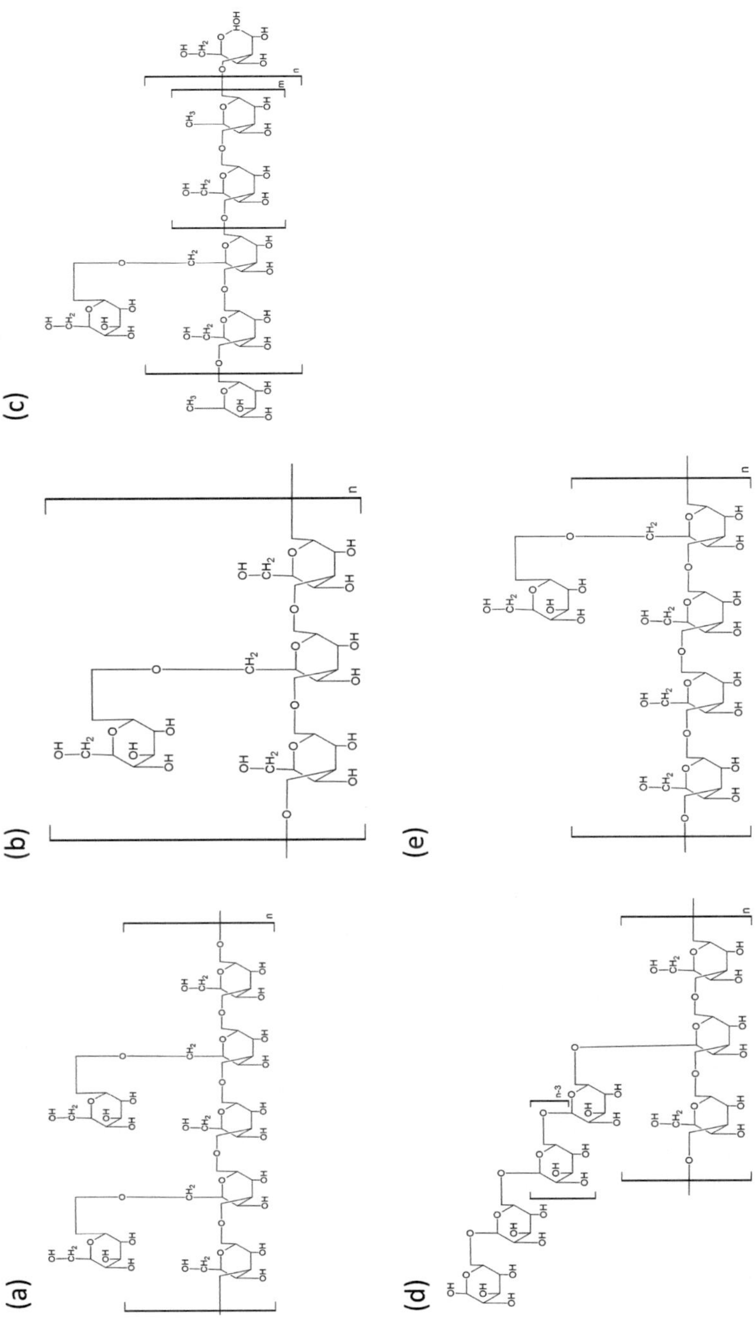

FIGURE 10.2 Beta-glucans found in basidiomycetes: chemical structure of β-glucan (a) Lentinan from *Lentinusedodes*, (b) Schizophyllan from *Schizophyllum commune*, (c) Krestin from *Trametes versicolor*, (d) Grifolan from *Grifolafondosa*, (e) Pleuran from *Pleurotus* spp. [6].

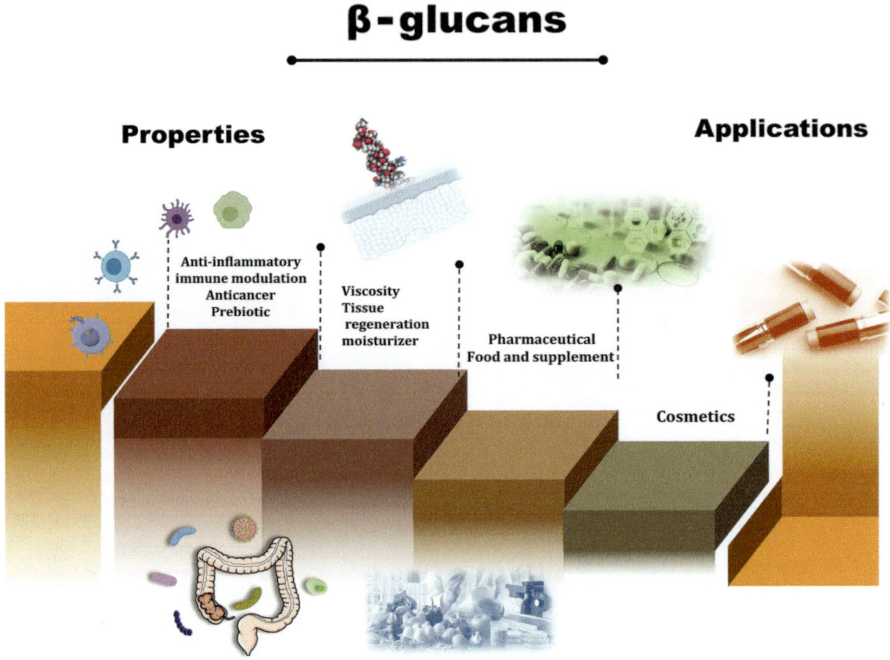

FIGURE 10.3 Properties and applications of β glucans.

10.3 INDUSTRIAL APPLICATIONS OF BETA-GLUCAN

Despite the natural use of β-glucan, they can also be chemically modified to enhance properties, bioactivities, and applications by acetylation, carboxymethylation, phosphorylation, and sulfonation. Each process may enhance one property [13].

Several beneficial properties led these molecules to be the target of multidisciplinary studies aiming at their potential industrial applications [14]. β-glucans present many health-promoting and prebiotic properties [2,15]; anti-aging [16,17]; anticancer [18]; antiviral [19]; anti-inflammatory [16] and immune-modulating effects [20, 21, 14]. Figure 10.3 summarizes some properties and applications.

10.3.1 Food Industry

The benefits of daily dietary fiber consumption are well documented, at approximately 21–25 g/day for women and 30–38 g/day for men. Soluble dietary fiber is known for its therapeutic potential for various diseases. The US Food and Drug Administration recommends 3 g of β-glucan-containing oats, recognized as a cholesterol-reducing food [22,23].

The benefit of consuming β-glucan-containing products is forming viscous solutions by retaining water in the digestive tract. This ability improves digestive health by increasing stool volume. It improves the homeostasis of blood glucose levels based on the impacts of hyperglycemia. Furthermore, the application of β-glucan may improve the textural characteristics of food, which has led to research initiatives to incorporate soluble dietary fiber, such as β-glucan, into commonly consumed food [23, 24].

Several patents about extracted or biosynthesized β-glucan incorporated into food and beverage manufacturing have been deposited. Some patents merge grounded β-glucan-containing plants, fruits, and vegetables into a liquid product added to the microbial strains to produce a functional fermented beverage. Shanghai Bright Dairy and Food Co and South China Institute of Collaborative Innovation patented the use of tomato juice fermented by *Leuconostoc mesenteroides* and ground fruits and vegetables fermented by *Lactobacillus* and *Bacillus*, respectively [25,26].

Even though cereal-extracted β-glucan may be used for food purposes, most of the production is done by microbial fermentation, as described in Table 10.1.

Microbial β-glucan can be produced using several carbon sources, including natural agro-industry residues, lowering the cost of the bioprocess. This theme is a vast area of research regarding the productivity, quality, and safety of these glucans for food application [13].

Cordyceps militaris, a recognized traditional Chinese medicine, contains polysaccharides with various biological activities. One polysaccharide, CM1, has been isolated and found to enhance cholesterol efflux. CM1 has a backbone of (1→4)-β-D-Glcp and (1→2)-α-D-Manp and presents a molecular weight of 700 kDa [45]. Lin *et al.* probed the anti-atherosclerotic impacts of CM1 on apolipoprotein E-deficient mice [46]. The findings demonstrate that CM1 notably diminished the levels of total cholesterol and triglycerides in mouse plasma and decreased lipid accumulation and atherosclerotic plaque formation in a dose-responsive manner. CM1 interacts with numerous signaling pathways, encompassing those pertinent to lipid metabolism, inflammatory response, oxidoreductase activity, and fluid shear stress, collectively contributing to its antiatherosclerotic properties. These insights accentuate the potential therapeutic utility of CM1 in addressing hyperlipidemia and atherosclerotic cardiovascular ailments.

Polygonum sibiricum polysaccharides (PSP), one of the primary active constituents of the traditional Chinese medicine *Polygonum sibiricum*, have demonstrated antidepressant-like properties, yet the exact mechanisms remain elusive. Zhang *et al.* [14] analyzed the antidepressant-like effects of PSP linked to the microbiota-gut-brain (MGB) axis. To this end, fecal microbiota transplantation (FMT) from mice treated with PSP was conducted on mice subjected to chronic unpredictable mild stress (CUMS), a model for inducing depressive symptoms. This investigation revealed that FMT significantly reversed depressive-like behaviors in these CUMS-induced mice across multiple behavioral tests. Furthermore, FMT elevated the concentrations of the neurotransmitters 5-hydroxytryptamine and norepinephrine while reducing proinflammatory cytokines in the hippocampus and the stress hormone corticosterone in the serum. The study suggests that the antidepressant-like effects exhibited by PSP are mediated through the MGB axis [14].

10.3.2 Pharmaceutical and Medical Industry

According to the report by Markets and Markets, the beta-glucan market size is estimated at USD 501 million in 2023 [47]. The projection to 2028 is estimated at

TABLE 10.1
Overview of Microbial β-Glucan Application in Food Manufacture

Microbial Strain	β-Glucan Application	Patent Applicant	Reference
Saccharomyces cerevisiae	Candy Pieces	Shanghai Acme Import And Export Co Ltd	[27]
Saccharomyces cerevisiae	Beverage Additive	Dongguan Natural Salutem Health Technology Co Ltd	[28]
Saccharomyces cerevisiae	Liquid And Food Additives	Glykos Finland Ltd. Hankkija Oy	[29]
Saccharomyces cerevisiae	Food Additive; Preparation Of Functional Food	Lesaffre Et Cie SA	[30]
Saccharomyces cerevisiae; *Kluyveromyces fragilis;* *Candida utilis* *Alkaligenes faecalis*	Functional Beverages	CUB Pty Ltd	[31]
Agaricus blazei *Pediococcus damnosus*	Margarine; Fat; Oil; Mayonnaise; Curry Sauce; Cookies; Bread; Candies	Adeka Corp	[32]
Aureobasidium sp.	Food And Drink Additives	Individual	[33]
Aureobasidium sp.	Yogurt; Cheese; Ice Cream; Soy Or Rice Milk	ONAKA YASUSHI	[34]
Aureobasidium pullulans	Production Of Beta-Glucan	SOOCHOW UNIVERSITY	[35]
Hericium erinaceus	Regulating Intestinal Flora	Shanghai Institute Of Technology	[36]
Aureobasidium pullulans	Production Of Beta-Glucan	DAISO	[37]
Aureobasidium sp.	Soft Drinks, Materials For Healthy Foods, Thickeners For Food And Medicines O	Daiso	[38]
Schizophyllum commune	Production Of Beta-Glucan	QUEGEN BIOTECH	[39]
Aureobasidium pullulans	Functional Food	BIOCOSMO	[40]
Saccharomyces pastorianus	Functional Food	JIANGNAN UNIVERSITY	[41]
Leuconostoc sp.	Feed Additive	Agricultural Technology Research Institute	[42]
Pholiota adiposa	Functional Food	WUXI ZANJIANG BIOLOGICAL TECHNOLOGY	[43]
Monascus pilosus	Functional Food	GUANGDONG ZHENHONG BIOTECHNOLOGY	[44]

USD 734 million, at a CAGR of 7.9% up to 2028. The pharmaceutical segment of beta-glucan applications is expected to reach USD 117.5 million by 2025. Most of this growth is led by new trends in drugs for cancer treatment, cholesterol-lowering drugs, immunomodulators, and weight reduction [18,47].The applications mentioned above have drawn the attention of several industries to novel drugs for cancer therapy, as discussed in Table 10.2.

10.3.2.1 β-Glucan Application in Cancer Therapy

β-glucan presents biological and immunological activities, such as activating immune cells as macrophages, neutrophils, and antigen-presenting cells. β-glucan's immunological potency and characteristics depend on its type, source, structure, and molecular weight. Therefore, it may enhance the host antitumor defense mechanism. The antitumor property is attributed to its structure, water solubility, high viscosity, composition, and high molecular weight. These characteristics make the type of production the critical parameter [62,63].

In vitro and animal tests have shown the benefits of using beta-glucan in cancer treatment; by Table 10.2, it is clear that the wide range of applications allows further studies on the matter, especially in biological functions and interactions with receptors, regarding the physical-chemical properties of glucan, its structure, and its origin [64,65].

β-glucans are recognized for their extensive immunomodulatory properties, activating innate and adaptive immunity. Historical research dating back to the 1940s analyzed their influence on the immune system, encompassing roles in cancer treatment, anti-infection immunity, and bone repair. These compounds activate innate immune cells [macrophages, dendritic cells, granulocytes, and natural killer (NK) cells], inhibiting tumor growth and metastasis through adaptive immune cells, specifically T and B cells. However, the exact mechanism of the antimetastatic actions of β-glucans remains partially understood [66].

β-glucans, primarily derived from yeast, have emerged as promising adjuvant candidates for cancer treatment, mainly due to their biocompatibility, safety profile, and distinctive biological response modifier (BRM) effects (more pronounced than β-glucans from oats or barley). According to Ikewaki *et al.* these glucans function as potent immunomodulators, molecules capable of enhancing or suppressing specific immune response components [67]. In the realm of cancer immunity, β-glucans have demonstrated capabilities such as increasing resistance to infections, enhancing antitumor effects by triggering both adaptive and innate immune systems, stimulating various immune cells, and exhibiting anticoagulant properties [68,67]. The antitumor action of β-glucans is primarily attributed to their role in augmenting the immune response against cancer cells and obstructing tumor progression through intricate modulation of apoptotic and angiogenic pathways. Specifically, β-glucans bind to receptors such as Dectin-1 on myeloid cells, transforming them into antigen-presenting cells [69]. Additionally, β-glucans can reshape the tumor microenvironment, reducing tumor growth and metastasis [67].

The efficacy of antitumor immune therapies is frequently challenged by an immune-suppressive microenvironment, where tumor-associated macrophages (TAMs) play a significant role. These TAMs are prevalent in various solid tumors (e.g., breast,

TABLE 10.2
Overview of Beta-Glucan Application in Cancer Treatment and Other Medical Applications

Glucan Application	Extracted/Synthesized	Applicant	Reference
Immunostimulatory Effect	Yeast	Biothera Inc	[48]
Increasing Nonspecific Immunity and Resistance to Infection	Yeast	Biopolymer Engineering Inc	[49]
Enhancing Anti-Proliferative Immunotherapy	Yeast	University of Louisville	[50]
Immuno-Enhancement And/or Immune-Balancing	*Aureobasidium pullulans*	Takashi Onaka; Samuel Jk Abraham	[51]
Enhances The C3-Antibody-Tumor Cell	*Saccharomyces cerevisiae*	University Of Louisville Research Foundation ULRF, Biothera Inc	[52]
Immuno-Enhancement And/or Immuno-Balancing, And For Adjuvant Use	*Aureobasidium pullulans*	Sophy Inc, GN Corp Co Ltd	[51]
Stimulate The Immune Response	*Coccidioides immitis, Trichophyton verrucosum, Saccharomyces Cerevisiae*	Innate Biotherapeutics LLC	[53]
Odor Removing	*Phellinus linteus*	University of Seoul	[54]
Treat And Prevent Avian Flu	*Saccharomyces cerevisiae*	En Bio Technology	[55]
Cancer Treatment	*Ganoderma lucidum*	Hayashibara Biochemical Laboratories	[56]
Immune-Boosting Compositions, Health Functional Foods	*Aureobasidium pullulans*	Korea Research Institute of Bioscience & Biotechnology	[57]
Immune Modulator	*Euglena* sp.	Kemin	[58]
Anticancer	*Sparassis crispa*	Sangcheol Yoon	[59]
Vaginal Mycose Treatment	*Saccharomyces cerevisiae*	Lesaffre	[60]
Immunostimulation	*Lactobacillus* sp.	DAISO	[61]

ovarian, bladder, and skin) that promote or suppress immune activation. Graaff *et al.* studied the role of β-glucans (oatβG, curdlan, grifolan, schizophyllan, lentinan, yeast whole glucan particles (yWGP), zymosan and two additional yeast-derived β-glucans a and b) and their potential antitumor effects. Notably, the receptor binding capacity of β-glucans, and thus their immunomodulatory activity, depends on their physicochemical structures, which vary according to their source. The investigation studied their physicochemical properties and their influence on macrophages. The findings indicate that β-glucans derived from *Alcaligenes faecalis* (curdlan) and yeast (yeast-b and zymosan) can be suggested as potential adjuvants to support immune therapies for various tumors [70].

The tumor microenvironment (TME) significantly influences cancer progression, highlighting the need for targeted therapies. β-glucan stands out as a crucial anticancer agent due to its immune-modulating properties. The structure of particulate β-glucan is also being harnessed for drug delivery. Although β-glucans, particularly those from fungi, have been used for centuries to stimulate immunity, new insights are emerging about their biology. This deepening knowledge is reshaping our comprehension of their immune interactions, especially regarding metabolic reprogramming and epigenetic shifts in specific immune cells. β-glucan's potential as an adjunct in antitumor vaccines and its combination with other therapies promise improved patient outcomes, but challenges such as molecule standardization remain [71].

10.3.2.2 β-Glucans Used in Drug Formulation

The basic structure of 1,3-β-glucan can differ depending on the production and application of an external modification. Such characteristics make it a promising drug delivery agent, and it can be used as gels, nano- and microparticles, and complexes [72].

β-glucans can be applied in drugs of oral administration. They protect the API encapsulated through the acid gastric, making it reach undamaged to the adsorption sites in the intestine [73]. The specific physicochemical properties of β-glucan, such as molecular weight, conformation in different solvents, and degree of branching, will be determined. It may be used in principles including electrostatic interaction, covalent binding, and encapsulation on the loading medicines with β-glucan [74].

The properties of beta-glucan have been researched, and several patents have been deposited describing its use conjugated to API. Fuzhou Qianzheng Pharmaceutical Co., Ltd. studied beta-glucan application with (−)-epigallocatechin gallate (EGCG). The authors discovered a synergetic function between EGCG and β-glucan, enhancing the drug effects on antitumor, antioxidation stress, and anti-tissue fibering [75].

Beta-glucans used as vehicles for drug component delivery may present sustained release [76]. They can also be used to produce microgels in drug delivery systems [77]; adjuvants in vaccines [78]; agents for controlled release [52]; particulate delivery systems [52]; and β-glucan-based delivery systems for sd-RNA [79].

Over the past decade, substantial research has been channeled toward creating fungal β-glucan-based nanomaterials, notably in biomedicine, because of the therapeutic capabilities of fungal β-glucan. Nanosystems containing β-glucan have been used in anticancer treatments, vaccinations, and anti-inflammatory therapies. Compared to their metal or inorganic counterparts, these carbohydrate-based nanomaterials are more likely to be biocompatible. Moreover, β-glucans stand out among other polysaccharides with pronounced immunological potentiating effects. The fact that they are biologically safe, nontoxic, and free from proteins and peptides that might activate nonspecific immune responses further amplifies their suitability for therapeutic applications [80]. The inherent structure of β-glucans, replete with hydroxyl groups, paves the way for structural modifications. This results in β-glucan-based nanoparticles with improved physiological attributes and the ability to carry immune-modulating agents effectively [80].

Schizophyllan (SPG) is a polysaccharide secreted by the fungus *Schizophyllum commune.* Its structure comprises linear β-1,3-d-glucose with a β-1,6-d-glucose side chain attached to every third main-chain glucose (Figure 10.4-b). SPG interacts with its receptor, Dectin-1, an essential receptor in anti-fungal innate immune responses. Uno *et al.* introduced a method using a complex of oligonucleotides and SPG selectively directed to cells with the Dectin-1 lectin receptor through SPG–Dectin-1 interactions. Dectin-1 is predominantly found in human monocytes and dendritic cell subsets and is vital for antigen presentation. This research highlights CD40, which is integral in regulating immune responses in antigen-presenting cells, as a potential target gene for small interfering RNA (siRNA). The results confirm that siCD40–SPG complex uptake was dose-dependent in Dectin-1 cells, and its gene-silencing efficacy was validated in vitro and in vivo. This receptor–ligand binding-mediated siRNA delivery method holds promise for drug development against various diseases, including autoimmune conditions and cancer [81].

10.3.3 β-Glucan in Immunotherapy

β-glucan generates an immune response, including stimulating cytokine production, oxidative burst, lymphocyte proliferation, and phagocytosis of opsonized tissues. Industries have employed those characteristics to deposit patents on the matter [72].

Biotherapeutics Co. deposited a patent based on the use of microbial beta-glucan conjugated to antibody to enhance its efficiency [82]. The immunostimulatory environment created by β-glucan can be used therapeutically to decrease suppressive mechanisms in chronic diseases to hasten full clearance of the infection [83], soluble yeast β-glucan can be used with CD40 agonist in cancer immunotherapy [83], a part of that, the proper dosage of beta-glucan optimize acute immune dynamic responses [84], it can also be used as biomarker-positive in immunotherapy [85].

Murphy *et al.* [86] analyzed the potential of β-glucan extracts from the *Lentinus edodes* mushroom in counteracting inflammatory responses related to COVID-19. Notably, the study compared two β-glucan extracts: IH-Lentinan and Carbosynth-Lentinan. Although derived from the same mushroom, the two extracts demonstrated distinct in vitro immunomodulatory activities. Specifically, IH-Lentinan more effectively inhibits NF-κB activation in human alveolar epithelial A549 cells, especially at lower doses. In contrast, Carbosynth-Lentinan showed superior attenuation of pro-inflammatory cytokine production in THP-1-derived macrophages.

Additionally, while Carbosynth-Lentinan mitigated early apoptosis induced by oxidative stress, IH-Lentinan was more effective against late apoptosis. These findings highlight the importance of recognizing the differential effects of similar β-glucan extracts and emphasize the need for further clinical research [5].

Recurrent respiratory infections (RRIs) in the pediatric population have profound ramifications on the infancy and childhood landscape and represent a challenge for pediatricians. Varricchio *et al.*[87] evaluated the effects of an aerosolized solution containing resveratrol plus carboxymethyl-β-glucan in preventing RRIs in children compared to a saline solution. A randomized study involved 82 children suffering from acute rhinopharyngitis and RRI.

The results indicated that the active compound notably decreased the number of days children experienced various respiratory symptoms, reduced medication use, medical visits, and school absences. The authors, however, acknowledge the preliminary nature of this study and its limitations, including the open design, lack of objective symptom assessments, and limited patient sample size. Considering these limitations, they recommend that subsequent, more rigorous research be conducted to address these issues. They conclude by suggesting the potential of resveratrol plus carboxymethyl-β-glucan aerosolized solution in prolonging the efficacy of anti-infective and anti-inflammatory treatments, thus possibly preventing the recurrence of RI in children with RRI [87].

Lin *et al.* studied the effects of a polysaccharide extracted from *Ganoderma lucidum* (PS-G) on human monocyte-derived dendritic cells (DC). The polysaccharide component from *G. lucidum*, specifically the branched (1→6)-β-D-glucan moiety known as PS-G, has garnered attention due to its potential antitumor activities and ability to activate natural killer cells. Upon treatment with PS-G, dendritic cells show enhanced cell-surface expression of several markers (CD80, CD86, CD83, CD40, CD54, and HLA-DR) and increased production of interleukins. Mechanistically, PS-G augments inhibitor of κB (IκB) kinase and nuclear factor (NF)-κB activity, as well as IκBα and p38 mitogen-activated protein kinase (MAPK) phosphorylation. Conclusively, the data suggest that PS-G can potentially promote the activation and maturation of immature dendritic cells, implying its possible role in modulating immune responses [88].

Euglena gracilis is a unicellular, photosynthesizing green microalga found in fresh water. It is renowned for its rich nutrient profile with a high concentration of vitamins, minerals, amino acids, and fatty acids. Notably, its lack of a cell wall makes it highly digestible. Several studies have demonstrated Euglena's potential health-enhancing benefits, making it a popular choice for human and animal dietary supplements. A significant constituent of *Euglena* is paramylon (PM), a β-1,3-glucan that constitutes approximately 20–70% of the alga's dry weight [89]. PM is an insoluble and unfermentable polysaccharide and exhibits several benefits [90], such as improvement in glycemic control [91], anti-obesity and anti-inflammatory effects [92], promotion of immune function [93], inhibition of atopic dermatitis development [94], anti-HIV effects [95], improvement of sleep quality [90,96] and promotion of immune defense and good health [97].

Founded in 2005, Euglena Corp. initiated the world's first large-scale outdoor cultivation of Euglena for food. After establishing itself in the market, the company launched Euglena-infused products, such as cookies and drinks, in the Japanese market. The company's portfolio has since expanded to nonfood items, including cosmetics. Their primary product is foods enhanced with dried *E. gracilis* powder [98].

There is a need for alternative treatment and prevention methods for invasive mycosis, such as cryptococcosis, due to the poor efficacy of current antifungal drugs and the lack of available vaccines. Basso *et al.* investigated the immunostimulatory properties of exopolysaccharides (EPS) containing β-(1,3)-glucan isolated from the edible mushroom *Auricularia auricula*. This EPS was found to activate macrophages and dendritic cells by binding to Dectin-1, a pattern recognition receptor of the C-type lectin receptor family. The EPS repolarized M2-like phenotype macrophages to an M1 pattern, associated with the efficient killing of *C. neoformans* through increased nitric oxide production. The administration of this β-glucan-containing EPS enhanced the survival rate of mice infected with *C. neoformans*. The study concludes that EPS from *A. auricula* could be a promising adjuvant for vaccines or an immunostimulant in antifungal therapy [99].

10.3.4 β-Glucan Application in Cosmetics

In contrast to the extensive literature about the application of β-glucan in the food and pharmaceutical industries, there are few studies regarding the skin health promotion promoted by beta-glucan, such as anti-wrinkle, wound healing, anti-ultraviolet effects, and moisturizing characteristics [16].

The use of β-glucan combined with a gelling agent in wound dressing has been proven to have the ability to heal [100]; β-1,3-glucan may be used in medicine for repairing skin injury, skin tissue regeneration, and anti-inflammatory drugs for skin [101], cosmetic composition for skin moisturizing, wrinkle improvement, and whitening [102], agents against skin aging [103], and skin rejuvenation [104].

10.4 MICROBIAL PIGMENTS

Microbial pigments are emerging as a potential source of industrial pigments. In addition to its primary use as a colorant, several applications are based on other properties, such as antioxidant, immunosuppressive, and anti-inflammatory activity [105], UV protection [106,107], antimicrobial action [108], anticancer effects [109], and several pharmacological effects. These characteristics allow the use of microbial pigments in vast segments of industry, including cosmetics, textiles, pharmaceuticals, and food [110,111]. Fungi, bacteria (Archaea and Bacteria domains), and macroalgae produce biopigments [112,113].

Synthetic pigments dominate the Mundial market and have been broadly utilized in several industries, such as food, pharmaceuticals, cosmetics, and textiles. However, some of them are potentially teratogenic and carcinogenic. In the current scenario worldwide, it is necessary to use safe pigments. In addition, consumers are more aware of the need and importance of safe products in food and other sectors. Natural plant pigments are included in this group. However, their use includes extraction, purification, sensitivity to light, climate, competition with other industrial segments, and long production times [114,115].

Environmentally friendly, biodegradable, nontoxic, and sustainable production could be produced on an industrial scale through bioprocesses using agro-industrial waste, reducing costs [116,117]. Another advantage is the great variety of structures. Table 10.3 shows examples of pigments found in microorganisms, and Figure 10.4 demonstrates some structures.

10.5 BIOPIGMENT APPLICATIONS

Synthesized colorants have been applied in different industrial sectors, such as food, pharmaceuticals, cosmetics, and textiles. The simplified production, cost efficiency, and finer coloring properties placed them in an advantageous position compared to biopigments. Nevertheless, the popularity of synthetic pigments waned late due to their toxicity and the generation of toxic effluents [128]. Figure 10.5 schematically represents some colors and structures of biopigments.

Biopigments are produced by several microorganisms, including bacteria *Pseudomonas aeruginosa*, *Streptomyces* sp. [129], fungus *Ashbya gossypii* [128], microalgae *Dunaliella, Chlorella, Hematococcus*, and *Arthrospira* [130]. Efforts to produce biopigments have been made due to their eco-friendly characteristics, nontoxicity to humans, and wide application in the food industry, pharmaceuticals, and cosmetics. Furthermore, biopigments promote color and health benefits [131].

10.5.1 Food Industry

The food industry places significant emphasis on the visual appeal of products, leading to the incorporation of either synthetic or natural colorants to enhance attractiveness. However, recent trends indicate a shift from synthetic to natural colorants, predominantly due to the health concerns linked to some synthetic dyes. Microbial pigments derived from sources such as *Monascus* sp., *Xanthophyllomyces dendrorhous*, and *Penicillium oxalicum,* among others, offer multiple benefits, including cost-effective production, straightforward extraction processes, abundant yields, and consistency in availability irrespective of seasonal changes. Moreover, specific pigments, such as canthaxanthin and riboflavin, find applications in a broad range of food products, from cheese to beverages. Furthermore, carotenoids, another category of pigments, serve a dual purpose by acting as a protective agent against intense light exposure, thereby maintaining food quality [132]. Table 10.4 summarizes some patented biopigments and the producer strain.

Microbial-derived pigments are assumed to be safe because of their nontoxic, nonallergic, and noncarcinogenic characteristics. Furthermore, they are biodegradable and can be produced by renewable sources [159]. Plants have a wide variety of natural colors that can be used in food manufacturing; however, seasonality and variability in color during extraction, the need for a land-use requirement, and high water for cultivation are in contrast to microbial colorant production. In addition, microbial colorants are supposed to be more stable and watersoluble [160]. Table 10.5 presents some microbial colorants that can be used in food manufacturing.

TABLE 10.3
Example of Microbial Pigments

Chemical Group	Family	Pigments	Microorganisms	References
Polyphenols	Flavonoids	Anthocyanins	Engineered *Escherichia coli*	[118]
		Naringenin	Engineered *Sacharomyces cerevisae*	[119]
Isoprenoids	Carotenoids	Astaxanthin	*Hematococcus pluvialis*	[120]
		Neurosporene	*Microbacterium* sp.	[107]
Phycobiliproteins	Phycocyanin	Phycocyanin C	Cyanobacterium *Arthrospira platensi*	[121]
Polyketide	Naphthoquinone	Bikaverin	*Fusarium* sp.	[122]
	Anthraquinones	Nigrodiquinone A Aspergiolide A	*Nigrospora* sp. *Aspergillus* sp.	[123]
	Azaphilones	Rubropunctamine monascorubramine	*Monascus purpureus* FAFU618 a	[124]
Indoles	Melanin	Eumelanins	*Auricularia auricula*	[125]
Pyrroles	Chlorophyll Bacteriochlorophylls	Chlorophyll F	Cyanobacteria *Acaryochloris marina*	[126]
Nitrogen-heterocyclic derivatives	Betalains	Betacyanin Betaxanthins	*Amanita muscaria*	[127]
Alkaloids	Prodiginine	Prodigiosin	*Serratia marcensens*	[105]

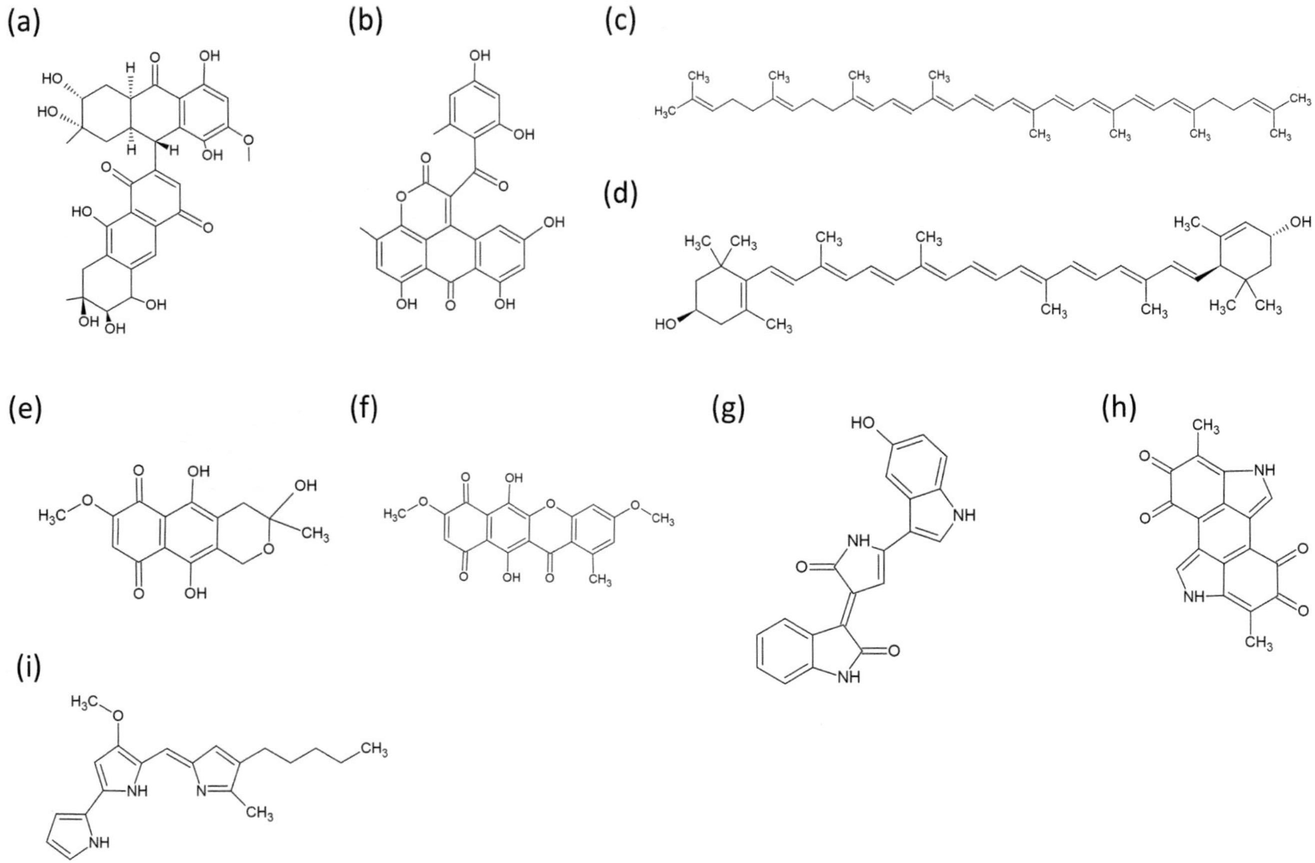

FIGURE 10.4 (a) Nigrodiquinone A, (b) Aspergiolide A, (c) Neurosporene, (d) Lutein, (e) Fusarubin, (f) Bikaverin, (g) Violacein, (h) Pyomelanim, (i) Prodigiosin.

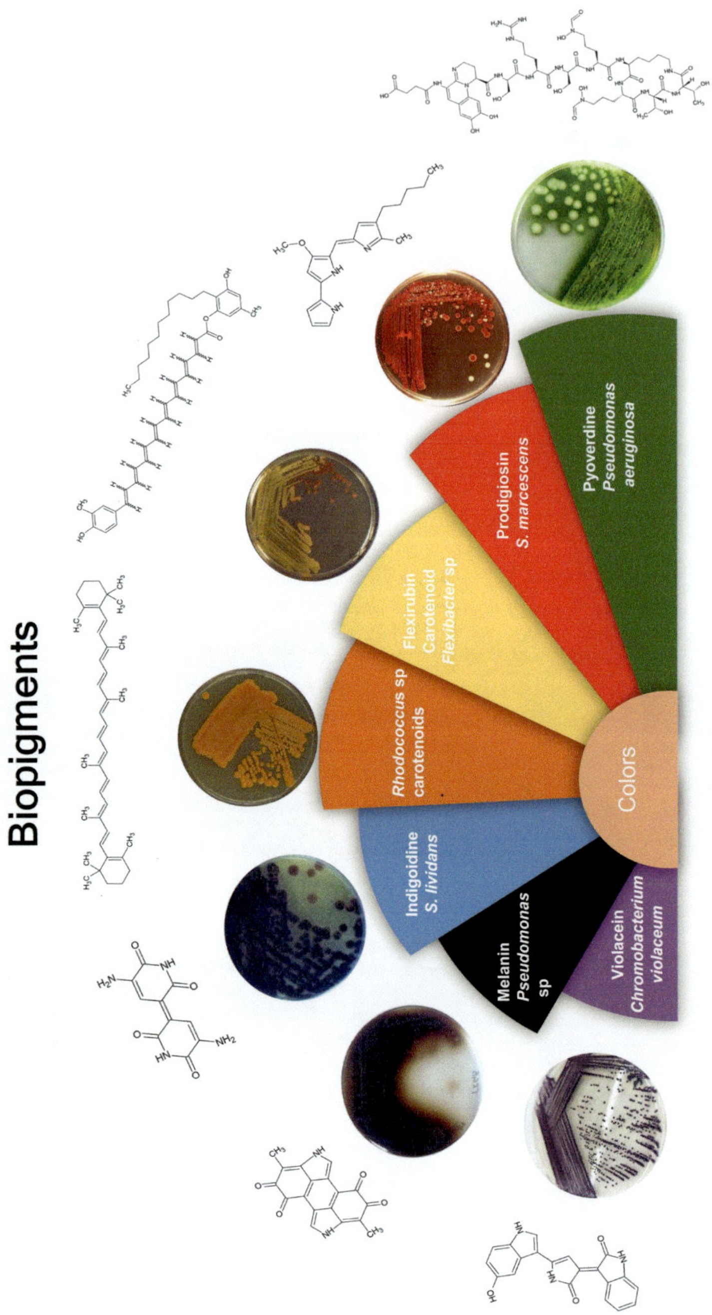

FIGURE 10.5 Biopigments in several microorganisms and colors.

TABLE 10.4
Patented Biopigments and Their Application

Biopigment	Microorganism	Applicant	Reference
Red pigment	*Rhodonellum psychrophilum*	Shoolini University	[133]
Natural dye	*Streptomyces cyaneofuscatus*	Zhejiang Sci-Tech University	[134]
Purple red	*Monascus purpureus*	YiwuDanxi Wine Industry	[135]
Xerochromic compound	*Geomyces* sp.	Wang Nengfei	[136]
Red pigment	*Streptomyces* sp.	Hunan University of Technology	[137]
Anthocyanin	*Aspergillus sabotei*	Sichuan University	[138]
Blue pigment	*Arthrobacter*	Northwest Institute of Eco Environment & Resources Chinese Academy Of Sciences	[139]
Monacolin K	*Monascuspilosus*	Korea University Industrial & Academic Collaboration Foundation	[140]
Blue pigment	*Rhodosporium toruloide*	Liaoning Marine Aquatic Product Science Research Institute	[141]
Natural pigments	*Lactobacillus brevis*	Food Industry Research & Development Institute	[142]
Red pigment	*Zooshikella* sp.	Institute For Industry-Academia Cooperation	[143]
Natural pigments	*Necrococcus niger*	Anhui Polytechnic University	[144]
Red pigment	*Endophytes*	Heze University	[145]
Red pigment	*Geomyces* sp.	Fang Long	[146]
Blue and red pigment	Lactic acid bacteria	Zhejiang University of Science & Technology	[147]
Lycopene	*Rhodospirillum sulfurexigens* *Rhodospirillum oryzae*	Eedera, ChintalapatiandChintalapati	[148]
Red pigment	*Colletotrichum Ichs*	Zhong NongKangzheng Technology Service	[149]
Carotenoids	*Deinococcus wulumuqiensis* R12M	Nanjing University of Technology	[150]
Blue pigment	*Streptomyces* sp.	Guangdong Institute of Microbiology	[151]
Violet blue pigment	*Streptomyces candelais*	Guangdong Institute of Microbiology	[152]

(continued)

TABLE 10.4 (Continued)
Patented Biopigments and Their Application

Biopigment	Microorganism	Applicant	Reference
Red pigment	*Monascus purpureus*	YiwuDanxi Wine Industry	[135]
Anthocyanins	*Fusarium oxysporum*	Beijing University of Technology	[153]
Violacein	*Bdellovibrio bacteriovorus*	CjCheiljedang	[154]
Zeaxanthin	*Gordonia melanodendrii*	Xihua University	[155]
Yellow pigment	*Monascus* sp.	South China University of Technology	[156]
Red yeast rice	*Monascus* sp.	Chenguang Biotech	[157]
Red pigment	*Cordyceps sinensis*	Foundation Of Daegu University	[158]

TABLE 10.5
Examples of Microbial Pigments

Species	Pigment Class	Color	Reference
Halococcus morrhuae *Halobacterium salinarum*	Carotenoids	Red–orange	[162]
Haloterrigena sp.	Carotenoids	Red	[163]
Haloterrigena thermotolerans	Carotenoids	Red	[164]
Erythrobacter sp.	Carotenoids	Yellow	[165]
Deinococcus radiodurans	Carotenoids	Red, red–orange, deep-red	[166]
Lysobacter oligotrophicus	Melanin	Dark-brown	[167]
Gluconobacter oxydans	Vitamin	yellow	[168]

Microorganisms and microalgae in nature frequently produce pigments such as carotenoids, melanins, flavins, quinones, and more specific compounds such as monascins, violacein, phycocyanin, and indigo. The oldest known application is red koji or angkak, a fermented rice pigment from the fungus *Monascus*, historically used in Asia for food coloring. Although synthetic pigments are still prevalent in the food industry, the rising consumer preference for natural additives has intensified research on plant-derived colorants. Nevertheless, transitioning from laboratory-based discoveries to industrial-scale production is challenging, with only a few achieving commercial success [161].

Carotenoids are mainly produced through chemical synthesis due to the inefficiencies associated with their production in organisms such as *Escherichia coli* or yeast. Historically, efforts to enhance microbial carotenoid production have centered mainly on *Escherichia coli* and *Saccharomyces cerevisiae,* wherein their metabolic networks were bolstered with foreign pathways for carotenoid synthesis. The genus *Paracoccus*, particularly species such as *Paracoccus marcusii, Paracoccus carotinifaciens,* and *Paracoccus haeundaensis*, stands out as potential candidates for this production due to their natural carotenoid-producing abilities [169].

Eneos Corp. has deposited a patent about *Paracoccus carotinifaciens* and the production of several natural colorants that can be applied in the food industry, such as astaxanthin, adonixanthin, beta-carotene, echinenone, canthaxanthin, zeaxanthin, beta-cryptoxanthin, 3-hydroxyechinenone, and adonirubin. They pointed out that the ratio of the carotenoid produced is dependent on the oxygen-dissolved concentration in the production media [170]. Congen Inc., Arch Innotek LLC., Conagen Inc., and Microbia Inc. worked on a patent for a recombinant microorganism for the production of carotenoids that may be used in food and beverages [171].

Carotenoids are natural pigments that can be used in food processing. Among them, astaxanthin has several applications. Although it has vast industrial value, it has few sources and high costs in production. These challenges led to studies for reducing production costs or developing a new astaxanthin-producing strain. Regarding the matter, Nippon Oil Co. Ltd. deposited a patent with two new strains of high-yield astaxanthin producers [172]. Naphthazarin red pigments can be produced by fungi and applied in food as described in the patent deposited by Finder Group Inc. [173].

Even though microorganisms produce a wide range of colors, their application in industry is challenging due to low productivity. There is a long path for biopigment from the laboratory to the market. It will demand profound studies and the development of ways to reduce costs [159].

10.5.2 Biopigment Application in the Medical and Pharmaceutical Fields

In addition to their colorant function, natural pigments have functional groups that provide pharmacological activity, such as antitumor, antimicrobial, antiviral, antifungal, antiparasitic, and immunosuppressors [174].

Several fungi, such as *Blakeslea trispora*, *Cantharellus cinnabarinus*, and *Phaffia rhodozyma*, have been reported as carotenoid producers. These carotenoids have applications in drug formulation due to their characteristics [128]. Not only can fungi produce pigments, but bacteria can also produce them. Among the pigment-producing bacterium *Bradyrhizobium* spp., *Staphylococcus aureus*, and *Rhodococcus maris* are known for the carotenoids they can produce [159]. Prodiginines, phycobiliproteins, and oxyindoles can also be synthesized by bacteria. They have applications in clinical uses as anticancer, anti-inflammatory, immunosuppressant, antimalarial, and antiplasmodialagents [175–177].

Violacein has previously been recognized for its potential therapeutic efficacy against Chagas disease and its effectiveness against leukemia and lymphoma cells in culture settings. Melo *et al.* studied the cytotoxic effects of violacein, a pigment derived from *Chromobacterium violaceum*, on V79 Chinese hamster (M-8) fibroblasts. The cytotoxicity of violacein in V79 fibroblasts was determined by observing changes in nuclear acid content, 3-(4,5-dimethylthiazole-2-yl)-2,5-biphenyl tetrazolium bromide reduction, and neutral red uptake. The results demonstrated that violacein is significantly cytotoxic to V79 fibroblasts. More intriguingly, at concentrations of 5 and 10 μM, violacein was found to induce apoptosis. Even at the highest concentration of violacein tested, necrosis was not detected. The research underscores the potential of violacein as a therapeutic agent, especially given its ability to induce apoptosis in V79 cells without leading to necrosis [178].

Bishit*et al.* isolated *Rhodonellum psychrophilum* from high-altitude Pangong Tso Lake in Leh Ladakh, India. This bacterium produces red pigments with significant antimicrobial activity against various pathogens, including *E. coli, S. aureus, C. albicans,* and *S. cerevisiae*. Notably, these red pigments enhanced the efficacy of several antifungal and antibacterial antibiotics. LC–MS/MS analysis of the red pigment extracts revealed a composition of various compounds, including prodigiosin, 2-methyl-3-butyl-prodigine, and others. These extracts produced noncytotoxic antimicrobial fabrics for the U87MG human glioblastoma cell line [179].

Due to the applications mentioned above of the microbial pigments, some industries have deposited patents regarding clinical uses of the pigments, as described in Table 10.6.

Diabetes is a chronic disorder characterized by persistent hyperglycemia, primarily due to inadequate insulin action. Central to the pathophysiology of both type 1 and type 2 diabetes is the death of pancreatic β cells. Astaxanthin is recognized for its superior antioxidant capacity. Hirakida *et al.* explored the effects of an

TABLE 10.6
Patented Biopigments and Their Clinical Application

Biopigment	Microorganism	Applicant	Clinical Uses	Reference
Carotenoid phytoene	*Blakeslea trispora*	Lycored Ltd	Use of the carotenoid in a lipid complex for the radioprotection	[180]
Azaphilones	*Penicillium* sp. HK1-6	Yangzhou University	Anti-MRSA drugs	[181]
Azaphilones	*Penicillium sclerotiorum* SJ0167	National Sun Yat Sen University	inhibitory effect on cancer cell proliferation	[182]
Iso-Penicillixanthone A (anthraquinone)	*Penicillium oxalicum*	Fuzhou University	Inhibitory effect on cervical cancer cells	[183]
Melanin	*Lachnum*YM156	Hefei University of Technology	Anti-renal-failure medicine with reduced side effects or toxicity	[184]
Monacolin, monascin, ankaflavin, rubropunctatin, monascorburin	*Monascus purpureus*	Sunway Biotech Co Ltd	Anti-inflammation and antioxidant agents, which can be used to treat patients with Alzheimer's disease	[185]

astaxanthin-rich extract (ARE) from *Paracoccus carotinifaciens* on rat pancreatic β-cell lines. The study assessed the insulin secretion capacity and protective effects of ARE. Their findings indicated that ARE from *P. carotinifaciens* enhances insulin secretion under specific glucose conditions and significantly protects against cell death. The mitigation of mitochondrial superoxide production was identified as a key underlying factor. The results suggest that ARE from *P. carotinifaciens* offers potential therapeutic benefits for diabetes, especially in augmenting glucose-stimulated insulin secretion and safeguarding pancreatic β cells. However, the authors emphasize the need for further research to conclusively determine the efficacy of ARE on pancreatic β cells [186].

Prospecting new secondary metabolites will open many possibilities in the pharmaceutical field. Natural pigments can be used in several biotechnological applications to increase the human lifespan in developing new drugs and treatments [187].

10.5.3 Biopigment Application in Cosmetics

The increasing consumer knowledge about environmental issues caused by synthetic coloring agents and the need for more environmentally friendly and safe, sustainable products have boosted the market for natural pigments in cosmetics [111].

In addition to the natural color of pigments, the antioxidant characteristics of some microbial pigments make them suitable for skin care products because the skin is an essential organ of the body, protecting us in many ways. However, skin exposure to ultraviolet radiation may lose its antioxidant capability, making it age faster. Therefore, skin care products are not only for appearance but also for skin protection and health [111,188].

The spectrum of colors produced by microbial pigments is enormous, including cream, yellow, orange, red, pink, gray, blue, brown, and black. Natural pigments can be produced either by bacteria, fungi, yeasts, or microalgae, and they are of several types, such as phycobiliproteins, carotenoids, melanin, prodigiosin, violacein, and rhodopsins [189,128].

Due to the great relevance of natural pigments in cosmetics, some patents have been deposited. The use of natural carotenoids from *Micrococcus luteus* to produce sunscreen was deposited by Promar As. [190], TerraVia Holdings Inc. filled an application with *Chlorella protothecoides* used in cosmetics, especially sunblock [191]; the use of *Parachlorella kessleri* HY1 to produce lutein that can be in the composition of cosmetics was patented by Mikrobiologickýústav AV ČR [192].

Cosmetics colorants can be produced in several strains, such as *Talaromyces atroroseus* [193], *Monascus purpureus* [194], *Serratia marcescens* [195], and *Chromobacterium violaceum* VP2 [196].

Nanoemulsions (NEs) have attracted the interest of the pharmaceutical domain due to their adaptability in integrating lipophilic substances into cosmetic formulations. Photoprotective nanoemulsions are crucial in mitigating skin damage from excessive sun exposure by counteracting the immediate effects of ultraviolet radiation. Mansur *et al.* developed and characterized oil-in-water photoprotective nanoemulsions enriched with microbial carotenoids, Buriti oil, and chemical filters. As a result, the nanoemulsion incorporated with microbial carotenoids and Buriti oil showed an

SPF of 36 ± 1.5. Furthermore, in vitro phototoxicity assessments ascertained that these nanoemulsions did not exhibit phototoxic traits and preserved cell viability postirradiation [197].

According to Kusmita *et al.* marine environments, particularly those undergoing environmental stress, can stimulate the generation of distinctive secondary metabolites such as carotenoids. Their study examined carotenoids isolated from the bacterial symbiont of *Sinularia* sp., a soft coral originating from Panjang Island in the North Java Sea. As a result, they isolated a *Virgibacillus* sp., which produced carotenoids that displayed antioxidant properties and powerful sun protection attributes (classifying them as UVA sunblocks). They exhibited antibacterial activities against bacteria such as methicillin-resistant *Staphylococcus aureus* (MRSA) and multidrug-resistant (MDR) *Escherichia coli* [198].

Carotenoids within cyanobacteria cells play a pivotal role in absorbing excess energy from photosynthesis, thereby preventing cellular damage. When extracted, these compounds have similar protective effects on human skin, particularly against potential damage from prolonged sunlight exposure. Phycobiliproteins, another type of pigment derived from cyanobacteria, exhibit multifarious bioactivities, including antioxidant, antiviral, and antimicrobial properties [199]. Pagels *et al.* evaluated the cosmetic potential of pigment-targeted extracts from *Cyanobium* sp., encompassing in vitro safety tests, cosmetic enzyme inhibition, ingredient stability, and serum formulation. The results show that no cytotoxic effects were observed in skin-related cell lines. The carotenoid extract exhibited antihyaluronidase capacity, while the phycobiliprotein extract demonstrated both antihyaluronidase and anticollagenase capacities [199].

There are still challenges in the field of microbial pigments to be solved. However, the increasing demand will accelerate research to improve yield and functionality. Induced metabolic engineering can be a powerful tool to achieve this goal [128].

ACKNOWLEDGMENTS

This research was funded in part by the postgraduate program of Science (Microbiology) at the Federal University of Rio de Janeiro (UFRJ) through the Coordination of Aperfeiçoamento Pessoal de Ensino Superior (CAPES) [grant number 001], Conselho National Council for Scientific and Technological Development (MCTI-CNPq) grant code [309461/2019–7] and Fundação de Amparo à Pesquisa do Estado do Rio de Janeiro (FAPERJ), grant 'Cientista do NossoEstado E26/200428/2023. We thank the slide model (https://slidemodel.com/) for the figure's schemes.

REFERENCES

[1] Bioreactors Market Size, Share, Competitive Landscape and Trend Analysis Report by Type, Usage, Scale and Control Type: Global Opportunity Analysis and Industry Forecast, 2021–2030. 2023. *Bioreactor Market*. October 10. www.alliedmarketresearch.com/bioreactors-market.

[2] H. Choi, and S. W. Kim. 2023. Characterization of β-Glucans from Cereal and Microbial Sources and Their Roles in Feeds for Intestinal Health and Growth of Nursery Pigs. *Animals* 13 (13): 2236. doi:10.3390/ani13132236.

[3] F. Zhu, B. Du, and B. Xu. 2016. A Critical Review on Production and Industrial Applications of Beta-Glucans. *Food Hydrocolloids* 52 (January): 275–88. doi:10.1016/j.foodhyd.2015.07.003.

[4] S. A. S. Aboushanab, D. V. Vyrova, I. S. Selezneva, and M. N. G. Ibrahim. 2019. The Potential Use of β-Glucan in the Industry, Medicine and Cosmetics, vol. 020198. Ekaterinburg, Russia. doi:10.1063/1.5134349.

[5] E. J. Murphy, E. Rezoagli, I. Major, N. J. Rowan, and J. G. Laffey. 2020. β-Glucan Metabolic and Immunomodulatory Properties and Potential for Clinical Application. *Journal of Fungi* 6 (4): 356. doi:10.3390/jof6040356.

[6] T. G. Timm, T. M. Costa, M. D. Alberton, C. V. Helm, and L. B. B. Tavares. 2023. Mushroom β-Glucans: Application and Innovation for Food Industry and Immunotherapy. *Applied Microbiology and Biotechnology* 107 (16): 5035–49. doi:10.1007/s00253-023-12656-4.

[7] K. Frick, T. Ebbing, Y. Yeh, U. Schmid-Staiger, and G. E. M. Tovar. 2023. Beta-Glucan Production of *Phaeodactylum tricornutum*, *Monodopsis subterranea* and *Cylindrotheca fusiformis* during Nitrogen Depletion. *Journal of Applied Phycology*, July. doi:10.1007/s10811-023-03026-8.

[8] M. E. Morales-Plasencia, L. Ibarra-Castro, J. M. Martínez-Brown, M. Nieves-Soto, J. F. Bermúdez-Lizárraga, and A. H. Rojo-Cebreros. 2023. The Effect of Nitrogen Limitation on Carbohydrates and β-Glucan Accumulation in *Nannochloro psisoculata*. *Algal Research* 72: 103125. https://doi.org/10.1016/j.algal.2023.103125.

[9] M. McIntosh, B. A. Stone, and V. A. Stanisich. 2005. Curdlan and Other Bacterial (1→3)-β-d-Glucans. *Applied Microbiology and Biotechnology* 68 (2): 163–73. doi:10.1007/s00253-005-1959-5.

[10] P. Ezhilarasi, and B. Vanavil. 2023. Curdlan Gum Production Using Marine Bacteria Isolated from Red Seaweeds: Screening and Optimization Studies. *Microbiology* 92 (5): 725–33. doi:10.1134/S0026261723600131.

[11] R. R. Dias, I. A. Severo, M. C. Deprá, M. M. Maroneze, L. Q. Zepka, and E. Jacob-Lopes. 2020. The Next-Generation of Microalgae-Based Products. In *Microbial Enzymes and Biotechniques*, edited by Pratyoosh Shukla, 15–42. Singapore: Springer Singapore. doi:10.1007/978-981-15-6895-4_2.

[12] M. G. Roth, N. M. Westrick, and T. T. Baldwin. 2023. Fungal Biotechnology: From Yesterday to Tomorrow. *Frontiers in Fungal Biology* 4 (March): 1135263. doi:10.3389/ffunb.2023.1135263.

[13] R. R Philippini, S. E. Martiniano, J. C. Santos, S. S. Silva, and A. K. Chandel. 2019. Fermentative Production of Beta-Glucan: Properties and Potential Applications. In *Bioprocessing for Biomolecules Production*, edited by Gustavo Molina, Vijai Gupta, Brahma Singh, and Nicholas Gathergood, 1st ed., 303–20. Wiley. Avaliable: https://onlinelibrary.wiley.com/doi/epdf/10.1002/9781119434436.ch15

[14] Y. Zhang, Y. Sun, Y. Liu, J. Liu, J. Sun, X. Liu, B. Fan, C. Lu, and F. Wang. 2023. *Polygonum sibiricum* Polysaccharides Exert the Antidepressant-like Effects in Chronic Unpredictable Mild Stress-induced Depressive Mice by Modulating Microbiota-gut-brain Axis. *Phytotherapy Research* 37 (8): 3408–23. doi:10.1002/ptr.7813.

[15] Y. Zhang, J. Hu, H. Tan, Y. Zhong, and S. Nie. 2022. *Akkermansia muciniphila*, an Important Link between Dietary Fiber and Host Health. *Current Opinion in Food Science* 47 (October): 100905. doi:10.1016/j.cofs.2022.100905.

[16] B. Du, Z. Bian, and B. Xu. 2014. Skin Health Promotion Effects of Natural Beta-Glucan Derived from Cereals and Microorganisms: A Review. *Phytotherapy Research* 28 (2): 159–66. doi:10.1002/ptr.4963.

[17] P. Sousa, D. Tavares-Valente, M. Amorim, J. Azevedo-Silva, M. Pintado, and J. Fernandes. 2023. β-Glucan Extracts as High-Value Multifunctional Ingredients for Skin Health: A Review. *Carbohydrate Polymers* 322 (December): 121329. doi:10.1016/j.carbpol.2023.121329.

[18] C. Caseiro, J. N. R. Dias, C. M. G. A. Fontes, and P. Bule. 2022. From Cancer Therapy to Winemaking: The Molecular Structure and Applications of β-Glucans and β-1,3-Glucanases. *International Journal of Molecular Sciences* 23 (6): 3156. doi:10.3390/ijms23063156.

[19] Z. Pérez-Bassart, I. Falcó, M. Martínez-Sanz, A. Martínez-Abad, G. Sánchez, A. López-Rubio, and M. J. Fabra. 2024. Antiviral and Technological Properties of β-Glucan-Rich Aqueous Fractions from *Pleurotus ostreatus* Waste Biomass. *Food Hydrocolloids* 146 (January): 109308. doi:10.1016/j.foodhyd.2023.109308.

[20] A. Rieder and A. B. Samuelsen. 2012. Do Cereal Mixed-linked B-glucans Possess Immune-modulating Activities?. *Molecular Nutrition & Food Research* 56 (4): 536–47. doi:10.1002/mnfr.201100723.

[21] L. J. Rather, S. S. Mir, S. A. Ganie, S. Islam, and Q. Li. 2023. Research Progress, Challenges, and Perspectives in Microbial Pigment Production for Industrial Applications – A Review. *Dyes and Pigments* 210 (February): 110989. doi:10.1016/j.dyepig.2022.110989.

[22] K. M. Andrzej, M. Małgorzata, K. Sabina, O. K. Horbańczuk, and E. Rodak. 2020. Application of Rich in β-Glucan Flours and Preparations in Bread Baked from Frozen Dough. *Food Science and Technology International* 26 (1): 53–64. doi:10.1177/1082013219865379.

[23] S. Mejía, M. Vásquez, A. Francisco, and B. M. Bohrer. 2020. A Comprehensive Review on Cereal β-Glucan: Extraction, Characterization, Causes of Degradation, and Food Application. *Critical Reviews in Food Science and Nutrition* 60 (21): 3693–3704. doi:10.1080/10408398.2019.1706444.

[24] G. Venkatachalam, S. Arumugam, and M. Doble. 2021. Industrial Production and Applications of α/β Linear and Branched Glucans. *Indian Chemical Engineer* 63 (5): 533–47. doi:10.1080/00194506.2020.1798820.

[25] J. Han, Z. Liu, Z. Wu, S. Wu, and X. Xu. 2016. Tomato Juice Fermentation Beverage Rich in Water-Insoluble Glucan and Preparation Method of Beverage. CN105581208A, filed December 21, 2015, and issued May 18, 2016.

[26] X. Xu, J. Yang, J. Ren, and W. Lou. 2017. Preparation Method and Application of Glucan Fruit and Vegetable Enzymes. CN106579115A, filed December 2, 2016, and issued April 26, 2017.

[27] L. Zhang, 2016. Beta-Glucan Candy Pieces and Preparing Method Thereof. CN105994873A, filed May 13, 2016, and issued October 12, 2016.

[28] D. Wu, L. Jin, M. Cao, Z. Li, and X. Gou. 2021. Product containing Plant Exosomes and Capable of Improving Oral Health. CN112425703A, filed November 30, 2020, and issued March 2, 2021.

[29] R. Niemelae, J. Saarinen, J. Helin, and J. Wakkinen. 2010. Nutritional Compositions. EP2167674A1, filed June 13, 2008, and issued March 31, 2010.

[30] P. Vandekerckove, G. Pignede, J. Simon, and P. Desreumaux. 2009. New *Saccharomyces cerevisiae* Strain Submitted to National Collection of Cultures of Microorganisms, useful e.g. for Preparation of Food Supplement, Probiotic, Functional

Food, Cosmeceutical and/or Active Pharmaceutical Ingredient. FR2928652A1, filed March 12, 2008, and issued September 18, 2009.

[31] R. Wheatcroft, W. H. Langeris, J. Kulandai, R. W. Gilbert, K. J. Sime, and C. G. Smith. 2003. Production of Beta-Glucan-Mannan Preparations by Autolysis of Cells Under Certain pH, Temperature and Time Conditions. EP0847446B1, filed June 28, 1996, and issued October 22, 2003.

[32] K. Tsubaki, H. Sugiyama, and Y. Shoji. 2014. Beta Glucan-Containing Fat and Oil Compositions and Novel Microorganism Capable of Producing Beta-Glucan. EP1533382B1, filed June 18, 2003, and issued December 24, 2014.

[33] H. Hattori, 1998. Production of Food and Beverage Containing Beta-1,3-1,6-Glucan or Its Alkali Salt. JPH10276739A, filed April 9, 1997, and issued October 20, 1998.

[34] N. Ikewaki, N. Fujii, and T. Onaka. 2005. Beta-1.3-1.6 Glucan (*Aureobasidium medium*). US6956120B2, filed November 9, 2001, and issued October 18, 2005.

[35] G. Wei, X. Chen, D. Wang, and C. Wang. 2021. *Aureobasidium pullulans f*or Improving Beta-Glucan Yield and Application Thereof. CN112175843A, filed October 12, 2020, and issued January 5, 2021.

[36] T. Feng, W. Li, H. Zhuang, L. Yao, S. Song, M. Sun, and H. Wang. 2020. Method for Evaluating Regulation Capability of *Hericium erinaceus* Beta-Glucan on Intestinal Flora. CN111560447A, filed April 17, 2020, and issued August 21, 2020.

[37] T. Suzuki, S. Nakamura, S. Nakayama, and K. Nishikawa. 2004. Method for Preparing Culture Solution Containing Water-Soluble Low Viscosity Beta-D-Glucan. JP2004321177A, filed March 26, 2004, and issued November 18, 2004.

[38] T. Suzuki, S. Nakamura, S. Nakayama, and K. Nishikawa. 2005. Beta-1, 3-1,6-D-Glucan and Its Use. US2005271613A1, filed March 28, 2005, and issued December 8, 2005.

[39] J. D. Lee, J. K. Kim, M. J. Kim, and Y. M. Jang. 2020. Method for Preparing Bioconversion Products of *Asparagus cochinchinensis* and Soybean Embryos by Means of Liquid Culturing of Basidiomycetes Mycelium, and Use Thereof. WO2020085735A1, filed October 21, 2019, and issued April 30, 2020.

[40] Y. Namba, 2016. Powder Functional Food and Manufacturing Method Therefor. JP2016049022A, filed August 28, 2014, and issued April 11, 2016.

[41] Q. Li, M. Zhang, X. Li, J. Wang, C. Liu, C. Niu, and F. Zheng. 2021. High-Temperature Sensitive *Saccharomyces pastorianus* and Application Thereof. US2021321654A1, filed June 30, 2021, and issued October 21, 2021.

[42] M. Huang, H. J. Joo, L. Zeng, X. Ye, and F. Zeng. 2022. Synbiotic-Containing Feed Additive for Aquatic Animals. CN114651913A, filed December 22, 2020, and issued June 24, 2022.

[43] B. Wang, L. Li, X. Fan, Q. Lu, and X. Gu. 2022. *Pholiota adipose* Strain Obtained through Ultraviolet Mutagenesis and Application Thereof. CN115181676A, filed July 1, 2022, and issued October 14, 2022.

[44] X. Yang, and H. Zheng. 2014. Method of Preparing Functional *Monascus* mycelia and Fermentation Liquor by Liquid Fermentation, and Products of the *Monascus* mycelia and the Fermentation Liquor. CN103602590A, filed September 29, 2013, and issued February 26, 2014.

[45] F. Yin, P. Lin, W. Yu, N. Shen, Y. Li, and S. Guo. 2021. The Cordyceps militaris-derived Polysaccharide CM1 Alleviates Atherosclerosis in LDLR (-/-) mice by Improving Hyperlipidemia." *Frontiers in Molecular Biosciences* 8 (December): 783807. doi:10.3389/fmolb.2021.783807.

[46] P. Lin, F. Yin, N. Shen, N. Liu, B. Zhang, Y. Li, and S. Guo. 2021. Integrated Bioinformatics Analysis of the Anti-Atherosclerotic Mechanisms of the

Polysaccharide CM1 from *Cordyceps Militaris*. *International Journal of Biological Macromolecules* 193 (December): 1274–85. doi:10.1016/j.ijbiomac.2021.10.175.

[47] Glucan Market. 2023. *Glucan Market*. www.marketsandmarkets.com/Market-Reports/beta-glucan-market5191796.html?gclid=Cj0KCQjw9rSoBhCiARIsAFOipllw4C-YX6lcC6nV05qoaIhbpxW7zy-1WRFzPhnJZOeKQGBpMdNkyhIaAti7EALw_wcB.

[48] N. Bose, K. Gorden, A. S. Chan, S. Leonardo, J. Graff, X. Qiu, T. Kangas, K. A. Fraser, J. A. Bykowski, and N. Ottoson. 2017. Beta-Glucan in Combination with anti-Cancer Agents Affecting the Tumor Microenvironment. EP3166616A1, filed July 10, 2015, and issued May 17.

[49] S. J. Karel, 2008. Combination of a Beta-Glucan and an Egf Receptor Antagonist for the Treatment of Cancer and Infection. EP1917019A2, filed May 2, 2006, and issued May 7, 2008.

[50] J. Yan, 2009. Therapeutic Beta-Glucan combinations. US2009074761A1, filed April 24, 2008, and issued March 19, 2009.

[51] T. Onaka and S. J. K. Abraham. 2021. Beta-Glucan for Immuno-Enhancement and/or Immuno-Balancing, and for Adjuvant Use. WO2021256470A1, filed June 15, 2021, and issued December 23, 2021.

[52] G. R. Ostroff, 2010. Drug Delivery Product and Methods. US2010221357A1, filed May 17, 2010, and issued September 2, 2010.

[53] I. Rubin-Bejerano and G. R. Fink. 2015. Immunomodulating Compositions and Methods of Use Thereof. US9005903B2, filed December 11, 2013, and issued April 14, 2015.

[54] H. W. Kim, 2016. Odor Reducing Composition Containing *Phellinus linteus* Extract or Beta-Glucan as Active Ingredient. WO2016159452A1, filed July 31, 2015, and issued October 6, 2016.

[55] W. K. Moon, D. W. Kim, B. H. Chung, and J. H. Park. 2005. Novel Use of Water Soluble Glucan Oligomer Isolated from *Saccharomyces cerevisiae* Is2 for Prevention and Treatment of Avian Flu. WO2005067940A1, filed January 19, 2005, and issued July 28, 2005.

[56] A. Misaki, Y. Sone, M. Yoshida, and K. Takeuchi. 1988. Beta-Glucan. US4769363A, filed March 4, 1985, and issued September 6, 1988.

[57] B. D. Yoon, K. H. Ahn, and H. S. Kim. 2016. *Aureobasidium pullulans* YK1 Produced by Using UV and a Producing Method for β-Glucan Using the Same. KR20160048393A, filed October 24, 2014, and issued May 4, 2016.

[58] R. B. Levine, J. R. Lebrun, and G. P. Horst. 2019. Multi-Stage Process for Production of Immune Modulator. EP2858652B1, filed May 7, 2013, and issued July 10, 2019.

[59] S. C. Yun, 2011. Health Care Food Using Extrcaton of *Sparassis crispa*. KR20110082758A, filed January 12, 2010, and issued July 20, 2011.

[60] P. Vandekerckove, 2019. *Saccharomyces cerevesiae* Yeast for the Prevention and/or Treatment of Vaginal Mycosis. EP2874636B1, filed July 9, 2013, and issued December 25, 2019.

[61] T. Suzuki, M. Kido, K. Yamamoto, and J. Wada. 2010. Effect Promoter for *Lactobacillus* Having Intestinal Immunostimulatory Activity. JP2010285421A, filed May 10, 2010, and issued December 24, 2010.

[62] A. Ahmad, Y. Gulraiz, S. Ilyas, and S. Bashir. 2022. Polysaccharide Based Nano Materials: Health Implications. *Food Hydrocolloids for Health* 2 (December): 100075. doi:10.1016/j.fhfh.2022.100075.

[63] S. M. Wani, A. Gani, S. A. Mir, F. A. Masoodi, and F. A. Khanday. 2021. β-Glucan: A Dual Regulator of Apoptosis and Cell Proliferation. *International Journal of Biological Macromolecules* 182 (July): 1229–37. doi:10.1016/j.ijbiomac.2021.05.065.

[64] S. Chaichian, B. Moazzami, F. Sadoughi, H. H. Kashani, M. Zaroudi, and Z. Asemi. 2020. Functional Activities of Beta-Glucans in the Prevention or Treatment of Cervical Cancer. *Journal of Ovarian Research* 13 (1): 24. doi:10.1186/s13048-020-00626-7.

[65] L. Steimbach, A. V. Borgmann, G. G. Gomar, L. V. Hoffmann, R. Rutckeviski, D. P. Andrade, and F. R. Smiderle. 2021. Fungal Beta-Glucans as Adjuvants for Treating Cancer Patients – A Systematic Review of Clinical Trials. *Clinical Nutrition* 40 (5): 3104–13. doi:10.1016/j.clnu.2020.11.029.

[66] V. Vetvicka, T. V. Teplyakova, A. B. Shintyapina, and T. A. Korolenko. 2021. Effects of Medicinal Fungi-Derived β-Glucan on Tumor Progression. *Journal of Fungi* 7 (4): 250. doi:10.3390/jof7040250.

[67] N. Ikewaki, V. Dedeepiya, K. Raghavan, K. Rao, S. Vaddi, H. Osawa, T. Kisaka, G. urosawa, S. Srinivasan, S, Ramasamy, B. Kumar, R. Senthilkumar, M. Iwasaki, S. Preethy, and S. J. K. Abraham. 2021. β-Glucan Vaccine Adjuvant Approach for Cancer Treatment through Immune Enhancement (B-VACCIEN) in Specific Immunocompromised Populations (Review). *Oncology Reports* 47 (1): 14. doi:10.3892/or.2021.8225.

[68] V. Cognigni, N. Ranallo, F. Tronconi, F. Morgese, and R. Berardi. 2021. Potential Benefit of β-Glucans as Adjuvant Therapy in Immuno-Oncology: A Review. *Exploration of Targeted Anti-Tumor Therapy*, 2: 122–138, March. doi:10.37349/etat.2021.00036.

[69] V. Vetvicka, L. Vannucci, P. Sima, and J. Richter. 2019. Beta Glucan: Supplement or Drug? From Laboratory to Clinical Trials. *Molecules* 24 (7): 1251. doi:10.3390/molecules24071251.

[70] P. Graaff, C. Berrevoets, C. Rösch, H. A. Schols, K. Verhoef, H. J. Wichers, R. Debets, and C. Govers. 2021. Curdlan, Zymosan and a Yeast-Derived β-Glucan Reshape Tumor-Associated Macrophages into Producers of Inflammatory Chemo-Attractants. *Cancer Immunology, Immunotherapy* 70 (2): 547–61. doi:10.1007/s00262-020-02707-4.

[71] A. Geller, R. Shrestha, and J. Yan. 2019. Yeast-Derived β-Glucan in Cancer: Novel Uses of a Traditional Therapeutic. *International Journal of Molecular Sciences* 20 (15): 3618. doi:10.3390/ijms20153618.

[72] M. Verma., and F. X. Gu. 2012. 1,3-Beta-Glucans: Drug Delivery and Pharmacology. In *The Complex World of Polysaccharides*, edited by Desiree Nedra Karunaratne. [Online], Available: www.intechopen.com/chapters/40585, doi:10.5772/50363.

[73] M. Salgado, F. Santos, S. Rodríguez-Rojo, R. L. Reis, A. R. C. Duarte, and M. J. Cocero. 2017. Development of Barley and Yeast β-Glucan Aerogels for Drug Delivery by Supercritical Fluids. *Journal of CO_2 Utilization* 22 (December): 262–69. doi:10.1016/j.jcou.2017.10.006.

[74] Y. Su, L. Chen, F. Yang, and P. C. K. Cheung. 2021. Beta-d-Glucan-Based Drug Delivery System and Its Potential Application in Targeting Tumor Associated Macrophages. *Carbohydrate Polymers* 253 (February): 117258. doi:10.1016/j.carbpol.2020.117258.

[75] G. Zhang, B. Wang, D. Bai, and X. Tang. 2015. EGCG and Beta-Glucan Composition, and Preparation Method, and Medical Application Thereof. CN104784197A, filed March 30, 2015, and issued July 22, 2015.

[76] S. Jamas, G. R. Ostroff, and D. D. Easson. 1994. Glucan Drug Delivery System and Adjuvant. EP0476063B1, filed June 14, 1990, and issued May 25, 1994.

[77] C. S. Ki, K. H. Lee, H. H. Tae, and S. Lee. 2019. Photo-Crosslinked Microgel for Protein Drug Delivery and Fabrication Method Thereof. KR20190082133A, filed December 28, 2018, and issued July 9, 2019.

[78] E. F. Fritsch, K. Nellaiappan, and I. Javeri. 2016. Formulations for Neoplasia Vaccines. US2016310584A1, filed December 5, 2014, and issued October 27, 2016.

[79] A. Khovorova, W. Salomon, J. Kamens, D. Samarsky, T. M. Woolf, and J. Cardia. 2017. Self-Delivered RNAi Compounds of Reduced Size. RU2615143C2, filed March 24, 2011, and issued April 4, 2017.

[80] F. Yang and P. C. K. Cheung. 2023. Fungal β-Glucan-Based Nanotherapeutics: From Fabrication to Application. *Journal of Fungi* 9 (4): 475. doi:10.3390/jof9040475.

[81] A. Uno, K. Arima, M. Shimazaki, M. Ushida, K. Amano, R. Namikawa, and K. Sakurai. 2021. A Novel β-Glucan–Oligonucleotide Complex Selectively Delivers siRNA to APCs via Dectin-1. *Journal of Controlled Release* 338 (October): 792–803. doi:10.1016/j.jconrel.2021.09.011.

[82] W. J. Grossman, M. A. Antonysamy, R. M. Walsh, M. I. Nelson, N. Bose, M. E. Danielson, and K. S. Michel. 2022. Methods for Identifying Beta-Glucan Binding to Immune Cells. US2022105181A1, filed December 15, 2021, and issued April 7, 2022.

[83] N. Bose, K. Gorden, R. Fulton, S. Leonardo, and J. Graff. 2023. Beta-Glucan Immunotherapies Affecting the Immune Microenvironment. US2023172967A1, filed August 12, 2022, and issued June 8, 2023.

[84] N. Bose, N. Ottoson, B. Harrison, J. Lowe, M. Uhlik, J. Graff, and R. Huhn. 2020. Beta Glucan Immunopharmacodynamics. US2020033362A1, filed February 23, 2018, and issued January 30, 2020.

[85] N. Bose, M. A. Antonysamy, K. B. Gorden, R. Walsh, M. E. Danielson, and P. Maimonis. 2018. Beta-Glucan Assay Methods. US10114027B2, filed December 18, 2017, and issued October 30, 2018.

[86] E. J. Murphy, C. Masterson, E. Rezoagli, D. O'Toole, I. Major, G. D. Stack, M. Lynch, J. G. Laffey, and N. J. Rowan. 2020. β-Glucan Extracts from the Same Edible Shiitake Mushroom *Lentinusedodes* Produce Differential in vitro Immunomodulatory and Pulmonary Cytoprotective Effects—Implications for Coronavirus Disease (COVID-19) Immunotherapies. *Science of The Total Environment* 732 (August): 139330. doi:10.1016/j.scitotenv.2020.139330.

[87] A. M. Varricchio, M. Capasso, A. D. Volpe, L. Malafronte, N. Mansi, A. Varricchio, and G. Ciprandi. 2014. Resveratrol plus Carboxymethyl-β-Glucan in Children with Recurrent Respiratory Infections: A Preliminary and Real-Life Experience. *Italian Journal of Pediatrics* 40 (1): 93. doi:10.1186/s13052-014-0093-3.

[88] Y. Lin, Y. Liang, S. Lee, and B. Chiang. 2005. Polysaccharide Purified from *Ganoderma lucidum* Induced Activation and Maturation of Human Monocyte-Derived Dendritic Cells by the NF-κB and P38 Mitogen-Activated Protein Kinase Pathways. *Journal of Leukocyte Biology* 78 (2): 533–43. doi:10.1189/jlb.0804481.

[89] M. Iida, M. J. Desamero, K. Yasuda, A. Nakashima, K. Suzuki, J. K. Chambers, K. Uchida, R. Ogawa, S. Hachimura, J. Nakayama, S. Kyuma, K. Miura, S. Kakuta and, K. Hirayama. 2021. Effects of Orally Administered *Euglena gracilis* and Its Reserve Polysaccharide, Paramylon, on Gastric Dysplasia in A4gnt Knockout Mice. *Scientific Reports* 11 (1): 13640. doi:10.1038/s41598-021-92013-5.

[90] A. Nakashima, K. Yasuda, A. Murata, K. Suzuki, and N. Miura. 2020. Effects of *Euglena gracilis* Intake on Mood and Autonomic Activity under Mental Workload, and Subjective Sleep Quality: A Randomized, Double-Blind, Placebo-Controlled Trial. *Nutrients* 12 (11): 3243. doi:10.3390/nu12113243.

[91] R. Shimada, M. Fujita, M. Yuasa, H. Sawamura, T. Watanabe, A. Nakashima, and K. Suzuki. 2016. Oral Administration of Green Algae, *Euglena gracilis*, Inhibits Hyperglycemia in OLETF Rats, a Model of Spontaneous Type 2 Diabetes. *Food & Function* 7 (11): 4655–59. doi:10.1039/C6FO00606J.

[92] R. Okouchi, S. E., K. Yamamoto, T. Ota, K. Seki, M. Imai, R. Ota, Y. Asayama, A. Nakashima, K. Suzuki, and T. Tsuduki. 2019. Simultaneous Intake of *Euglena gracilis* and Vegetables Exerts Synergistic Anti-Obesity and Anti-Inflammatory Effects by Modulating the Gut Microbiota in Diet-Induced Obese Mice. *Nutrients* 11 (1): 204. doi:10.3390/nu11010204.

[93] K. Ishibashi, M. Nishioka, N. Onaka, M. Takahashi, D. Yamanaka, Y. Adachi, and N. Ohno. 2019. Effects of *Euglena gracilis* EOD-1 Ingestion on Salivary IgA Reactivity and Health-Related Quality of Life in Humans. *Nutrients* 11 (5): 1144. doi:10.3390/nu11051144.

[94] A. Sugiyama, S. Hata, K. Suzuki, E. Yoshida, R. Nakano, S. Mitra, R. Arashida, Y. Asayama, Y. Yabuta, and T. Takeuchi. 2010. Oral Administration of Paramylon, a beta-1,3-D-Glucan Isolated from *Euglena gracilis* Z Inhibits Development of Atopic Dermatitis-Like Skin Lesions in NC/Nga Mice. *Journal of Veterinary Medical Science* 72 (6): 755–63. doi:10.1292/jvms.09-0526.

[95] N. Koizumi, H. Sakagami, A. Utsumi, S. Fujinaga, M. Takeda, K. Asano, I. Sugawara, S. Ichikawa, H. Kondo, S. Mori, K. Miyatake, Y. Nakano, H. Nakashima, T. Murakami, N. Miyano and, N. Yamamoto. 1993. Anti-HIV (Human Immunodeficiency Virus) Activity of Sulfated Paramylon. *Antiviral Research* 21 (1): 1–14. doi:10.1016/0166-3542(93)90063-O.

[96] C. Ryan, S. Cao, M. Sekiguchi, A. Haraguchi, A. Murata, A. Nakashima, Ke. Suzuki, and S. Shibata. 2023. *Euglena gracilis*-Derived β-Glucan Paramylon Entrains the Peripheral Circadian Clocks in Mice. *Frontiers in Nutrition* 10 (March): 1113118. doi:10.3389/fnut.2023.1113118.

[97] T. Kawano, A. Miura, J. Naito, N. Nishida, K. Ishibashi, Y. Adachi, N. Ohno, and Y. Naito. 2023. High-Parameter Immune Profiling and Subjective Health Assessment of the Immunomodulatory Effects of Paramylon-Rich *Euglena gracilis* EOD-1: A Randomized, Double-Blind, Placebo-Controlled, Parallel-Group Study. *Journal of Functional Foods* 109 (October): 105804. doi:10.1016/j.jff.2023.105804.

[98] K. Suzuki, 2017. Large-Scale Cultivation of *Euglena*. In *Euglena: Biochemistry, Cell and Molecular Biology*, edited by Steven D. Schwartzbach and Shigeru Shigeoka, 979: 285–93. Advances in Experimental Medicine and Biology. Cham: Springer International Publishing. doi:10.1007/978-3-319-54910-1_14.

[99] A. M. M. Basso, R. J. A. De Castro, T. B. de Castro, H. I. Guimarães, V. L. P. Polez, E. R. Carbonero, and V. H. Pomin. 2019. Immunomodulatory Activity of β-Glucan-Containing Exopolysaccharides from Auricularia auricular in Phagocytes and Mice Infected with *Cryptococcus Neoformans. Medical Mycology,* 58(2):227-39, myz042. doi:10.1093/mmy/myz042.

[100] R. Engstad, D. Ramsøy, and E. Steene. 2018. Glucan Compositions. US9956245B2, filed November 29, 2011, and issued May 1, 2018.

[101] Y. Du and L. Du. 2020. Application of Beta-1,3-Glucan in Preparation of Medicines for Repairing Skin Damage and Daily Chemical Products. CN111773239A, filed July 31, 2020, and issued October 16, 2020.

[102]. Y. H. Kho, E. S. Lee, B. W. Kang, D. W. Kim, J. J. Choung, J. O. Lee, and B. S. Seo. 2019. – *Aureobasidium pullulans* for Highly Producing Melanin Free Beta-Glucan and Method for Culturing. KR20190104956A, filed August 27, 2019, and issued September 11, 2019.

[103] U. Griesbach, R. Wachter, A. Ansmann, B. Fabry, and R. E. Engstad. 2005. Use of Water-Soluble Beta-(1,3) Glucans as Agents for Producing Therapeutic Skin Treatment Agents. US6875754B1, filed February 4, 2002, and issued April 5, 2005.

[104] R. Gammelsaeter and J. Remmereit. 2018. Use of Cellular Extracts for Skin Rejuvenation. US10092504B2, filed November 4, 2016, and issued October 9, 2018.

[105] V. Srilekha, G. Krishn, B. Sreelatha, E. J. Kumar, and K. V. N. Rajeshwari. 2023. Prodigiosin: A Fascinating and the Most Versatile Bioactive Pigment with Diverse Applications. *Systems Microbiology and Biomanufacturing*, 4(June): 66–76. doi:10.1007/s43393-023-00192-1.

[106] D. Coppola, D., C. Lauritano, G. Zazo, G. Nuzzo, A. Fontana, A. Ianora, M. Costantini, C. Verde, and D. Giordano. 2023. Biodiversity of UV-Resistant Bacteria in Antarctic Aquatic Environments. *Journal of Marine Science and Engineering* 11 (5): 968. doi:10.3390/jmse11050968.

[107] M. C. P. P. Reis-Mansur, J. S. Cardoso-Rurr, J. V. M. A. Silva, G. R. Souza, V. S. Cardoso, F. R. P. Mansoldo, Y. Pinheiro, *et al.* 2019. Carotenoids from UV-Resistant Antarctic *Microbacterium* sp. LEMMJ01. *Scientific Reports* 9 (1): 9554. doi:10.1038/s41598-019-45840-6.

[108] S. Naisi, M. Bayat, T. Z. Salehi, B. R. Zarif, and R. Yahyaraeyat. 2023. Antimicrobial and Anti-Biofilm Effects of Carotenoid Pigment Extracted from *Rhodotorula glutinis* Strain on Food-Borne Bacteria. *Iranian Journal of Microbiology*, 15(1):79–88 February. doi:10.18502/ijm.v15i1.11922.

[109] S. Sinha, S. Das, B. Saha, D. Paul, and B. Basu. 2023. Anti-Microbial, Anti-Oxidant, and Anti-Breast Cancer Properties Unraveled in Yeast Carotenoids Produced via Cost-Effective Fermentation Technique Utilizing Waste Hydrolysate. *Frontiers in Microbiology* 13 (January): 1088477. doi:10.3389/fmicb.2022.1088477.

[110] C. Ramesh, K., N. V. Vinithkumar, R. Kirubagaran, C. K. Venil, and L. Dufossé. 2019. Multifaceted Applications of Microbial Pigments: Current Knowledge, Challenges and Future Directions for Public Health Implications. *Microorganisms* 7 (7): 186. doi:10.3390/microorganisms7070186.

[111] M. J. Kiki, 2023. Biopigments of Microbial Origin and Their Application in the Cosmetic Industry. *Cosmetics* 10 (2): 47. doi:10.3390/cosmetics10020047.

[112] M. E. Paillié-Jiménez, P. Stincone, and A. Brandelli. 2020. Natural Pigments of Microbial Origin. *Frontiers in Sustainable Food Systems* 4 (September): 590439. doi:10.3389/fsufs.2020.590439.

[113] W. Sajjad, G. Din, M. Rafiq, A. Iqbal, S. Khan, S. Zada, B. Ali, and S. Kang. 2020. Pigment Production by Cold-Adapted Bacteria and Fungi: Colorful Tale of Cryosphere with Wide Range Applications. *Extremophiles* 24 (4): 447–73. doi:10.1007/s00792-020-01180-2.

[114] M. Aman Mohammadi, H. Ahangari, S. Mousazadeh, S. M. Hosseini, and L. Dufossé. 2022. Microbial Pigments as an Alternative to Synthetic Dyes and Food Additives: A Brief Review of Recent Studies. *Bioprocess and Biosystems Engineering* 45 (1): 1–12. doi:10.1007/s00449-021-02621-8.

[115]. J. V. O. Barreto, L. M. Casanova, A. N. Junior, M. C. P. P. R. Mansur, and A. B. Vermelho. 2023. Biotechnological Applications of Microbial Pigments. *Biology and Life Sciences*. [Online] Avaliable: www.preprints.org/manuscript/202310.0121/v1, doi:10.20944/preprints202310.0121.v1.

[116] F. C. Lopes and R. Ligabue-Braun. 2021. Agro-Industrial Residues: Eco-Friendly and Inexpensive Substrates for Microbial Pigments Production. *Frontiers in Sustainable Food Systems* 5 (March): 589414. doi:10.3389/fsufs.2021.589414.

[117] R. Panesar, S. Kaur, and P. S. Panesar. 2015. Production of Microbial Pigments Utilizing Agro-Industrial Waste: A Review. *Current Opinion in Food Science* 1 (February): 70–76. doi:10.1016/j.cofs.2014.12.002.

[118] S. Yang , L. Mi, J. Wu, X. Liao, and Z. Xu. 2022. Strategy for Anthocyanins Production: From Efficient Green Extraction to Novel Microbial Biosynthesis. *Critical Reviews in Food Science and Nutrition*, April, 63 (28): 9409–24 doi:10.1080/10408398.2022.2067117.

[119] Q. Zhang, S. Yu, Y. Lyu, W. Zeng, and J. Zhou. 2021. Systematically Engineered Fatty Acid Catabolite Pathway for the Production of (2 *S*)-Naringenin in *Saccharomyces cerevisiae*. *ACS Synthetic Biology* 10 (5): 1166–75. doi:10.1021/acssynbio.1c00002.

[120] A. K. Patel, V. S. Tambat, C. Chen, A. S. Chauhan, P. Kumar, A. P. Vadrale, C. Huang, C. Dong, and R. R. Singhania. 2022. Recent Advancements in Astaxanthin Production from *Microalgae*: A Review. *Bioresource Technology* 364 (November): 128030. doi:10.1016/j.biortech.2022.128030.

[121] N. T. Eriksen, 2008. Production of Phycocyanin—A Pigment with Applications in Biology, Biotechnology, Foods and Medicine. *Applied Microbiology and Biotechnology* 80 (1): 1–14. doi:10.1007/s00253-008-1542-y.

[122] A. G. Medentsev, A. lu. Arinbasarova, and V. K. Akimenko. 2005. Biosynthesis of Naphthoquinone Pigments by Fungi of the Genus *Fusarium*. *Applied Biochemistry and Microbiology* 41 (5): 503–7. doi:10.1007/s10438-005-0091-8.

[123] M. Sebak, F. Molham, C. Greco, M. A. Tammam, M. Sobeh, and A. El-Demerdash. 2022. Chemical Diversity, Medicinal Potentialities, Biosynthesis, and Pharmacokinetics of Anthraquinones and Their Congeners Derived from Marine Fungi: A Comprehensive Update. *RSC Advances* 12 (38): 24887–921. doi:10.1039/D2RA03610J.

[124] Z. Huang, W. Zhou, X. Yang, A. Tong, J. Hong, W. Guo, T. Li, R. Jia, Y. Pan, J. Lin, X. C. Lv and, B. Liu. 2018. The Regulatory Mechanisms of Soluble Starch and Glycerol for Production of Azaphilone Pigments in *Monascus purpureus* FAFU618 as Revealed by Comparative Proteomic and Transcriptional Analyses. *Food Research International* 106 (April): 626–35. doi:10.1016/j.foodres.2018.01.037.

[125] Z. Wu, M. Zhang, H. Yang, H. Zhou, and H. Yang. 2018. Production, Physico-Chemical Characterization and Antioxidant Activity of Natural Melanin from Submerged Cultures of the Mushroom *Auricularia auricula*. *Food Bioscience* 26 (December): 49–56. doi:10.1016/j.fbio.2018.09.008.

[126] S. Ohkubo and H. Miyashita. 2017. A Niche for Cyanobacteria Producing Chlorophyll f within a Microbial Mat. *The ISME Journal* 11 (10): 2368–78. doi:10.1038/ismej.2017.98.

[127] F. Stintzing, and W. Schliemann. 2007. Pigments of Fly Agaric (*Amanita Muscaria*). *ZeitschriftFürNaturforschung C* 62 (11–12): 779–85. doi:10.1515/znc-2007-11-1201.

[128] H. Meruvu and J. C. Santos. 2021. Colors of Life: A Review on Fungal Pigments. *Critical Reviews in Biotechnology* 41 (8): 1153–77. doi:10.1080/07388551.2021.1901647.

[129] R. Chatragadda and L. Dufossé. 2021. Ecological and Biotechnological Aspects of Pigmented Microbes: A Way Forward in Development of Food and Pharmaceutical Grade Pigments. *Microorganisms* 9 (3): 637. doi:10.3390/microorganisms9030637.

[130] A. L. Morocho-Jácome, N. Ruscinc, R. M. Martinez, J. C. M. Carvalho, T. S. Almeida, C. Rosado, J. G. Costa, M. V. R. Velasco, and A. R. Baby. 2020. (Bio) Technological Aspects of Microalgae Pigments for Cosmetics. *Applied Microbiology and Biotechnology* 104 (22): 9513–22. doi:10.1007/s00253-020-10936-x.

[131] A. Mohammad, S. N. Huda, N. S. Sulaiman, M. S. M. Nor, P. S. Abdullah, Z. A. Kari, and S. Pati. 2022. A Review on Recent Advances on Natural Plant Pigments in

Foods: Functions, Extraction, Importance and Challenges. *Applied Biochemistry and Biotechnology* 194 (10): 4655–72. doi:10.1007/s12010-022-04050-z.

[132] M. P. N. Rao, M. Xiao, and W-J. Li. 2017. Fungal and Bacterial Pigments: Secondary Metabolites with Wide Applications. *Frontiers in Microbiology* 8 (June): 1113. doi:10.3389/fmicb.2017.01113.

[133] K. Dev, 2018. *A Psychro-Halophilic Rhodonellum psychrophilum Strain Gl8 And Pigments Thereof.* www.istem.gov.in/patents/patent_details/64266c984528cd373891d3ce.

[134] B. Han, Y. K. Xiong, Z. Fanzhong, Y. Zhicheng, S. Kaili, and Y. Xinyue. 2020. Application of actinomycin. CN110983816A, filed November 7, 2019, and issued April 10, 2020.

[135] Z. Lanqin and C. Haofeng. 2008. Purple *Monascus* strain Danxi-2. CN100443584C, filed November 9, 2005, and issued December 17, 2008.

[136] N. Wang, Q. Zhao, and P. Zhang. 2015. *Geomyces* sp. wnf-15A(phy) Coloring Compound. CN104804007A, filed March 14, 2015, and issued July 29, 2015.

[137] L. Ma, L. Wang, X. Guo, M. Zhu, C. Hu, and S. Lyu. 2023. *Streptomyces hematochrome* as Well as Method for Extracting Pigmen*t of Streptomyces hematochrome* and Application of *Streptomyces hematochrome*. CN115612639A, filed September 5, 2022, and issued January 17, 2023.

[138] C. Yi, Q. Dairong, and X. Hui. 2021. *Aspergillus sydowii H-1* Producing *anthocyanin* and Application of *Aspergillus sydowii*. CN110438012B, filed August 5, 2019, and issued October 26, 2021.

[139] X. Yu, W. Zhang, G. Zhang, G. Liu, T. Chen, and Y. Wu. 2022. Screening Method for Obtaining High-Yield *Arthrobacter* Blue by Using Heavy Ion Irradiation. CN115141821A, filed June 8, 2022, and issued October 4, 2022.

[140] H. J. Hwang and S. H. Suh. 2006. *Monascus* sp. Mutants and Cholesterol Inhibitors Comprising Extracts Thereof. KR100635452B1, filed September 4, 2004, and issued October 17, 2006.

[141] B. Chen, Z. Zhou, and Y. Li. 2023. Pigment and Production Engineering Bacterium Construction Method Thereof. CN115991761A, filed November 29, 2022, and issued April 21, 2023.

[142] L. Yu, Y. Chen, K. Chen, M. Cai, M. Wu, Z. Wang, and S. Xu. 2019. *Lactobacillus*, Method of Preparing Pigment by Using *Lactobacillus, Lactobacillus* Culture, Pigment Composition Having Lactobacillus Culture. CN109136121A, filed July 5, 2017, and issued January 4, 2019.

[143] J. S. Park, J. M. Park, and K. H. Lee. 2019. KCTC 13387BP New Bacterial Species *Zooshikella* sp. KCTC 13387BP Producing Red Pigment and Method for Culturing the Same. KR20190054518A, filed November 13, 2017, and issued May 22, 2019.

[144] S. Li, K. Zhang, F. Ge, Y. Tao, and S. Wei. 2018. *Epicoccum nigrum* LS10H and Application Thereof. CN108004150A, filed December 23, 2017, and issued May 8, 2018.

[145] S. Zhu, 2022. Method for Increasing Content of *Paeonia lactiflora hematochrome* through Exogenous Inoculated Strain. CN114532096A, filed January 24, 2022, and issued May 27, 2022.

[146] N. Wang and L. Fang. 2021. *Geomyces* Mutant Strain and Application Thereof. US2021363549A1, filed October 14, 2019, and issued November 25, 2021.

[147] L. Li, J. Fan, J. Gong, G. Song, D. Wang, and T. Yuan. 2022. Method for Synthesizing Gardenia Blue Pigment and Gardenia Red Pigment through High-Yield Beta-Glucosidase Lactic Acid bacteria Immobilized Cells. CN115011506A, filed May 5, 2022, and issued September 6, 2022.

[148] E. Ramaprasad, C. Sasikala, and C. Ramana. 2012. A Process for the Isolation and Purification of Lycopene from *Rhodospirillum sulfurexigens* JA143T and *Rhodospirillum oryzae* JA318T and Utilization Thereof. 49/2012, filed 2012, and issued 2012. https://iprsearch.ipindia.gov.in/PublicSearch/PublicationSearch/PatentDetails.
[149] G. Fan, C. Wu, S. Fang, J. Wang, T. Li, and N. Xie. 2015. *Cyclocarya paliurus* Endophyte Producing Hematochrome, and Method for Fermenting *Cyclocarya paliurus* Endophyte to Produce Hematochrome. CN103436451B, filed July 22, 2013, and issued April 1, 2015.
[150] H. Huang, X. Xu, W. Jin, L. Jiang, S. Li, and P. Song. 2014. Carotenoid High Yield Strain and Application Thereof. CN104073457A, filed July 21, 2014, and issued October 1, 2014.
[151] L. Zhikun, D. Mingrong, Y. Qing, Y. Songzhen, S. Xiaotang, Z. Honghui, W. Yan, and J. Guo. 2010. *Streptomyces caeruleatus* GIMN4.002 and Application Thereof. CN101525586B, filed January 22, 2009, and issued December 1, 2010.
[152] Z. Honghui, J. Guo, S. Xiaotang, Y. Songzhen, L. Yanxuan, and C. Meibiao. 2009. *Streptomyces vietnamensis*. CN100448984C, filed March 7, 2006, and issued January 7, 2009.
[153] Y. Li, Z. Yang, Y. Wu, and A. Zhang. 2022. Application of Strain CH18 in Preparation of Anthocyanin and Other Flavonoid Compounds. CN114480532A, filed January 26, 2022, and issued May 13, 2022.
[154] G. R. Ryu, H. A. Bae, A. J. Jeon, and J. W. Jang. 2021. Method for Preparing Violacein or Deoxyviolacein Using *Bdellovibrio bacteriovorus*. KR20210079825A, filed December 20, 2019, and issued June 30, 2021.
[155] Q. Zhang, J. Tang, W. Xiang, J. Wang, X. Sun, M. Wu, C. Li, M. Zheng, and H. Yao. 2021. *Gordonia melanodendri*, Application and Pigment Production Method. CN113249248A, filed April 25, 2021, and issued September6, 2022.
[156] Z. Wu, G. Chen, and M. Wang. 2015. *Monascus ruber* Strain with High Extracellular Yellow Pigment Yields, Method for Breeding *Monascus ruber* Strain and Application Thereof. CN105062895A, filed July 28, 2015, and issued November 18, 2015.
[157] H. Pu, X. Wang, and J. Wang. 2014. Method for Quickly Producing Red Yeast Rice with High Pigment Value. CN102550940B, filed December 27, 2011, and issued November 12, 2014.
[158] Y. J. Cho, C. H. Song, and J. W. Yoon. 2005. Red Pigment by Submerged Culture of *Paecilomyces sinclairii* and Method for Preparing the Same. KR100473886B1, filed March 22, 2004, and issued March 14, 2005.
[159] T. Sen, C. J. Barrow, and S. K. Deshmukh. 2019. Microbial Pigments in the Food Industry—Challenges and the Way Forward. *Frontiers in Nutrition* 6 (March): 7. doi:10.3389/fnut.2019.00007.
[160] M. Thakur, and V. K. Modi. 2022. Biocolorants in Food: Sources, Extraction, Applications and Future Prospects. *Critical Reviews in Food Science and Nutrition*, December 11: 1–40. [online] doi:10.1080/10408398.2022.2144997.
[161] L. Dufossé, P. Galaup, A. Yaron, S. M. Arad, P. Blanc, K. N. C. Murthy, and G. A. Ravishankar. 2005. Microorganisms and Microalgae as Sources of Pigments for Food Use: A Scientific Oddity or an Industrial Reality?. *Trends in Food Science & Technology* 16 (9): 389–406. doi:10.1016/j.tifs.2005.02.006.
[162] F. Mandelli, V. S. Miranda, E. Rodrigues, and A. Z. Mercadante. 2012. Identification of Carotenoids with High Antioxidant Capacity Produced by Extremophile Microorganisms. *World Journal of Microbiology and Biotechnology* 28 (4): 1781–90. doi:10.1007/s11274-011-0993-y.

[163] N. Flores, S. Hoyos, M. Venegas, A. Galetović, L. M. Zúñiga, F. Fábrega, B. Paredes, C. S. Ardiles, C. Vilo, C. Ascaso, J. Wierzchos, V. S. Egipsy, J. E. Araya, R. A. B. García and, B. G. Silva. 2020. *Haloterrigena* sp. Strain SGH1, a Bacterioruberin-Rich, Perchlorate-Tolerant Halophilic Archaeon Isolated From Halite Microbial Communities, Atacama Desert, Chile. *Frontiers in Microbiology* 11 (March): 324. doi:10.3389/fmicb.2020.00324.

[164] F. I. Kesbiç and N. Gültepe. 2022. C50 Carotenoids Extracted from *Haloterrigena thermotolerans* Strain K15: Antioxidant Potential and Identification. *Folia Microbiologica* 67 (1): 71–79. doi:10.1007/s12223-021-00905-w.

[165] E. Setiyono, T. H. P. Brotosudarmo, D. Pringgenies, Heriyanto, M. N. U. Prihastyanti, and Y. Shioi. 2019. Analysis of β-Cryptoxanthin from Yellow Pigmented Marine Bacterium *Erythrobacter* sp. Kj5. *IOP Conference Series: Earth and Environmental Science* 246 (May): 012004. doi:10.1088/1755-1315/246/1/012004.

[166] S. Jinendiran, H. Dahms, B. S. D. Kumar, V. K. Ponnusamy, and N. Sivakumar. 2020. Diapolycopenedioic-Acid-Diglucosyl Ester and Keto-Myxocoxanthin Glucoside Ester: Novel Carotenoids Derived from *Exiguobacterium acetylicum* S01 and Evaluation of Their Anticancer and Anti-Inflammatory Activities. *Bioorganic Chemistry* 103 (October): 104149. doi:10.1016/j.bioorg.2020.104149.

[167] T. Kimura, W. Fukuda, T. Sanada, and T. Imanaka. 2015. Characterization of Water-Soluble Dark-Brown Pigment from Antarctic Bacterium, *Lysobacter oligotrophicus*. *Journal of Bioscience and Bioengineering* 120 (1): 58–61. doi:10.1016/j.jbiosc.2014.11.020.

[168] A. E. Noman, N. S. Al-Barha, A. M. Sharaf, Q. A. Al-Maqtari, A. Mohedein, H. H. H. Mohammed, and F. Chen. 2020. A Novel Strain of Acetic Acid Bacteria *Gluconobacter oxydans* FBFS97 Involved in Riboflavin Production. *Scientific Reports* 10 (1): 13527. doi:10.1038/s41598-020-70404-4.

[169] A. Maj, L. Dziewit, L. Drewniak, M. Garstka, T. Krucon, K. Piatkowska, K. Gieczewska, J. Czarnecki, E. Furmanczyk, R. Lasek, J. Baj, and D. Bartosik. 2020. *In Vivo* Creation of Plasmid pCRT01 and Its Use for the Construction of Carotenoid-Producing *Paracoccus* spp. Strains That Grow Efficiently on Industrial Wastes. *Microbial Cell Factories* 19 (1): 141. doi:10.1186/s12934-020-01396-z.

[170] T. Akira and M. Haruyoshi. 2003. Process for Producing Carotenoid Pigments. CN1388830A, filed June 8, 2001, and issued January 1, 2003.

[171] Y. Wang and O. Yu. 2021. Compositions and Methods of Biosynthesizing Carotenoids and Their Derivatives. US2021115454A1, filed December 16, 2020, and issued April 22, 2021.

[172] K. Hirasawa, A. Tsubokura, H. Satou, and T. Yata. 2010. Carotenoid Fermentation Method. WO2010044469A1, filed October 16, 2009, and issued April 22, 2010.

[173] M. A. Kozubal, R. E. Macur, and W. P. Inskeep. 2017. Acidophilic *Fusarium oxysporum* Strains, Methods of Their Production and Methods of Their Use. US9796989B2, filed July 2, 2015, and issued October 24, 2017.

[174] H. Agarwal, S. Bajpai, A. Mishra, I. Kohli, A. Varma, M. Fouillaud, L. Dufossé, and N. C. J. 2023. Bacterial Pigments and Their Multifaceted Roles in Contemporary Biotechnology and Pharmacological Applications. *Microorganisms* 11 (3): 614. doi:10.3390/microorganisms11030614.

[175] Y. Song, G. Liu, J. Li, H. Huang, X. Zhang, H. Zhang, and J. Ju. 2015. Cytotoxic and Antibacterial Angucycline- and Prodigiosin- Analogs from the Deep-Sea Derived *Streptomyces* sp. SCSIO 11594. *Marine Drugs* 13 (3): 1304–16. doi:10.3390/md13031304.

[176] B. Tian, Z. Sun, S. Shen, H. Wang, J. Jiao, L. Wang, Y. Hu, and Y. Hua. 2009. Effects of Carotenoids from *Deinococcus radiodurans* on Protein Oxidation. *Letters in Applied Microbiology* 49 (6): 689–94. doi:10.1111/j.1472-765X.2009.02727.x.

[177] D. C. Woodhams, B. C. LaBumbard, K. L. Barnhart, M. H. Becker, M. C. Bletz, L. A. Escobar, S. V. Flechas, *et al.*, 2018. Prodigiosin, Violacein, and Volatile Organic Compounds Produced by Widespread Cutaneous Bacteria of Amphibians Can Inhibit Two *Batrachochytrium* Fungal Pathogens. *Microbial Ecology* 75 (4): 1049–62. doi:10.1007/s00248-017-1095-7.

[178] P. S. Melo, S. S. Maria., B. C. Vidal, M. Haun, and N. Durán. 2000. Violacein cytotoxicity and induction of apoptosis in V79 cells. *In Vitro Cellular & Developmental Biology – Animal* 36 (8): 539. doi:10.1290/1071-2690(2000)036<0539:VCAIOA>2.0.CO;2.

[179] G. Bisht, S. Srivastava, R. Kulshreshtha, A. Sourirajan, D. J. Baumler, and K. Dev. 2020. Applications of Red Pigments from psychrophilic *Rhodonellum psychrophilum* GL8 in Health, Food and Antimicrobial Finishes on Textiles. *Process Biochemistry* 94 (July): 15–29. doi:10.1016/j.procbio.2020.03.021.

[180] V. Tyurenkov, A. Tyurenkov, and Y. D. Turiyansky. 2015. Preparation for Radio Protection. US2015328269A1, filed December 19, 2012, and issued November 19, 2015.

[181] M. Chen, N. Shen, and Z. Chen. 2018. Azaphilones Compounds in Marine Fungus HK1-6 and Application of azaphilones Compounds as MRSA-Resistant Drug. CN108101878A, filed December 15, 2017, and issued June 1, 2018.

[182] J. Li, X. Yang, L. Liu, Y. Lin, Y. Cen, Y. Lu, L. He, M. Li, J. Yuan, and J. He. 2014. Marine Fungus-Derived Azaphilone Dimer Compound, and Preparation Method and Application Thereof. CN103911407A, filed February 28, 2014, and issued July 9, 2014.

[183] L. Chen, Y. Li, X. Li, S. Wang, Q. Liu, M. Cheng, and Q. Zhang. 2019. Iso-Penicillixanthone A Derived from *Penicillium oxalicum* and Application to Cervical Cancer. CN109776478A, filed December 4, 2018, and issued May 21, 2019.

[184] M.Ye, S. Li, F. Shi, L. Yang, and J. Li. 2017. Application of *Lachnum melanin* as Anti-Renal-Failure Medicine. CN106389482A, filed September 22, 2016, and issued February 15, 2017.

[185] T. Pan and C. Lee. 2016. Composition and Method for Prevention and Treatment of Alzheimer's Disease. EP2072055B1, filed December 21, 2007, and issued March 23, 2016.

[186] H. Hirakida, S. Nakamura, S. Inagaki, S. Tsuji, M. Hayashi, M. Shimazawa, and H. Hara. 2022. Anti-Diabetic Effects of Astaxanthin-Rich Extract Derived from *Paracoccus arotinifaciens* on Pancreatic β Cells. *Journal of Functional Foods* 97 (October): 105252. doi:10.1016/j.jff.2022.105252.

[187] R. S. Celedón and L. B. Díaz. 2021. Natural Pigments of Bacterial Origin and Their Possible Biomedical Applications. *Microorganisms* 9 (4): 739. doi:10.3390/microorganisms9040739.

[188] B. Nowruzi, G. Sarvari., and S. Blanco. 2020. The Cosmetic Application of Cyanobacterial Secondary Metabolites. *Algal Research* 49 (August): 101959. doi:10.1016/j.algal.2020.101959.

[189] C. Lourenço-Lopes, M. Fraga-Corral, C. Jimenez-Lopez, M. Carpena, A. G. Pereira, P. Garcia-Oliveira, M. A. Prieto, and J. Simal-Gandara. 2021. Biological Action Mechanisms of Fucoxanthin Extracted from Algae for Application in Food and Cosmetic Industries. *Trends in Food Science & Technology* 117 (November): 163–81. doi:10.1016/j.tifs.2021.03.012.

[190] A. Goksoeyr, 2011. Carotenoid Sunscreen. WO2011151426A2, filed June 1, 2011, and issued December 8, 2011.

[191] G. Brooks and S. Franklin. 2017. Cosmetic Compositions Comprising Microalgal Components. US9668966B2, filed November 3, 2015, and issued June 6, 2017.

[192] J. Kopecký, J. Masojídek, D. Kubáč, P. Hrouzek, R. Lhotský, K. Ranglová, and S. Pekařová. 2021. *Parachlorella kessleri* HY1 Microalgae Strain for Lutein Production and Use of This Strain for Industrial Lutein Production. CZ308652B6, filed December 16, 2019, and issued January 27, 2021.

[193] G. Tolborg, T. I. Petersen, T. O. Larsen, and M. Workman. 2018. Process for Producing an Azaphilone in *Talaromyces atroroseus*. WO2018206590A1, filed May 8, 2018, and issued November 15, 2018.

[194] G. Lakshmi, 2004. A process for preparation of water soluble pigments. www.allindianpatents.com/patents/248547-a-process-for-preparation-of-water-soluble-pigments.

[195] A. Abudureheman, W. Mijiti, and G. Zuonuni. 2011. *Bacillus Prodigiosus* and Prodigiosin Produced Thereby. CN101392227B, filed October 23, 2008, and issued June 1, 2011.

[196] K. Hidachi, M. Ueno, H. Kishi., K. Toda, H. Yuki, Y. Takiguchi, O. Sasaki, S. Hirano, K. Kokusho, and T. Yuba. 2015. Cosmetics, Pharmaceuticals, and Quasi Drugs Containing Violacein. JP2015105264A, filed December 2, 2013, and issued June 8, 2015.

[197] M. C. P. P. R. Mansur, C. Campos, A. B. Vermelho, J. Nobrega, L. C. Boldrini, L. Balottin., C. Lage, A. S. Rosado, E. Ricci-Júnior, and E. P. Santos. 2020. Photoprotective Nanoemulsions Containing Microbial Carotenoids and Buriti Oil: Efficacy and Safety Study. *Arabian Journal of Chemistry* 13 (8): 6741–52. doi:10.1016/j.arabjc.2020.06.028.

[198] L. Kusmita, A. N. P. Edi, Y. D. Franyoto, M, S. Haryanti, and A. D. R. Nurcahyanti. 2023. Sun Protection and Antibacterial Activities of Carotenoids from the Soft Coral *Sinularia* sp. Symbiotic Bacteria from Panjang Island, North Java Sea. *Saudi Pharmaceutical Journal* 31 (8): 101680. doi:10.1016/j.jsps.2023.06.013.

[199] F. Pagels, C. Almeida, V. Vasconcelos., and A. C. Guedes. 2022. Cosmetic Potential of Pigments Extracts from the Marine *Cyanobacterium cyanobium* sp. *Marine Drugs* 20 (8): 481. doi:10.3390/md20080481.

11 Menthol

The Story of a Plant-Derived Molecule Through the Patents

Reda El Boukhari, Hanane Nahid, and Ahmed Fatimi

11.1 INTRODUCTION

Plants refer to living organisms belonging to the plant kingdom. Different legislations have adopted different concepts, giving rise to a more in-depth debate on their patentability. In general, it is recognized that plants preexisting in nature are not patentable since they would constitute simple discoveries [1]. Patents are granted only for inventions and not for discoveries. "Discovery differs from invention in that it is the perception by observation of a natural phenomenon preexisting any human intervention, whereas invention is characterized by the fact that it is the voluntary coordination by man of material means" [2]. Discovery is merely the uncovering of previously unknown elements. Invention, on the other hand, is a technical solution to technical problems. Nevertheless, the question of patentability is subject to the laws and regulations of each country as well as international agreements, which play an important role in regulating the patentability of plants and plant derivatives. Two main aspects have been generated: plants as such and specific uses of plants.

For the first aspect, under Moroccan legislation, the following are not considered patentable in themselves: "plants and derivatives, plant varieties, and essentially biological processes for obtaining plants" [3], such as traditional selection and crossing methods. However, there are exceptions to this article where specific genetic modifications or unique characteristics can be patented. Accordingly, "inventions relating to plant varieties are patentable when they relate to:

- biological material isolated from the plant variety… produced by means of a technical process;
- plants… if the technical feasibility of the invention is not limited to a plant variety…;
- a new plant variety which has not been the subject of an application for a plant variety certificate…" [4].

DOI: 10.1201/9781003405665-11

Therefore, biotechnology, which is the application of scientific technologies to modify, improve, or add value to plants and microorganisms, has the capacity to intervene in plants and plant varieties and produce a result different from that which exists in nature. In this respect, many patent offices consider, for example, that a biological entity is patentable if man's technical intervention (manufacturing) has produced an artificial state of affairs that does not exist in nature [5].

In the same context, the World Trade Organization's Agreement on Trade-Related Aspects of Intellectual Property Rights (TRIPS) authorizes countries to exclude from patentability higher living organisms, animals, or plants and "essentially biological processes for the production of plants and animals". However, TRIPS members are needed to provide general patent protection for microorganisms and "nonbiological and microbiological processes" [6]. Furthermore, the TRIPS Agreement [Article 27.3(b)] has widely recognized the patentability of plant varieties, whether through a patent system, a sui generis system, or a combination of the two. In this respect, it is important to distinguish between plants and plant varieties. In this regard, the Agreement does not define the sui generis system. Article 27.3(b), in particular, has been interpreted in many different ways, but it is generally accepted that it requires the extension of intellectual property rights to plant varieties [7]. In the context of plant variety protection, the sui generis system may be extended to the Plant Variety Certificate.

In addition, some countries have chosen to recognize the patentability of plants. In China, plants are conceived, in the context of patent law, as a form of living organism whose life is sustained by the synthesis of carbohydrates and proteins from inorganic materials such as water, carbon dioxide, and inorganic salts and which is generally fixed [8]. The Japanese Patent Office also specifies in its examination guidelines that the term "plant" refers to one of the three categories used for classifying organisms, namely, microorganisms, plants, and animals. Japanese national legislation contains no provisions relating to plants or plant varieties, which are therefore patentable provided the invention meets the criteria of novelty, industrial application, and inventive step and is adequately disclosed. The Japanese Patent Office has adopted guidelines for inventions in particular fields, such as genetic engineering and certain types of biological inventions [9].

Moreover, the United States does not expressly exclude the patentability of living organisms. In fact, several forms of protection are available in the United States for plants. A patent may be granted to anyone who invents, discovers, and asexually reproduces a distinctive and novel plant variety other than a tuber-propagated plant or a plant found in the wild. Plants can be protected by a utility patent (35 U.S.C. § 101), a plant patent (35 U.S.C. § 161), or a plant variety certificate (7 U.S.C. § 2321). By analyzing the US Plant Patent Law of 1930, the Supreme Court stated: "In enacting the Plant Patent Law, Congress took into account these two concerns [that plants, even those obtained by artificial means, were products of nature under patent law and that plants did not lend themselves to written description]." [9].

For the second aspect, specific uses of a plant, such as medicinal, cosmetic, food, industrial, or any other innovative application, can potentially be patented in certain cases. This depends on the regulations in force in the country concerned and on patentability criteria such as novelty, inventiveness, and industrial application. It is important to note that patents are designed to protect new, nonobvious inventions. In the case of plants and their uses, this generally means that there must be a significant innovation or inventive solution that goes beyond what is already known in nature. Moreover, the TRIPS Agreement does not define invention, nor does the Paris Convention for the Protection of Industrial Property. The answer to this question is therefore to be found in legislation and judicial decisions. The dividing line between products of nature and those resulting from human intervention has changed thanks to technology, judicial decisions, and changes in the laws governing intellectual property [7].

This agreement establishes minimum standards for the patentability of inventions, including specific uses of plants, in member countries. Article 27.1 of TRIPS specifies that an invention shall be patentable "provided that it is new, involves an inventive step, and is capable of industrial application". However, the interpretation and implementation of these standards may vary from one country to another, depending on national legislation. According to the Implementing Regulations of the European Patent Convention (EPC), biotechnological inventions are "inventions relating to a product consisting of or containing biological material or to a process for producing, treating, or using biological material". Biological material is defined as "any material that contains genetic information and can reproduce itself or be reproduced in a biological system". The European Union (EU) Directive 98/44/EC on the legal protection of biotechnological inventions established guidelines for the patentability of biotechnological inventions, including plants and their uses. However, the interpretation of this directive has given rise to debate, and the national laws of each member state have played a role in how patents are granted.

In short, the patentability of plants and their derivatives, as well as plant varieties, is a complex area where international agreements, conventions, and national laws interact to define the rules and criteria for protection, with the involvement of other intellectual property titles.

In this chapter, we provide an in-depth analysis of the patents related to menthol and its derivatives and their applications. By analyzing the patents related to menthol as a plant-derived molecule, we will provide hereinafter an overview of the current state of the art and identify emerging trends in the field. First, a scientific overview of menthol is presented. The second part presents an extensive review of the market analysis of menthol-based products. Part 3 focuses on the analysis of the menthol patent data. Finally, the last part of this chapter is proposed in the form of future trends and challenges associated with patenting menthol and its derivatives, as well as various recommendations that could help one plan and innovate research strategies.

11.2 A SCIENTIFIC OVERVIEW OF MENTHOL

Menthol is a covalent organic molecule that is naturally present in a variety of plants. It was first isolated in 1771 by Gaubius [10]. This organic compound is obtained either by synthesis or by extraction from mint essential oils. The most common stereoisomer of menthol is the configuration (–)-menthol (1R,2S,5R) [11]. It belongs

FIGURE 11.1 (a) Photograph of menthol crystals at room temperature. (Photo obtained from Wikimedia Commons and licensed under the Creative Commons Attribution-Share Alike 3.0 Unported license [32].) (b) Photograph of mint plant leaves (Photo obtained from Pexels Free Images with no attribution needed, and free use under a Pexels license [33].)

to the monoterpenol family. At room temperature, it is in the form of crystals, waxy white in color (Figure 11.1a). It melts if the temperature is slightly increased (Table 11.1).

The primary sources of menthol production are especially plants in the mint family, particularly peppermint (*Mentha* x *piperita*) and spearmint (*Mentha spicata*). Menthol is produced in glandular trichomes, which are small hair-like structures on the surface of mint plant leaves (Figure 11.1b). Menthol can be extracted from the leaves of the mint plant by a process called steam distillation [12]. During this process, steam passes through the leaves of the mint plant, causing the glandular trichomes to burst and release the essential oils containing menthol [13]. The resulting mixture of steam and essential oil is then cooled, causing the menthol to solidify and separate from the water [12,13]. The menthol crystals can be further purified and processed to obtain a high-purity menthol product [14].

In addition to mint plants, menthol can also be synthesized from other natural sources, such as cornmint oil, as well as through chemical synthesis using various starting materials. However, mint plants remain the most important and reliable source of menthol production [11].

Menthol has anti-inflammatory and antiviral properties [15]. It is a local anesthetic, and it is also used to relieve minor throat irritation [16]. It is well known for its relaxing and cooling effects, and its significance is felt in several applications, including health and wellness and industrial applications. In pain-relieving lotions and balms, their cooling impacts calms muscles, joints, and small injuries in health and wellness [17,18]. It facilitates breathing in respiratory products such as cough drops and inhalers as a decongestant and bronchodilator [19]. The revitalizing flavor and sensation of menthol are used in toothpaste and mouthwash for dental care [20,21]. Regarding vascular effects, recent investigations have revealed that menthol triggers an indirect vascular response through sensory fibers, in addition to its long-established direct effects on the endothelium and vascular smooth muscle of the vasculature [22]. Industrially, it increases the potency of medications and imparts a minty flavor to food, beverages, tobacco, and cosmetics [23–26]. The broad use of menthol is also influenced by its cultural relevance in conventional medicine and its energizing

TABLE 11.1
Chemical Characteristics of the Menthol Molecule (Adapted From Silva, 2020 [22]), Copyright© 2020 Frontiers Media S.A. Under the Terms of the Creative Commons Attribution License (CC BY).

Appearance	White or colorless crystalline solid
InChI	1S/C10H2°c1-7(2)9-5-4-8(3)6-10(9)11/h7-11H,4-6H2,1-3H3
InChiKey	NOOLISFMXDJSKH-UHFFFAOYSA-N
Molecular Formula	$C_{10}H_{20}O$
Molecular Weight	156.2674 g.mol^{-1}
Synonym	– 5-methyl-2-(1-methylethyl)-cyclohexan-1-ol – (1R,2S,5R)-2-isopropyl-5-methylcyclohexanol – 2-Isopropyl-5-methylcyclohexan-1-ol – 3-p-Menthanol – Hexahydrothymol – Menthomenthol – Peppermint camphor
Structures of menthol isomers	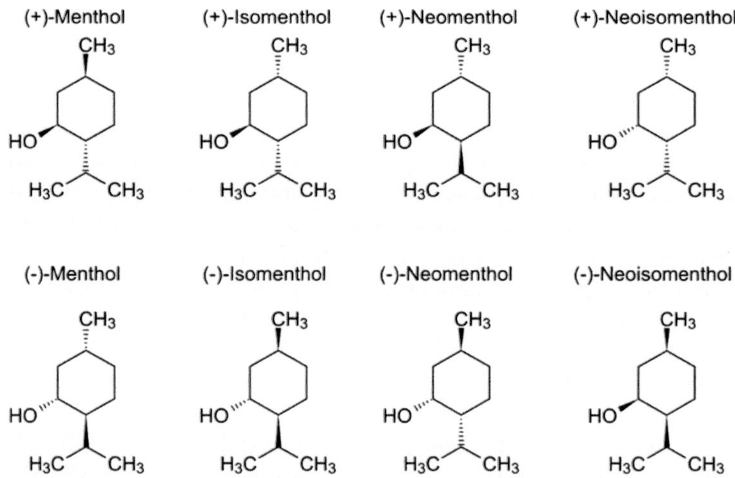

fragrance in aromatherapy [27–30]. Furthermore, menthol fits perfectly with consumer demands for sustainability as a natural substitute for synthetic compounds [31].

11.3 THE COUNTRY-WISE MARKET ANALYSIS OF MENTHOL-BASED PRODUCTS

The global menthol market is projected to experience significant growth during the forecast period from 2020 to 2030, as indicated by Transparency Market Research's latest report [34]. The rise in the usage of menthol in the flavoring and additives sector, along with its increased application in oral care products, is expected to drive the market's expansion. Additionally, the preference for sweet-tasting flavoring, particularly among young adults, is boosting the demand for menthol in the food and beverage sector. Based on revenue, the global menthol market is estimated to surpass USD 1.2 billion by the year 2030, with increasing usage in aromatherapy products such as candles and essential oils playing a significant role in its expansion, as well as a steady compound annual growth rate of approximately 4% throughout the forecast period [34].

On the other hand, the global menthol market is witnessing growth, driven by increasing demand from tobacco companies for menthol cigarettes. These companies market menthol cigarettes as "cool" and "refreshing", targeting new consumers who find them more appealing than regular, nonflavored cigarettes. Engaging advertisements featuring water, nature, and refreshing elements further contribute to the popularity of menthol cigarettes worldwide. As a result, menthol cigarettes currently hold a substantial market share in the combustible cigarette space. The preference for menthol cigarettes among beginner smokers is expected to maintain steady market growth. However, the rising awareness of the harmful effects of tobacco and tobacco products may hinder the demand for menthol cigarettes in the latter half of the forecast period, potentially influencing the overall growth of the global menthol market [34].

Due to the COVID-19 pandemic, the global menthol market is experiencing a moderate impact. The utilization of menthol as a vital component in essential oils for aromatherapy is also supported. The use of menthol in cough drops, nasal sprays, and inhalers has contributed to steady growth in the market throughout 2020. Despite challenges posed by the pandemic, the increasing adoption of menthol in aromatherapy products is expected to sustain demand in the menthol market during this period [34].

Competition in the global menthol market was highly fragmented in 2019, with leading manufacturers collectively holding a 65.7% market share. Major players operating in the market include BASF SE (Ludwigshafen, Germany), Archer-Daniels-Midland Company (Chicago, IL, United States), Symrise AG (Holzminden, Germany), Takasago International Corporation (Tokyo, Japan), and Nectar Lifesciences Limited (Chandigarh, India). Among them, BASF SE held a prominent share of 19.6% with its menthol products such as L-Menthol Flakes (CAS Number: 2216-51-5) that satisfy the "Food Chemical Codex" criteria and are appropriate for all food, beverage, and nutritional supplement applications. Archer-Daniels-Midland Company also stood as a global leader, with a 15.7% share of the global menthol market in 2019 [34].

11.4 ANALYSIS OF THE PATENT DATA FOR MENTHOL AND ITS DERIVATIVES

11.4.1 First Menthol and its Derivative-Related Patent

Although menthol was first isolated in 1771 by Gaubius [10], the first menthol-related invention covered by a patent was claimed 125 years later. Through the patent application (filed in 1896 and then published in 1897), Helbing has proposed some improvements in the preparation and application of menthol [35]. Based on the earliest priority date of this patent, it is possible to precisely assume that 1896 is considered the starting year of the patenting of menthol and its derivatives, as well as the preparation and application of menthol.

11.4.2 Patent Documents and Publication Date

Patent documents are official records that provide detailed descriptions of inventions. These documents comprise technical information about the invention, including claims, specifications, and eventually drawings, and are submitted to government authorities (i.e., the patent office or intellectual property office) for patent protection [36]. On the other hand, the publication date represents the date when, among others, a patent application or granted patent becomes publicly accessible, enabling anyone to access the specificities of the innovation. After the publication date, the information enters the public domain and becomes part of the prior art, potentially affecting the patentability of similar inventions [37,38].

According to the search database used in this study and based on "title", "abstract", and "claims", 22335 patent documents related to menthol were found (between January 1896 and July 2023) [39]. The found patent documents include different types, as shown in Table 11.2. Furthermore, Figure 11.2 presents the patent document count evolution as a function of filed, granted, and published years. As a maximum, 1330 patent applications were filed in 2016, and 1367 patent documents were published in 2018. However, 555 patents were granted in 2021.

The time evolution of patent application filing rates demonstrated noticeable growth until 2008, with a decrease in 2009, followed by another decline in 2011 and a subsequent drop in 2017. Additionally, there was a consistent decrease in filings from 2019 to 2022. However, a notable growth trend was observed between 2011 and 2016. On the other hand, the time evolution of the granted patents showed noticeable growth until 2021, with a dip between 2007 and 2009 followed by another slight decrease in 2019 and 2022. In the publication state, the time evolution of the published

TABLE 11.2
Patent Document Count Related to Menthol as a Function of Patent Document Type.

Patent Document Type	Patent Document Count
Patent application	17005
Granted patent	4716
Limited patent	295
Patent of Addition	72
Amended patent	21
Amended application	13
Plant patent	8
Other patent documents	205

Source: The Lens Patent Data Set (until July 2023) [39].

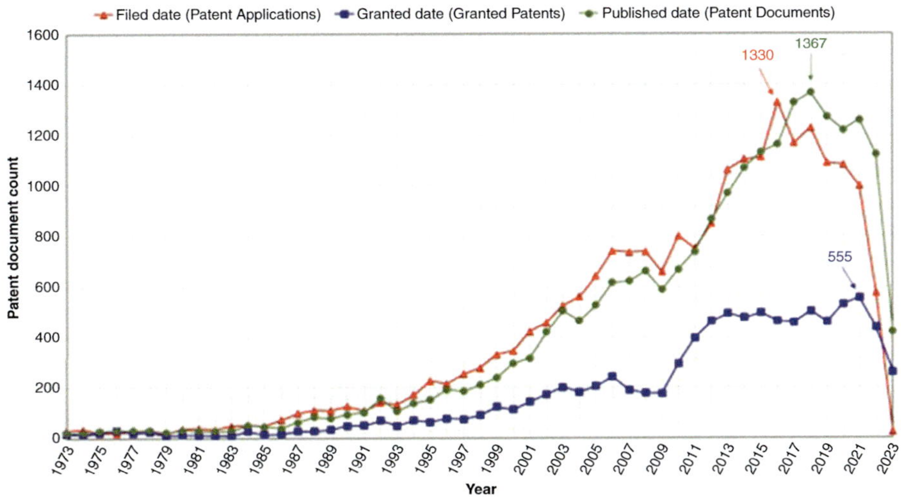

FIGURE 11.2 Patent document count evolution as a function of filed, granted, and published years (data include only the last 50 years of patenting activity in this area). (The Lens Patent Data Set (till July 2023) [39].)

patent documents showed noticeable growth until 2003. However, it showed a noticeable decrease in 2004 and 2009, as well as a dip between 2019 and 2020, before light rebounding in 2021 and finally a dip in 2022. Finally, it should be noted that data in 2023 are nonrepresentative since the search was performed until July 2023.

11.4.3 Inventors and Applicants

An inventor is a creative individual or group recognized for conceiving and creating inventions. They are the creative minds behind the new solution or technology, and they hold intellectual property rights to inventive work [38,40]. On the other hand, an applicant is an individual or an entity seeking legal protection for the invention by filing a patent application or other intellectual property forms. The applicant may be the inventor or a person or entity that is in possession of the invention's rights. The applicant oversees the application procedure and, if approved, assumes ownership of the invention's sole rights for the duration provided by the applicable patent laws [38,41].

Figure 11.3 presents the top ten inventors of menthol-related patent documents. First, inventors from China with waivers of the right to be mentioned as inventors have recorded 97 patent documents. In some cases, the inventor has the possibility of waiving the right to be mentioned as an inventor in the published patent application and the patent specification. A written waiver of their right to be mentioned as inventors is a declaration signed and filed by the inventor with the relevant patent authority [42]. Second, inventor Mishra Munmaya registered 89 patent documents in which the American tobacco company Philip Morris USA Inc. (Richmond, VA, United States) was an applicant. Third, inventor Gao Feng has recorded 58 patent

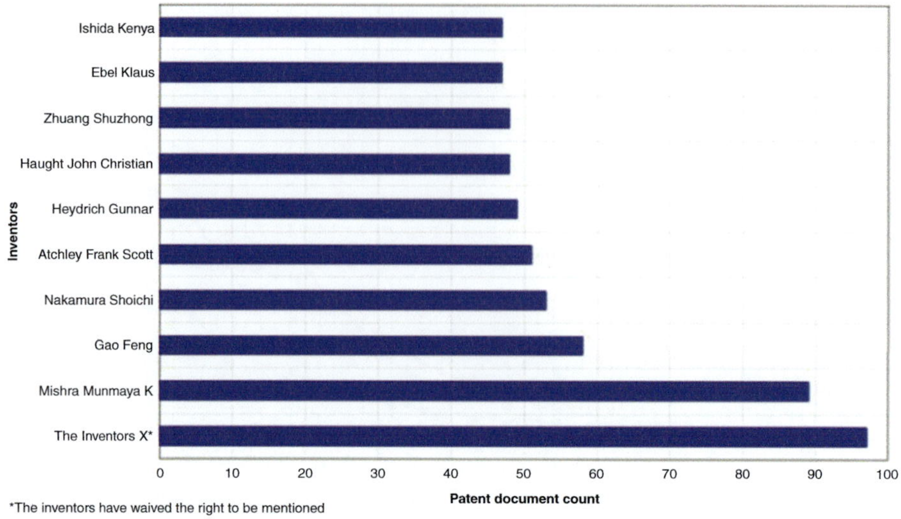

FIGURE 11.3 Top ten inventors of menthol-related patent documents as a function of patent document count. (The Lens Patent Data Set (till July 2023) [39].)

documents for which the American company Altria Client Services LLC (Richmond, VA, United States) was an applicant. It should be noted that both previous applicants are considered subsidiaries of the American tobacco corporation Altria Group (Richmond, VA, United States).

Table 11.3 presents the top ten applicants for menthol-related patent documents and their profiles, as well as the patent document count of each applicant. Regarding this top ten, all applicants are considered organizations (i.e., legal entities). The corporation Procter & Gamble (Cincinnati, OH, United States), as a legal entity, is ranked as the first applicant to record 549 patent documents. Second, Colgate Palmolive Co. (New York, NY, United States), as a legal entity, has recorded 229 patent documents. Third, the company Takasago Perfumery Co., Ltd. (Tokyo, Japan), as a legal entity, has recorded 212 patent documents. Through these top ten applicants, we can confirm the high number of menthol applications, including but not limited to health and wellness as well as industrial applications such as:

- Foods, beverages, and aroma ingredients
- Detergents, soaps, and other toiletries
- Medications and oral hygiene products
- Flavors, fragrances, and cosmetic ingredients
- Fine chemicals
- Tobacco and vaping

11.4.4 Patent Jurisdictions

Patent jurisdictions are certain geographical regions or territories where applicants for or owners of patents may seek or acquire protection for their inventions. Patent

TABLE 11.3
Top Ten Applicants of Menthol-Related Patent Documents as a Function of Patent Document Count.

Applicant (Headquarters)	Profile	Patent Document Count
Procter & Gamble (Cincinnati, OH, United States)	American multinational consumer goods corporation.	549
Colgate Palmolive Co (New York, NY, United States)	American multinational consumer products company.	229
Takasago Perfumery Co LTD (Tokyo, Japan)	Major international producer of flavors and fragrances.	212
BASF SE (Ludwigshafen, Germany)	European multinational company and the largest chemical producer in the world.	202
Lion Corp (Tokyo, Japan)	Japanese multinational manufacturer of detergent, soap, medications, and oral hygiene products and other toiletries.	190
Kao Corp (Tokyo, Japan)	Chemical and cosmetics company.	183
Symrise AG (Holzminden, Germany)	German chemicals company that is a major producer of flavors and fragrances.	171
Japan Tobacco INC (Tokyo, Japan)	Japanese diversified tobacco and vaping company.	155
Warner Lambert Co (Morris Plains, NJ, United States)	American pharmaceutical company.	144
Philip Morris USA INC (Richmond, VA, United States)	American tobacco division of the American tobacco corporation Altria Group.	133

Source: The Lens Patent Data Set (until July 2023) [39].

applications must be submitted to the appropriate national or regional patent office to obtain legal protection within that particular area. Each jurisdiction, whether a nation or a regional body, has its own set of patent laws and regulations. A patent only offers protection within the limits of the country or region in which it is granted [38,43]. Based on their business plans and the market potential for their inventions in various areas, inventors, or applicants may decide to apply for patent protection in many jurisdictions. Through a centralized filing process, international patent applications, such as those submitted under the Patent Cooperation Treaty (PCT), can make it easier to secure protection in several jurisdictions. It is important to remember that each jurisdiction handles patent issues independently and that patent rights and obligations are specific to the jurisdiction in which the patent is granted [44].

The repartition of patent jurisdictions for menthol-related patent documents is presented in Figure 11.4. The top ten jurisdictions represent 96.09% of patenting activity related to menthol and its derivatives. However, the rest of the other jurisdictions include 47 patent offices with 874 patent documents (3.91% per total). Regarding data, China, through the China National Intellectual Property Administration

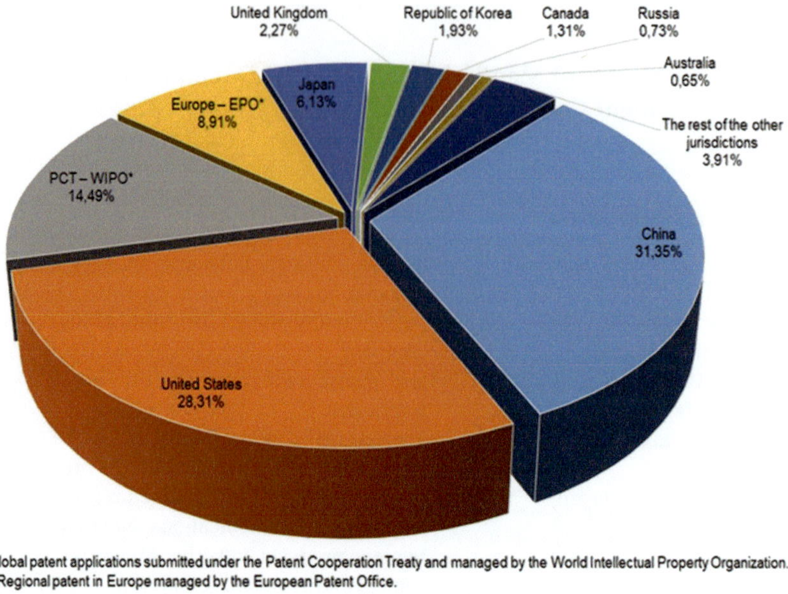

FIGURE 11.4 Repartition (%) of patent jurisdictions of menthol-related patent documents. (The Lens Patent Data Set (till July 2023) [39].)

(CNIPA), encompasses 7003 patent documents, with a higher patent contribution per total of 31.35%. On the other hand, the United States, through the United States Patent and Trademark Office (USPTO), encompasses 6322 patent documents with a higher patent contribution per total of 28.31%, as well as the global system for filing patent applications, known as PCT and administered by the World Intellectual Property Organization (WIPO), which encompasses 3237 patent documents with a patent contribution per total of 14.49%. Finally, the European Patent Office (EPO), through which patent applications are filed regionally (Europe), encompasses 1991 patent documents with a total patent contribution of 8.91%.

11.4.5 International Patent Classifications

A standardized approach used internationally to classify and arrange patents according to their technical subject matter is called International Patent Classification (IPC). The IPC system, which is maintained by the WIPO, provides alphanumeric codes to particular technological sectors and subfields to make it simple to find and analyze pertinent patent documents. It is arranged in the form of a section (e.g., A), class (e.g., A61), subclass (e.g., A61K), group (e.g., A61K31), and subgroup (e.g., A61K31/045). This classification system helps researchers, inventors, and patent examiners comprehend and look for inventions within particular technological fields. The IPC serves a crucial role in fostering effective patent examination, allowing the interchange of technical information and evaluating the uniqueness and

TABLE 11.4
Resulted International Patent Classifications (Top Ten) of Menthol-Related Patent Documents and Their Descriptions (the Description of Each IPC Code is Adapted from the IPC Publication of the WIPO [46]).

IPC code	Description	Patent Document Count
A61K31/045	Medicinal preparations containing organic active ingredients such as hydroxy compounds (e.g., alcohols salts, alcoholates hydroperoxides, etc.).	3577
A61K8/34	Cosmetics or similar toiletry preparations characterized by the composition and containing organic compounds with oxygen (e.g., alcohols).	2450
A61K9/00	Medicinal preparations characterized by special physical form.	2402
A61Q11/00	Preparations for care of the teeth, of the oral cavity, or of dentures (e.g., dentifrices, toothpaste mouth rinses, etc.).	2048
A61K47/10	Medicinal preparations characterized by nonactive ingredients with organic compounds containing oxygen, such as alcohols and phenol salts (e.g., glycerol, polyethylene glycols, poloxamers, alkyl ethers, etc.).	1639
A61P29/00	Specific therapeutic activity of chemical compounds or medicinal preparations such as noncentral analgesic, antipyretic, or anti-inflammatory agents (e.g., antirheumatic agents, nonsteroidal anti-inflammatory drugs, etc.).	1589
A61K9/70	Medicinal preparations characterized by special physical form such as web, sheet, or filament bases.	1349
A61Q19/00	Preparations for care of the skin.	1292
A61K8/49	Cosmetics or similar toiletry preparations characterized by the composition and containing organic compounds such as heterocyclic compounds.	1179
A61K9/06	Medicinal preparations characterized by special physical form such as ointments bases and apparatus for manufacturing them.	1149

creativity of new discoveries. It is regularly updated to take into account technological improvements [45,46].

The resulting top ten international patent classifications for menthol-related patent documents and their descriptions are presented in Table 11.4. Regarding this top ten, there are four groups (i.e., A61K9/00, A61Q11/00, A61P29/00, and A61Q19/00) and six subgroups (i.e., A61K31/045, A61K8/34, A61K47/10, A61K9/70, A61K8/49, and A61K9/06). The most common IPC codes correspond to subclasses A61K and A61Q, which are dedicated to preparations for medical, dental, or toiletry purposes using techniques or methods specifically adapted for bringing pharmacy products into particular physical environments, as well as specific use of cosmeceuticals or comparable toiletries.

11.5 FUTURE TRENDS AND RECOMMENDATIONS

Future innovations in the field of menthol and its derivatives are anticipated, driven by a variety of factors, including shifting consumer tastes, technological advancements, regulatory changes, and sustainability requirements. The growth of natural and organic products is one of the most significant trends. A growing interest in menthol derivatives made from natural, plant-based sources is being seen as customers look for cleaner and more environmentally friendly solutions. This is in line with more general societal tendencies that value products with clear origins and little environmental impact [47,48]. The use of menthol compounds in functional foods and beverages is another interesting trend. The refreshing qualities of menthol and its capacity to improve flavors will be advantageous for these products, which include enhanced water, energy beverages, and those with wellness-focused characteristics. This will satisfy consumer desires for both taste and functional advantages [49–52]. Continuous study in the medical and pharmaceutical fields could lead to the discovery of new uses for menthol and its derivatives. This could result in novel pharmacological treatments, such as enhanced pain management techniques, advanced wound care products, and customized drug delivery systems [53–55]. The use of menthol may increase dramatically in the personal care and cosmetics industries. With the increasing consumer focus on self-care and personal cleaning, its cooling and calming qualities make it a desirable ingredient in skincare, haircare, and dental care products [56,57]. The usage of menthol in aromatherapy and relaxation products is anticipated to increase as awareness of mental health and general wellbeing rises. Menthol has a stimulating aroma and may have therapeutic benefits, making it a potential aid in reducing stress and improving mood [58]. Alternative nicotine delivery systems could be yet another important trend. Innovative menthol-containing solutions for safer nicotine delivery, such as electronic cigarettes with menthol-infused formulations, may become available as a result of the continued push toward reduced-risk nicotine products and tobacco harm reduction [59–61]. Future sustainability is crucial, and the menthol sector may adopt green business strategies. This entails looking into alternative menthol-producing plants using eco-friendly extraction techniques and reducing the environmental impact of the production process [62–65].

To drive the field of menthol and its derivatives forward, scientists and regulatory bodies ought to implement a multifaceted approach that includes research and innovation, safety and efficacy studies, regulation and labeling, collaboration, environmental impact, international cooperation, and adaptation to new trends. First, research and development should investigate a wide range of applications, including medicinal, therapeutic, and industrial applications, while encouraging innovation to unleash untapped potential. Comprehensive safety and efficacy research must be encouraged across applications to ensure strict quality standards while minimizing health concerns. Clear regulatory criteria should be established for a variety of companies, such as pharmaceuticals, food, cosmetics, and vaping products, and should be accompanied by transparent labeling procedures to inform customers. Collaboration between scientific organizations, industry stakeholders, and regulatory bodies is crucial for knowledge sharing, best practices, and addressing emerging challenges.

Environmental considerations are critical, with a focus on sustainable sourcing and environmentally friendly extraction processes for menthol. International cooperation helps to align rules, share research, and promote responsible worldwide use. Finally, remaining flexible to trends and changes in consumer preferences ensures that the sector remains responsive and relevant.

The most important of all these future trends and recommendations is that when an invention related to menthol and its derivatives meets all the specific criteria for patentability, the inventor should not hesitate to patent it. This applies to all menthol-related areas of research, development, and innovation in raw material sourcing, extract preparation processes, methods for extraction, formulations, production technology, and applications. Knowledge that, to be patentable, an invention must be new or novel, involve an inventive step, and be industrially applicable or useful. This will not only protect the invention of the mind from unfair exploitation but also accelerate technological development by rewarding the intellectual creativity of the inventor and, at the same time, by facilitating the spread of new technological knowledge to the public.

11.6 CONCLUSIONS

Based on the country-wide market analysis of menthol-based products, BASF SE held a prominent share of 19.6% with its menthol products, and Archer-Daniels-Midland Company also stood as a global leader with a 15.7% share of the global menthol market in 2019. Moreover, through global market analysis, the significance of menthol is felt in a number of applications, including health and wellness and industrial applications such as medications (e.g., cough drops, nasal sprays, and inhalers), the flavoring and additives sector, oral care products, the food and beverage sector, cigarettes (i.e., electronic nicotine delivery system flavor use), and essential oils for aromatherapy.

Patent analysis for menthol confirmed that 1896 is considered the starting year of the patenting of menthol and its derivatives, as well as the preparation and application of menthol. Through this research study, 22335 patent documents related to menthol were found. Furthermore, China leads the patent race in the menthol sector as the first patenting jurisdiction, with a higher patent contribution per total of 31.35%. On the other hand, the corporation Procter & Gamble leads the patenting process as an applicant with 549 patent documents. Finally, the classification of patents reveals that most inventions are intended for preparations for medical, dental, or toiletry purposes using techniques or methods specifically adapted for bringing pharmacy products into particular physical environments, as well as specific use of cosmeceuticals or comparable toiletries.

The future trends in the area of menthol and its derivatives reflect a combination of evolving consumer preferences, health considerations, technological innovations, and regulatory dynamics, shaping the industry's direction in the years to come. By following the cited recommendations, scientists and regulatory bodies can create a supportive environment that fosters the growth and advancement of the field of menthol and its derivatives while ensuring consumer safety and responsible use.

REFERENCES

[1] Moroccan Office of Industrial and Commercial Property. *Industrial Property Law in Morocco, Article 23, Law No. 17–97 on the protection of industrial property*; OMPIC: Casablanca, Morocco, 2000.

[2] Borges, R.M. Patents and plants: What are the issues? *International Journal of Economic Intelligence* 2013, *5*, 9–23.

[3] Moroccan Office of Industrial and Commercial Property. *Industrial Property Law in Morocco, Article 24, Law No. 17–97 on the protection of industrial property*; OMPIC: Casablanca, Morocco, 2000.

[4] Moroccan Office of Industrial and Commercial Property. *Industrial Property Law in Morocco, Article 25, Law No. 17–97 on the protection of industrial property*; OMPIC: Casablanca, Morocco, 2000.

[5] Committee on Development and Intellectual Property. Patent-Related Flexibilities in the Multilateral Legal Framework and their Legislative Implementation at National and Regional Levels – Part III, p.4, Committee on Development and Intellectual Property (CDIP) Thirteenth Session Geneva, May 19 – 23, 2014, in Australian Patent Office – Guide to Practice and Procedures, 2.9.2.14 (last revised June 3, 2013).

[6] Food and Agriculture Organization of the United Nations. Available online: www.fao.org (accessed on July 25, 2023).

[7] Ravi Srinivas, K. Intellectual property and the common domain in biology: Open source and beyond. *International Social Science Journal* 2006, *188*, 339–355.

[8] China National Intellectual Property Administration. *Chinese Patent Office Examination Guidelines: Chapter 1, Article 25(1)(4)*; CNIPA: Beijing, China, 2010.

[9] Office, J.P. *Examination Guidelines for Inventions in Certain Fields: Biological Inventions. Part VII, Chapter 2, Section 3*; JPO: Tokyo, Japan, 2012.

[10] Knochel, P. Natural products in the chemical industry. *Synthesis* 2015, *47*, 1356–1356, doi:10.1055/s-0034-1378701.

[11] Dylong, D.; Hausoul, P.J.C.; Palkovits, R.; Eisenacher, M. Synthesis of (−)-menthol: Industrial synthesis routes and recent development. *Flavour and Fragrance Journal* 2022, *37*, 195–209, doi:10.1002/ffj.3699.

[12] Wijaya, B.K.; Hardjo, P.H.; Emantoko, S. Menthol from the stem and leaf in-vitro Mentha piperita Linn. *IOP Conference Series: Earth and Environmental Science* 2019, *293*, 012009, doi:10.1088/1755-1315/293/1/012009.

[13] Turner, G.W.; Gershenzon, J.; Croteau, R.B. Development of peltate glandular trichomes of Peppermint1. *Plant Physiology* 2000, *124*, 665–680, doi:10.1104/pp.124.2.665.

[14] Mehmood, M.; Mushtaq, A. Methods for crystal production of natural compounds: A review of recent advancements. *International Journal of Chemical and Biochemical Sciences* 2023, *23*, 258–267.

[15] Cheng, H.; An, X. Cold stimuli, hot topic: An updated review on the biological activity of menthol in relation to inflammation. *Frontiers in Immunology* 2022, *13*, 1023746. doi:10.3389/fimmu.2022.1023746.

[16] Zhao, H.; Ren, S.; Yang, H.; Tang, S.; Guo, C.; Liu, M.; Tao, Q.; Ming, T.; Xu, H. Peppermint essential oil: Its phytochemistry, biological activity, pharmacological effect and application. *Biomedicine & Pharmacotherapy* 2022, *154*, 113559, doi:10.1016/j.biopha.2022.113559.

[17] Jeffries, O.; Waldron, M. The effects of menthol on exercise performance and thermal sensation: A meta-analysis. *Journal of Science and Medicine in Sport* 2019, *22*, 707–715, doi:10.1016/j.jsams.2018.12.002.

[18] Stevens, C.J.; Best, R. Menthol: A fresh ergogenic aid for athletic performance. *Sports Medicine* 2017, *47*, 1035–1042, doi:10.1007/s40279-016-0652-4.

[19] Pereira, E.J.; Sim, L.; Driver, H.S.; Parker, C.M.; Fitzpatrick, M.F. The effect of inhaled menthol on upper airway resistance in humans: A randomized controlled crossover study. *Canadian Respiratory Journal* 2013, *20*, 383019, doi:10.1155/2013/383019.

[20] Fatmawati, F.; Putri Wachyuni, A.; Puriyani Azhary, D. Review of menthol in oral care formulation as freshner and preventing halitosis. *International Journal of Pharmaceutical Research and Applications* 2022, *7*, 1085–1091, doi:10.35629/7781-070210851091.

[21] Vranić, E.; Lačević, A.; Mehmedagić, A.; Uzunović, A. Formulation ingredients for toothpastes and mouthwashes. *Biomolecules and Biomedicine* 2004, *4*, 51–58, doi:10.17305/bjbms.2004.3362.

[22] Silva, H. Current knowledge on the vascular effects of menthol. *Frontiers in Physiology* 2020, 11, 298. doi:10.3389/fphys.2020.00298.

[23] Crespi, E.; Hardesty, J.J.; Nian, Q.; Sinamo, J.; Welding, K.; Cohen, J.E.; David Kennedy, R. Device and liquid characteristics used with sweet, menthol/mint, and tobacco ENDS liquid flavors: The population-based VAPER study. *Addictive Behaviors* 2023, *144*, 107727, doi:10.1016/j.addbeh.2023.107727.

[24] Abdulsalaam, S.R.; Akdo, H.J. Evaluation of addition peppermint oil on the microbial and sensory properties of a soft milk drink (Shenina). *IOP Conference Series: Earth and Environmental Science* 2023, *1213*, 012078, doi:10.1088/1755-1315/1213/1/012078.

[25] de Aquino, D.A.O.; Oliveira, A.S.; Amorim, M.V.P.; Gomes, A.P.B.; Veríssimo, L.M.; Ferrari, M. Thermal behavior of cetylpyridinium hydrochloride and its association with sugar alcohols and flavoring agents: a preformulation study. *Journal of Thermal Analysis and Calorimetry* 2023, *148*(3). doi:10.1007/s10973-023-12320-z.

[26] Nisar, S. The production of water-soluble derivatives of natural menthol for commercial applications: A comprehensive review. *International Journal of Chemical and Biochemical Sciences* 2022, *21*, 335–345.

[27] Ali, B.; Al-Wabel, N.A.; Shams, S.; Ahamad, A.; Khan, S.A.; Anwar, F. Essential oils used in aromatherapy: A systemic review. *Asian Pacific Journal of Tropical Biomedicine* 2015, *5*, 601–611, doi:10.1016/j.apjtb.2015.05.007.

[28] Sowndhararajan, K.; Kim, S. Influence of fragrances on human psychophysiological activity: With special reference to human electroencephalographic response. *Scientia Pharmaceutica* 2016, *84*, 724–751, doi:10.3390/scipharm84040724.

[29] Božović, M.; Pirolli, A.; Ragno, R. Mentha suaveolens Ehrh. (Lamiaceae) essential oil and its main constituent piperitenone oxide: Biological activities and chemistry. *Molecules* 2015, *20*, 8605–8633, doi:10.3390/molecules20058605.

30] Mikaili, P.; Mojaverrostami, S.; Moloudizargari, M.; Aghajanshakeri, S. Pharmacological and therapeutic effects of *Mentha Longifolia* L. and its main constituent, menthol. *Ancient Science of Life* 2013, *33*, 131–138, doi:10.4103/0257-7941.139059.

[31] Ribeiro, B.; Shapira, P. Anticipating governance challenges in synthetic biology: Insights from biosynthetic menthol. *Technological Forecasting and Social Change* 2019, *139*, 311–320, doi:10.1016/j.techfore.2018.11.020.

[32] Wikimedia Commons. Photograph of menthol crystals at room temperature. Available online: https://en.m.wikipedia.org/wiki/File:Menthol_Crystals_close_up.jpg (accessed on October 23, 2023).

[33] Pexels Free Images. Green leaves of a spearmint plant. Available online: www.pexels.com/photo/green-leaves-of-a-spearmint-plant-4599920 (accessed on October 23, 2023).

[34] Transparency Market Research. Menthol Market Report. Available online: www.transparencymarketresearch.com/menthol-market.html (accessed on August 08, 2023).

[35] Helbing, H. Improvements in the preparation and application of menthol. Patent Application: GB189625778A, Great Britain, October 9, 1897.

[36] World Intellectual Property Organization. *Handbook on industrial property information and documentation*; WIPO: Geneva, Switzerland, 2013; p. 43.

[37] Hachimi Alaoui, C.; Fatimi, A. A 20-year patent review and innovation trends on hydrogel-based coatings used for medical device biofabrication. *Journal of Biomaterials Science, Polymer Edition* 2023, *34*, 1255–1273, doi:10.1080/09205063.2022.2161777.

[38] European Patent Office. Espacenet Glossary. Available online: https://worldwide.espacenet.com/patent/help/espacenet-glossary (accessed on August 25, 2023).

[39] Cambia Institute. The Lens Patent Data Set. Available online: www.lens.org (accessed on July 30, 2023).

[40] Fatimi, A. A patent data analysis of the innovation trends in biological control agent formulations. *Recent Advances in Food, Nutrition & Agriculture* 2022, *13*, 59–69, doi:10.2174/2772574X13666220831122154.

[41] Fatimi, A. Exploring the patent landscape and innovation of hydrogel-based bioinks used for 3D bioprinting. *Recent Advances in Drug Delivery and Formulation* 2022, *16*, 145–163, doi:10.2174/2667387816666220429095834.

[42] El Boukhari, R.; Fatimi, A. A review of the patentability of rosemary-derived drugs and bioactive compounds. *Drugs and Drug Candidates* 2023, *2*, 172–188, doi:10.3390/ddc2010011.

[43] Fatimi, A. Seaweed-based biofertilizers: A patent analysis. *Recent patents on biotechnology* 2022, *16*, 144–154, doi:10.2174/1872208316666220128105056.

[44] World Intellectual Property Organization. Summary of the Patent Cooperation Treaty (PCT). Available online: www.wipo.int/treaties/en/registration/pct/summary_pct.html (accessed on July 30, 2023).

[45] World Intellectual Property Organization. *Guide to the International Patent Classification (IPC)*; WIPO: Geneva, Switzerland, 2020; p. 51.

[46] World Intellectual Property Organization. IPC Publication. Available online: www.wipo.int/classifications/ipc/ipcpub (accessed on July 30, 2023).

[47] Ghalili, B.; Goldberg, A.; Borja, J. Stabilized menthol and other volatile compound compositions and methods. Patent Application: US20230226189A1, United States, July 20, 2023

[48] Wang, L.; Wang, X. Water-soluble menthol and preparation method thereof. Patent Application: CN114917140A, China, August 19, 2022.

[49] Ke, J.; Xu, H.; Ma, M.; Cui, F.; Dai, P. Production method of menthol with adjustable bitter taste. Patent Application: CN 115124406 A, China, September 30, 2022.

[50] Yang, J.H.; Park, G.D. Menthol gummy gelly. Patent Application: KR20220089918A, Republic of Korea, December 22, 2020.

[51] Han, C.; Luan, Z. Preparation method and use method of natural menthol spice. Patent Application: CN114621055A, China, June 14, 2022.

[52] Wang, Y.; Luo, J.; Zhang, L.; Wang, Y. Menthol-containing beverage. Patent Application: CN109757636A China, May 17, 2019.

[53] Chen, Y.-H.; Huang, S.-S.; Hung, S.-Y.; Su, H.-H.; Wang, Y.-H.; Chung, H.-Y.; Luo, S.-T.; Chen, C.-J.; Chu, Y.-T.; Macdonald, I.J. The use of menthol for the preparation

of a topical composition to improve neurodegenerative disease and stroke. Patent Application: US20230133657A1, United States, May 4, 2023.

[54] Yang, G.; Zhou, P. Menthol monoterpenoid dimer pilea alcohol B and application thereof in preparation of antitumor drugs. Patent Application: CN115611796A, China, January 17, 2023.

[55] Camacho, S.; Michlig, G.S.; Le Coutre, J.; Markram, H.; Pezzoli, M. Methods of treating or preventing autism disorders using menthol, linalool and/or icine. Patent Application: CN115006399A, China, September 6, 2022.

[56] Kalem, C.; Stuhlmann, D.; Trunet, A.; Brouard, D. Compositions comprising menthol compounds as smoothing agents. Granted Patent: US11696884B2, United States, July 11, 2023.

[57] Choi, S.H.; Park, C.H.; Oh, M.S. A shampoo composition comprising menthol microcapsule. Patent Application: KR20230089726A, Republic of Korea, June 21, 2023

[58] Jang, Y.J.; Gwak, H.M.; Lee, J.H.; Jin, S.J. Cartridge containing menthol and flavoring agent, and aerosol generating system comprising same. Patent Application: EP4197363A1, Europe, June 21, 2023.

[59] Becker, U.; Howell, T.; Xue, L.; Karles, G.D.; Fournier, J.A. Menthol cigarette. Granted Patent: US11700877B2, United States, July 18, 2023.

[60] Han, S.; Chen, H.; Liu, T.; Wu, S.; Hou, H.; Hu, Q. Method for measuring content of menthol in electronic cigarette liquid. Granted Patent: CN103808849B, China, January 7, 2015.

[61] Jiang, W.; Liu, C.; Tang, S.; Zhu, Z.; Si, X.; Li, Z.; Zhang, F.; Wang, H.; Zhou, G.; He, P.; Guan, Y.; Gui, Q.; et al. Synthesis method of menthol amine compound and application of menthol amine compound in cigarettes. Patent Application: CN115872881A, China, March 31, 2023.

[62] Liu, Z.; Shen, Z.; Liu, W. Novel menthol extraction device capable of rapidly realizing condensing and crystallizing. Patent Application: CN114042333A, China, February 15, 2022.

[63] Liu, W. Menthol production crystallization refrigerating chamber. Limited Patent: CN218506872U, China, February 21, 2023.

[64] Liu, W. Efficient drying device for menthol production. Limited Patent: CN218511324U United States, February 21, 2023.

[65] Zhang, G.; Xie, F.; Zhang, Q.; Han, H.; Yang, P.; Lu, X. System and process for removing acetone after crystallization of L-menthol synthesis solvent. Patent Application: CN115779459A, China, March 14, 2023.

12 Technology Transfer in the Field of Biotechnology

Concepts, Actors, Models, and Processes

Tamara Arizbe Virgilio-Virgilio and Alejandro Barragán-Ocaña

12.1 INTRODUCTION

Developing technology involves a continuous process taking place between tangible and intangible elements. Therefore, the foundations of such a development must be found in different information sources, ranging from patents to industrial secrets, which, together with the sum of an organization's capacities and abilities, can be translated into the production of goods and the delivery of services [1,2]. Technology transfer (TT) plays a fundamental role in this context, and all the elements that converge for its adequate operation must be taken into account. Among these are the agent that transfers the technology, the means to execute the transfer, what is transferred, and the final recipient [3]. Due to its complexity, an approach to the TT process requires a solid conceptualization and construction of a robust theoretical framework [4].

Synthetically, TT can be defined as a series of complex stages that start from the dissemination of results emanating from basic science to the implementation of inventions in the market, although it is necessary to consider the interrelationship of all the elements and intermediate actors that emerge during this process [5]. The main purpose of TT is to obtain economic benefits and it can involve research and development (R&D) under academia-company collaboration or other types of agreements, such as the acquisition of technology developed by academia for coping with technological change. TT can also be arranged as a service provided to a third party that cannot carry out certain technical activities due to a lack of specialization or because it chooses not to carry them out [6,7].

The generation of social capital has been observed to produce collective ownership, which can be used to stimulate innovation dynamics, organizational development, employment, and the socialization of knowledge, and intellectual capital [8], where knowledge represents a strategic and intensively used resource

DOI: 10.1201/9781003405665-12

to create value in the contemporary world [9]. To measure the impact and effectiveness of TT, different models integrate aspects such as the achievement of the transfer itself and its impact at various levels, including the market, the economy, politics, scientific progress, human resources training, and, more recently, public value [10].

In this way, the approaches to understanding the theoretical body of TT include a macro perspective, which includes an analysis of the mechanisms that comprise it, the definition of policies, and the methods to measure and assess its effectiveness. At the meso level, the central actor under study is constituted by the institutions that promote the transfer, such as technology transfer offices, business incubators and accelerators, and science parks. Finally, the micro level focuses on case studies and multiple topics such as culture, collaboration, and motivations, among other elements [11]. TT has particular characteristics relevant to its operation. Some of them are the type of technology or knowledge involved, technical capacities, communication among all actors (consultants, intermediaries or facilitators, scientists, and risk investors), and the cultural context [12–15].

In Europe, small, and medium enterprises (SMEs) have been shown to present technological management limitations due to their lack of financial capital and specialized human resources [16]. In this regard, organizations focused on TT and knowledge transfer can play the role of a link between different actors and be an essential source of tangible and intangible assets to support innovation in the companies they serve. However, sometimes the provision of these services is limited, and the theoretical background and tools employed are limited [17]. At the industrial level, biotechnology is attractive due to its intensive use of knowledge since many of these ventures require the participation of academia due to its strategic role in offering technological developments and licensing schemes, in addition to other sources of investment to obtain venture capital, government participation, and the collaboration of corporate partners [18]. This branch of knowledge, which makes intensive use of R&D and creates state-of-the-art technology, is often associated with the foundation of SMEs, for example, spin-offs; therefore, given its multidisciplinary nature, technology management is one of the fundamental elements to adequately exploit biotechnology [19].

Thus, biotechnology creates technologically advanced inventions, often based on the contributions of basic research, which are strategic for the industrial sector [20]. As a result of its importance, sizeable innovation networks have formed around biotechnology, exerting a positive impact on different industries due to collaboration and shared resources [21]. In fact, biotechnology is one of venture capital's most interesting fields of knowledge due to its capacity to deliver innovative solutions. Within this dynamic, academia has been able to launch into the market many start-ups that were created in their facilities [22]. The role of biotechnology has been decisive in the current industrial innovation process, predominantly in economies working on the frontier of knowledge, although more and more developing countries are showing interest [23]. As a result, the different courses of action offered by TT will play a fundamental role in current and future biotechnological applications that will significantly affect the well-being and lifestyles of human populations.

In this way, the research questions addressed by the present study are the following: 1. How can TT be conceptualized in the field of biotechnology?; 2. What are its main actors, models, and processes?; and 3. What is the technological potential that frames the global biotechnology market? To answer these questions, a bibliometric and network analysis was carried out focused on research related to technology transfer and biotechnology to identify the most addressed thematic clusters and topics. Additionally, granted patents were identified to analyze the advances in this field of knowledge. These analyses provided a general vision of biotechnology and its technological transfer status. In addition, the main institutions, countries, publications, and areas of knowledge in which this topic is addressed were identified, and it was possible to outline the most important technological sectors where biotechnology is being developed, the applicants, and the leading countries where the protection of these inventions is being sought.

12.2 ELEMENTS, MODELS, AND TECHNOLOGY TRANSFER PROCESSES

Defining TT using a unique concept is complicated due to its multidimensional and transdisciplinary nature. One of the most frequent approaches is TT oriented toward the novel application of scientific knowledge to address specific problems [24]. The traditional type includes the presence of an offeror who facilitates the transfer of know-how to an assignee who has the capabilities or is willing to develop them to make appropriate use of the technology; other relevant elements include processes, products, or even methodologies [25]. On the other hand, the indirect effects of producing knowledge and technology in developed countries and disseminating those advances have been shown to positively impact their economic advance [26]. There are diverse areas of knowledge where different forms of TT are carried out, such as those oriented to the application of technologies and the use of global-interest knowledge. To cite one case, some fields focus on environmentally friendly solutions and combating climate change [27,28].

Advanced economies lead R&D and have implemented intellectual property protection systems to safeguard their technological developments, and they can use different routes for their diffusion [29]. Furthermore, the concept of proximity, which moves in the geographical and technological directions between the different economic agents, can help facilitate the technology or knowledge transfer process, although it is essential to consider other aspects in order not to hinder this process, for example, the social, financial, normative, and production-related, among others [30]. All these considerations take on a very different dimension in developing economies, and consequently, it is essential to consider the specific context of this group of countries. In this regard, experiences in this technological field highlight that the transfer process from developed to developing countries needs to be more inclusive; in addition, there is a need to recognize the technological efforts made by developing economies and to seize their existing capacities by exporting them to countries around the world [31].

Technology is an intangible asset, often recognized through different forms of intellectual property, including patents or utility models. In fact, value technologies that cannot be classified among the legal options is a complicated matter. However, by definition, the valuation of a technology constitutes a heterogeneous and complex task, which can be based on different approaches that use different approaches, such as technological foresight, organizational competence, social, and environmental and economic impact studies [32]. From an economic point of view, TT involves the diffusion of knowledge associated with different technology groups and difficulty levels. In turn, knowledge is comprised of information, experience, and a learning process that reflects in the construction of tacit or explicit knowledge depending on each user's needs and adoption capabilities [33].

12.3 ACTORS AND DYNAMICS IN TECHNOLOGY TRANSFER

Foreign direct investment (FDI) has traditionally been seen as beneficial by recipient economies because it promotes technology transfer, improves business management, and opens different opportunities for export markets. It also promotes learning through imitation and observation, although its success is affected by the restrictions set by transnational corporations to protect their know-how. There are other obstacles to the benefits of FDI in local companies because the technological gap and the differences in organizational maturity and target markets between the involved actors are often considerable [34]. Thus, TT, international trade, and FDI catalyze the flows and dynamics of knowledge use and dissemination [35].

TT is a way to provide companies with innovative technology, scientific infrastructure, and technical information [36]. The role of universities in the field of TT is relevant for the generation of knowledge that, as well as the protection and promotion of innovation processes, has been associated with the construction of competitive features in different industries. Thus, its main transaction mechanisms in connection with the industrial sector include products developed using sponsored research, patents, consultancies, collaboration, licensing, publications, start-up companies, support by students focused on the research within production activities, and direct contact with researchers [4, 37,38].

Other ways must be addressed to promote a more significant economic impact on a region, for example, science parks or business incubators [39]. Business incubators are associated not only with the promotion of innovation and business activity but can also be a channel for technology and knowledge transfer and establishing connections with the market [40]. Additionally, the discussion concerning the industry-university-government trinomial has pointed out the relevance of technology transfer offices (TTO) and the features that promote their success, namely, organizational maturity, experience, prestige, technology provided by the associated institution, and technological demand in the target region [41].

University TTOs allow for dissemination activities to showcase the innovations and facilitate the compatibility of scientific activity and commercial actions carried out in this context, such as the creation of *spin-offs* and licensing efforts. Nevertheless, certain cases show that commercial effect is not always easy to achieve, and TTOs function instead as an intermediary mechanism for collaboration with the industry and the venture capital investors, which nurtures the university's academic and social activity, the role of faculty is essential in this context, as they can relay their inventions to commercial agents through this mechanism [42–46].

In their regular role, universities promote the disclosure and commercialization of inventions, search for financing and collaboration, and promote intellectual property mechanisms to allow knowledge to be used in the development of organizations and their role in the market through licensing is gaining relevance [47,48]. An interesting case of TT in the university context are innovation centers in science parks, which carry out a series of actions that enable the operation of technology-based companies; among these actions are producing knowledge and ideas, granting licenses, searching for financing and venture capital investors, and protecting inventions under legal terms, as well as conducting technological assessments and designing business plans oriented to business development [49].

Despite their comparable capacities to patent technology, it has been pointed out that American universities are more efficient when it comes to transferring technology than European and even Asian ones, which is why the two latter need to increase their dissemination and linkage efforts and improve their management, public policies, and other features conducive to their improved operation [50,51]. In fact, in specific contexts, patents can serve as an information supply for the construction of TT indicators at the international level [52]. Thus, it is important to emphasize that universities must have the capability to carry out TT, which can be divided into two stages. The first is the search for innovations based on research; this stage involves funding attraction and technology accumulation capacities (intellectual property protection mechanisms are used, and technologies created are marketable). The second stage is aimed at creating value; therefore, it is necessary to engage in external cooperation, understand the technological market, and have the ability to market the final products [53].

12.4 TECHNOLOGY TRANSFER AND BIOTECHNOLOGY

Innovation in the biotechnology area can be a viable mechanism for economic advancement and to address poverty. On the other hand, health problems represent a significant burden that requires investment and infrastructure. In this regard, technology transferred by advanced economies to developing ones can help mitigate health problems and promote the population's well-being [54]. However, despite the relevance of biotechnology, its transfer is controversial, and public perception influences whether it is accepted, as illustrated by the case of genetically modified foods [55], although other areas of biotechnology are accepted; therefore, each case must be approached in a particular way, taking into account the context where these applications are used.

Patenting and licensing are both relevant for TT. In the area of biological sciences, the literature reports on cases in which technology derived from stem cell research was transferred to the industry. The issues around its management are complex, including those related to patenting, regulations, ethics, licensing opportunities, and tough competition in the path to progress [56]. Likewise, there are relevant cases in medical biotechnology, such as the development achieved by Cuba in addressing health problems among its population using vaccines, formulations, and other applications strongly tied to the international context, in which patents have played an essential role for certain technologies [57]. However, there are also institutions, for example, in Spain, that prove the dynamics of biotechnology to be more strongly focused on scientific article production and less on patent production and knowledge transfer [58].

In Universities, agricultural biotechnology patents represent a fundamental input for TTOs and have a central role in the sector [59]. Public financing for agricultural R&D has been a fundamental engine for generating new inventions. However, it is also worth noting that, on the one hand, in specific contexts, public financing conditions have changed and, on the other, its link with the private sector has strengthened, and the presence of TTOs has favored the new technological management schemes in public institutions and the creation of firms based on all this know-how [60]. Unlike other technological sectors, biotechnology is quite close to the university context because it requires highly specialized human capital and innovation to guarantee its competitiveness [61].

Cutting-edge science and technology increase the opportunities for development in a competitive industry such as biotechnology; for instance, there are multiple applications to address not only health and agricultural issues but also other areas. In this context, TT can improve organizational performance and exert positive impacts on society, the economy, and other spheres of action. However, there are obstacles to tackle, especially in developing economies where the scenario is unfavorable [62]. In some cases, genomics and biotechnology have contributed to the solution of health issues in developing countries [63]; biotechnological agriculture has been associated with benefits derived from the use of genetically modified crops, although there is an extensive debate between supporters and detractors, which calls for a regulatory and technical agenda to guarantee its safe use [64]. Consequently, cooperation between developing economies is necessary, as demonstrated by success stories; for example, vaccine formulation has produced benefits not only for the participants but also for other stakeholders [65].

12.5 METHODOLOGICAL APPROACH

Academic indicators describe the progress of knowledge in a particular field, for example, the scientific publications used for drafting patents point at know-how relevant to the biotechnology industry [66]. Industrial biotechnology is an emerging and technologically advanced field in which knowledge is used intensively; therefore, network analysis can reveal the dynamics and characteristics of different actors in this sector [67]. At the same time, patent studies allow us to understand different aspects of management and intellectual property in the field of biotechnology [68,69]. The combination of bibliometric studies and network analysis has been used to address TT-related issues [70,71], but none have focused on TT and biotechnology as objects of study. Thus, the purpose of the present study was to define the concept of TT and its relevant actors, models, and processes; in addition, we sought to explain and contextualize its dimensions and importance in the field of biotechnology and, finally, to explore the technological potential to be transferred in this area of knowledge.

Therefore, we conducted a search in the Scopus and Web of Science (WoS) databases, considered the most important globally [72–74], to carry out a comparative bibliometric study using the data obtained from these high-academic prestige databases on documents related to TT and biotechnology. In the first case, the search criterion within the title, abstract, and keywords was as follows: (TITLE-ABS-KEY ("technology transfer") AND TITLE-ABS-KEY (biotechnology)). With this, 925 documents published from 1985 to 2022 were obtained. In the second case, the criterion was established within the field titled "topic" (includes the search in the title, abstract, author keywords, keywords plus) as: "technology transfer" (Topic) and biotechnology (Topic). In this case, the search retrieved considerably fewer results, for a total of 315 documents from 1986 to 2022 (Figure 12.1).

The first elements of interest obtained from the bibliometric analysis were related to academic document production, depending on their place of origin and the authors' institutional affiliations. These results pointed out the main countries and institutions where TT is being studied in connection with biotechnology. When analyzing the first ten positions in Scopus, it was clear that most of this production was conducted in the United States and Europe, with significant participation from Canada and India. A similar trend could be observed in WoS, except that Japan, which belongs to the Asia region, replaces India. Regarding institutions, organizations from the United

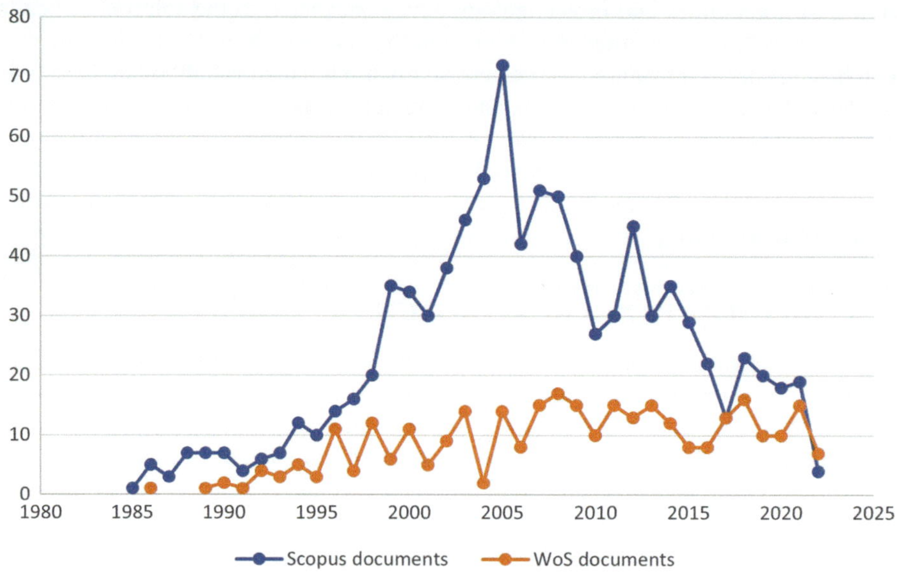

FIGURE 12.1 Scientific documents indexed in Scopus and WoS (1985–2022). (Elaborated by the authors based on Scopus [73] and WoS [74].)

States lead the group; in the search carried out in Scopus, these represent more than 50%, while the rest includes three European institutions and one Canadian. In the search carried out in WoS, all the identified institutions are from the United States, except for Ku Leuven University, a European institution located in the sixth position (see Table 12.1).

Regarding the thematic or research areas, it is interesting to observe that the search in both databases produced documents related to TT studies in biotechnology in technical academic areas, not only in those of a social or economic nature. However, the first position in Scopus corresponds to a technical area, while in WoS, this position corresponds to the economics and business area, although the results show consolidated journals from the technology management area (Table 12.2). When analyzing the most important journals concerned with the object of study in each database, we observed that the measurement given to *Nature Biotechnology* given by the SCImago Journal Rank (SJR) for 2021 was 20.12, and its best quartile is Q1 in Applied Microbiology and Biotechnology. In the case of *Research Policy*, the Journal Citation Reports (JCR) gave it an impact factor of 9.473 for 2021, and next in the journal impact factor ranking, this publication is in the 24th position out of 226 in the management category, which corresponds to quartile Q1. That is, both journals are highly influential in their respective areas.

Subsequently, two co-occurrence analyses were carried out using the data from Scopus and WoS to outline the most important thematic clusters and nodes around these two bibliographic searches. In both cases, the network's conformation followed these three parameters: (1) Type of analysis: co-occurrence; (2) Unit of analysis: all keywords; and (3) Counting method: full counting. A total of 5,917 words were identified in Scopus. Given that this number was too high, we decided to include only the keywords whose frequency or minimum occurrence was ten, which resulted in 208 words. These data resulted in a network comprising six clusters and this was normalized under the Association Strength method. This network comprised 208 nodes and 10,046 links, with a total link strength of 42,919. The same parameters were used for WoS, including the normalization method, but the number of words identified was significantly lower: a total of 1,358. Therefore, we decided to leave only the words having a minimum occurrence of three, which resulted in 143 items. The obtained network distributed these 143 nodes in nine clusters and 2,233 links, with a total link strength of 4,377.

Four clear thematic clusters can be observed in the first network. Cluster 2 (green) addresses topics related to technology management and its actors and medical research, but there are many nodes to do with technology transfer in biotechnology, specifically with patents, licensing, legal matters, and trade. In cluster 1 (red), the two most important nodes correspond to the search terms, that is, technology transfer and biotechnology, which are grouped with other items also related to technology management, in addition to other elements such as intellectual property, public policy, venture capital, information technology, international trade, and investment, as well as other sectors of interest, such as nanotechnology, chemical, biomedical engineering, and sustainable development. For its part, cluster 4 (yellow) focuses on agricultural biotechnology, which includes genetics, biosafety, and developing countries. Cluster

TABLE 12.1
Documents by Country, Territory, Region, and Affiliation in Scopus (a) and WoS (b)

(a)

No.	Country/Territory	Documents	Affiliation	Documents
1	United States	328	University of California, Berkeley	19
2	United Kingdom	96	University of Toronto	13
3	Germany	50	National Institutes of Health NIH	12
4	Canada	46	The George Washington University	9
5	India	33	University of Cambridge	9
6	Switzerland	32	University of Sussex	8
7	France	29	Massachusetts Institute of Technology	8
8	Belgium	26	Stanford University	7
9	Netherlands	26	University of California, Davis	7
10	Italy	24	ETH Zürich	7

(b)

No.	Countries/Regions	Documents	Affiliations	Documents
1	USA	121	University of California System	20
2	England	35	University of California Berkeley	11
3	Germany	20	United States Department of Agriculture USDA	8
4	France	15	Harvard University	7
5	Belgium	14	Cornell University	6
6	The Netherlands	13	Ku Leuven	6
7	Italy	12	State University of New York Suny System	5
8	Canada	11	University of California Davis	5
9	Switzerland	11	University of North Carolina	5
10	Japan	10	University of Sussex	5

Source: Elaborated by the authors based on Scopus [73] and WoS [74].

3 (blue) illustrates specific medical technologies and tools such as vaccines, drug design, bioprocesses, and fermentation, among others (Figure 12.1 and Table 12.3).

Although the second network has a smaller number of terms, it is possible to appreciate a larger number of clusters. The item "technology-transfer," one of our

TABLE 12.2
Documents by Thematic Area and Publication in Scopus (a) and WoS (b)

(a)

No.	Subject area	Documents	Source title	Documents
1	Biochemistry, Genetics and Molecular Biology	273	Nature Biotechnology	60
2	Engineering	239	International Journal of Biotechnology	38
3	Business, Management and Accounting	238	Research Policy	25
4	Chemical Engineering	175	Science	24
5	Immunology and Microbiology	123	Journal of Commercial Biotechnology	23
6	Medicine	118	Journal of Technology Transfer	21
7	Social Sciences	105	Nature	20
8	Economics, Econometrics and Finance	80	International Journal of Technology Management	13
9	Agricultural and Biological Sciences	76	Technological Forecasting and Social Change	10
10	Pharmacology, Toxicology and Pharmaceutics	69	Technovation	10

(b)

No.	Research Areas	Documents	Publication Titles	Documents
1	Business Economics	145	Research Policy	28
2	Engineering	44	Journal of Technology Transfer	16
3	Biotechnology Applied Microbiology	42	Technovation	9
4	Agriculture	27	Chimia	7
5	Environmental Sciences Ecology	20	Technological Forecasting and Social Change	6
6	Public Administration	20	Agricultural Biotechnology in International Development	4
7	Science Technology other Topics	19	Biotechnology in Agriculture	4

(continued)

TABLE 12.2 (Continued)
Documents by Thematic Area and Publication in Scopus (a) and WoS (b)

(b)

No.	Research Areas	Documents	Publication Titles	Documents
8	Operations Research Management Science	15	Small Business Economics	4
9	Chemistry	14	Agricultural Biotechnology in Developing Countries towards Optimizing the Benefits for the Poor	3
10	Government Law	11	American Journal of Agricultural Economics	3

Source: Elaborated by the authors based on Scopus [73] and WoS [74].

search terms, is located in cluster 1; this node is linked to other nodes referring to TT attributes, including items such as knowledge, performance, commercialization, university, firm, scientists, start-ups, university spin-offs, strategies, academic entrepreneurship, networks, alliances, and spillover. The same node is also related to terms associated with innovation sources, for example, biotechnology, industry, research and development, intellectual property, and patents, among others. Clusters 5 (purple) and 6 (light blue) stand out; they include intellectual property topics such as licensing and patents, but also different areas of biotechnological applications such as health and agriculture and other fundamental topics related to the development of the field, for example, education and research (Figure 12.2 and Table 12.4).

In the case of patent analysis, the search was carried out in the Lens [76] database, which integrates data from other databases such as the United States Patent and Trademark Office (USPTO). A structured search in English was carried out within the field "Title, Abstract or Claim" using the term "biotechnology" for a ten-year period (January 1, 2012, to December 31, 2021) and including filters to obtain only granted patents and IPC ratings. Thus, the search criterion was as follows: (Title:(biotechnology) OR (Abstract:(biotechnology) OR Claims (biotechnology)), which produced 5,219 documents grouped into 4,748 simple families and 4,752 extended families, which allowed the principal applicants, jurisdictions, and technological classes to be analyzed.

Concerning where inventions were registered and their applicants, Russian academic institutions rank first, third, sixth, and 10th; the rest of the positions belong to Korean, Chinese, and European organizations and are mainly oriented to the commercial part. Concerning jurisdictions, developed European countries and the United States are the main markets of interest, reflected by the significant number of inventions from different technological fields registered typically in these countries.

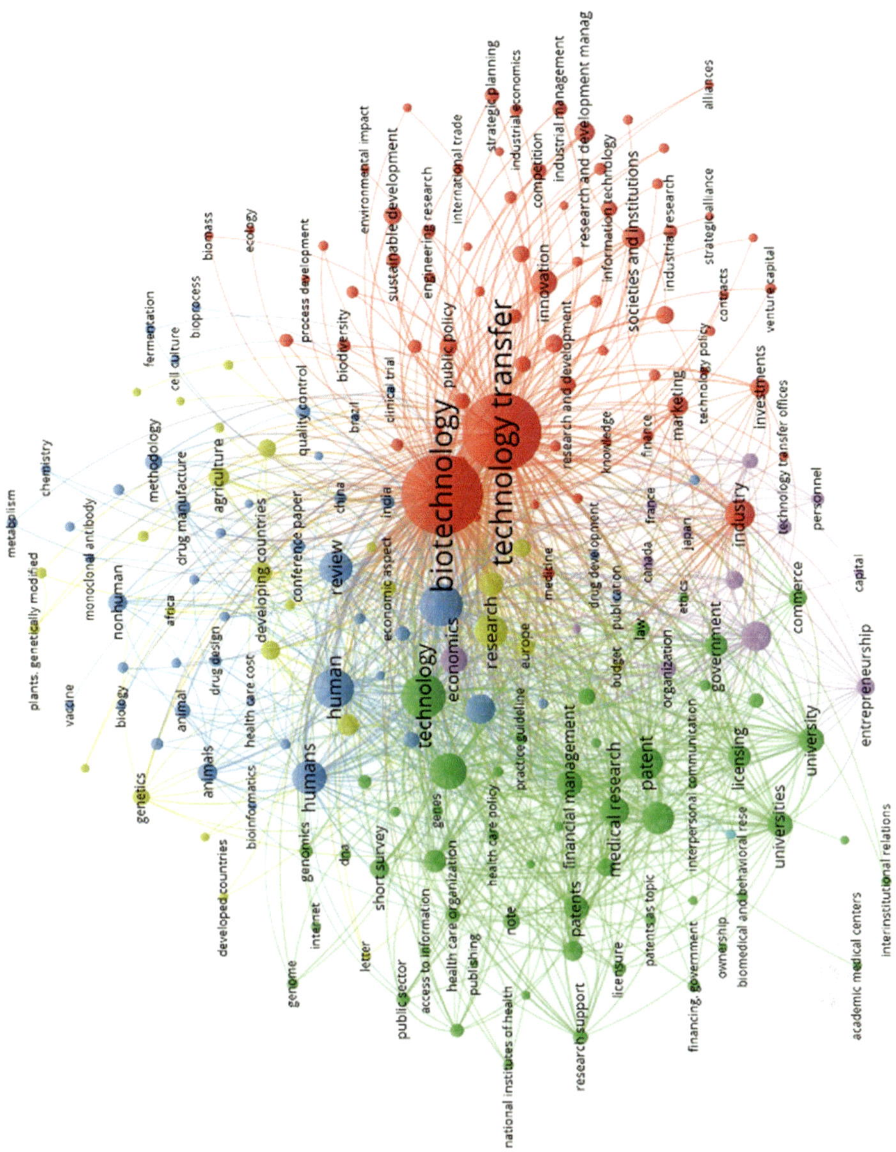

FIGURE 12.2 Network map (co-occurrence) – technology transfer and biotechnology – Scopus. (Elaborated by the authors based on Scopus [73] and VOSviewer [75].)

However, despite the economic importance and technological leadership of these regions, this study shows the cases of two countries considered as developing that stand out for their being in the first positions, that is, Russia and China, and with a smaller number of patents, Moldova and Mexico (Table 12.5).

TABLE 12.3
Network Detail – Technology Transfer and Biotechnology – Scopus

Cluster 1				Cluster 2			
69 items				53 items			
	links	link strength	occurrences		links	link strength	occurrences
alliances	20	43	10	academic medical centers	48	138	10
biodiversity	94	239	23	access to information	83	230	12
biomass	42	78	10	biomedical and behavioral research	78	230	13
biomedical engineering	91	167	18	biomedical research	136	557	37
biotechnology	207	6135	717	commerce	108	422	36
biotechnology industry	67	127	15	diffusion of innovation	73	138	11
chemical industry	55	93	14	DNA	76	176	15
commercialization	99	230	26	editorial	65	130	10
competition	99	249	28	ethics	85	193	14
contracts	61	121	13	financial management	161	1160	75
decision making	94	215	23	financing, government	86	264	17
drug products	103	254	28	forecasting	85	163	11
drug products plants	47	109	18	funding	106	325	24
ecology	51	94	10	genes	88	197	15
economic and social effects	115	321	34	genome	87	260	16
economic development	125	359	30	genomics	114	308	22
economic growth	45	81	10	Government	163	873	59
engineering education	54	97	13	health care organization	122	398	25
engineering research	79	162	22	health care policy	98	254	19
environmental impact	50	93	11	human genome project	63	141	10
environmental protection	74	148	14	information processing	89	188	14

Cluster 3				Cluster 4			
46 items				26 items			
	links	link strength	occurrences		links	link strength	occurrences
animal	109	363	29	Africa	67	166	14
animals	143	552	41	agricultural biotechnology	47	79	12
article	195	2414	209	agriculture	130	507	49
bioinformatics	74	128	10	biosafety	73	131	12
biology	77	155	14	budget	90	184	12
biomedical technology assessment	70	117	11	crop production	63	106	10
biomedicine	124	408	28	crops, agricultural	64	171	11
bioprocess	33	67	11	developed countries	70	169	10
Brazil	72	115	10	developed country	66	151	10
cell culture	61	119	13	developing countries	153	677	55
chemistry	64	142	14	developing country	147	618	42
China	90	218	19	economic aspect	119	365	28
clinical research	89	173	12	Europe	125	401	34
clinical trial	57	128	11	genetic engineering	131	358	37
conference paper	114	271	25	genetics	113	466	33
controlled study	73	109	10	health	92	185	13
drug approval	80	189	12	intellectual property	176	831	63
drug design	99	228	17	international cooperation	144	674	55
drug development	78	147	12	letter	58	121	10
drug industry	173	1325	97	plant biotechnology	73	153	15
drug manufacture	99	298	28	plants, genetically modified	74	202	16

(continued)

TABLE 12.3 (Continued)
Network Detail – Technology Transfer and Biotechnology – Scopus

Cluster 1				Cluster 2			
69 items				53 items			
	links	link strength	occurrences		links	link strength	occurrences
finance	107	233	20	interinstitutional relations	61	173	13
health care	97	211	20	internet	74	149	12
industrial economics	61	135	19	interpersonal communication	70	156	10
industrial management	57	156	26	law	132	405	28
industrial research	89	198	23	legal aspect	138	702	46
industry	182	1214	99	licensing	155	839	59
information dissemination	71	126	11	licensure	99	346	21
information technology	78	195	29	medical research	171	1365	94
innovation	140	517	77	medical technology	106	262	21
intellectual property rights	67	122	15	National Institute of Health (U.S.)	69	219	13
international trade	59	123	16	note	119	505	35
investments	129	500	41	organization and management	155	880	59
knowledge	86	169	14	ownership	65	152	10
knowledge based systems	54	86	11	patent	183	1852	124
knowledge management	46	82	12	patents	149	940	68
laws and legislation	73	147	15	patents as topic	85	231	15
management	86	160	12	practice guideline	105	268	18
marketing	144	530	49	priority journal	191	2022	148
mathematical models	48	79	10	private sector	109	416	22
medicine	84	167	17	public health	92	204	14
mergers and acquisitions	55	103	12	public sector	102	371	22
nanotechnology	108	294	37	publishing	78	172	11
new technologies	47	85	10	research support	101	419	27
patents and inventions	110	318	36	science	133	419	34

Cluster 3			
46 items			
	links	link strength	occurrences
drug marketing	80	189	15
drug research	115	331	25
European Union	74	156	14
fermentation	52	87	13
good manufacturing practice	70	166	15
health care cost	53	109	10
human	196	2275	178
humans	189	1795	133
India	106	267	23
instrumentation	55	132	14
medical education	90	189	13
metabolism	64	160	15
methodology	117	343	33
molecular biology	76	179	18
monoclonal antibody	58	122	12
nonhuman	132	501	44
procedures	81	170	13
publication	96	213	14
quality control	98	238	24
review	193	1436	120
technology, pharmaceutical	88	225	16
translational research	73	126	10
unclassified drug	60	133	12
vaccine	76	175	13

Cluster 4			
26 items			
	links	link strength	occurrences
policy	145	549	37
politics	85	184	12
research	194	1491	118
tissue culture	71	109	10
transgenic plant	68	162	16

(*continued*)

TABLE 12.3 (Continued)
Network Detail – Technology Transfer and Biotechnology – Scopus

Cluster 1				Cluster 2			
69 items				53 items			
	links	link strength	occurrences		links	link strength	occurrences
pharmaceutical industry	79	176	15	scientist	85	192	16
process development	67	116	10	short survey	122	466	42
product development	97	217	26	technology	201	2666	218
productivity	121	283	25	trends	90	223	13
professional aspects	58	99	10	United States	178	1559	116
public policy	131	414	37	Universities	147	1052	71
research and development	111	289	32	University	159	1158	76
research and development management	95	346	48	university hospital	54	140	10
risk assessment	96	206	21				
risk management	56	106	12				
science and technology	75	137	14				
social aspects	66	130	17				
societies and institutions	126	511	57				
South Africa	67	121	11				
strategic alliance	48	91	11				
strategic planning	58	159	25				
sustainable development	118	352	36				
technological change	49	74	10				
technological development	86	183	19				
technological forecasting	92	171	18				
technology policy	52	89	11				
technology transfer	207	5690	675				
technology transfer offices	53	90	11				
venture capital	51	105	15				

	Cluster 3			Cluster 4		
	46 items			26 items		
	links	link strength	occurrences	links	link strength	occurrences
vaccines	82	180	13			

(continued)

TABLE 12.3 (Continued)
Network Detail – Technology Transfer and Biotechnology – Scopus

Cluster 5				Cluster 6			
13 items				1 items			
	links	link strength	occurrences		links	link strength	occurrences
Canada	86	207	20	cooperative behavior	75	176	12
capital	80	199	14				
commercial phenomena	181	1365	98				
economics	166	1051	81				
education	135	453	38				
entrepreneurship	106	466	39				
France	77	137	15				
investment	116	448	31				
Japan	71	123	14				
leadership	69	157	11				
organization	136	464	33				
personnel	105	304	21				
United Kingdom	102	327	26				

Source: Elaborated by the authors based on Scopus [73] and VOSviewer [75].

Concerning technological sectors as defined by the International Patent Classification (IPC), the two technological classes with the highest number of patents granted belong to C12 N1/20, which deals with culture media (bacteria) and is, in turn, a subclass of C12 1/00, focused on microbiology and the preparation of medicinal compositions that can integrate antigens or bacterial antibodies, among other technical elements, and second, with C12Q 1/68, which refers to inventions based on nucleic acids and is a subcategory of C12Q 1/00, which considers enzymes and microorganisms with an approach to composition measurement and preparation. The rest of the categories also have a strong focus on medical applications, highlighting the strong dynamics of biotechnology within this sector. Inventions oriented to the use of genetic engineering in topics as controversial as transgenic crops are also present (Table 12.6).

Data from the 1,000 most relevant patent documents obtained from Lens was downloaded, and a network based on text data was created under the following criteria: 1. The terms were extracted from the title and the abstract; 2. The method used was full counting; 3. A minimum occurrence of ten was set, which produced 249 items out of 8,073; and 4. Of these items, 60% were selected considering their relevance. The network map was finally comprised of 149 nodes distributed in seven clusters and 2,498 links, with a total link strength of 22,695. The network shows profuse data confirming the importance of biotechnology in pharmacy. Cluster 2 (green) is characteristic because it has nodes associated with high occurrences, such as antibody (193),

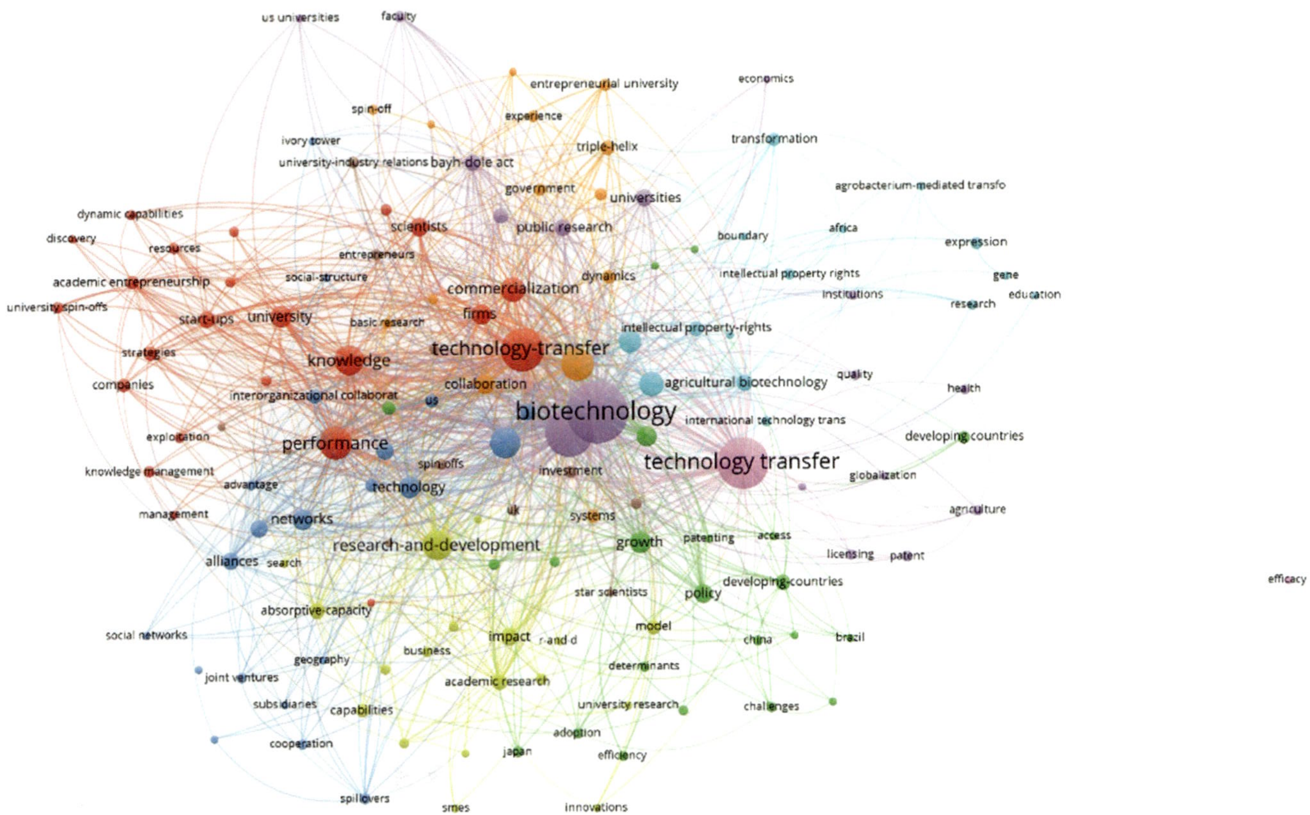

FIGURE 12.3 Network map (co-occurrence) – technology transfer and biotechnology – WoS. (Elaborated by the authors based on Web of Science [74] and VOSviewer [75].)

TABLE 12.4
Network Detail – Technology Transfer and Biotechnology – WoS

Cluster 1				Cluster 2			
16 items				13 items			
	links	link strength	occurrences		links	link strength	occurrences
academic entrepreneurship	30	48	7	access	16	18	3
commercialization	73	177	23	adoption	14	19	4
companies	32	50	7	Brazil	16	20	3
creation	27	34	5	challenges	22	28	4
discovery	19	24	3	China	18	20	4
dynamic capabilities	24	33	4	determinants	18	21	4
exploitation	25	33	4	developing countries	14	20	6
firm	21	29	4	developing-countries	32	49	9
firms	58	125	16	efficiency	14	17	4
innovation policy	24	34	4	entrepreneurship	60	123	16
knowledge	83	230	34	evolution	30	40	7
knowledge management	22	29	5	foreign direct-investment	23	26	4
management	20	25	3	growth	55	95	17
performance	102	311	40	institutional change	24	29	4
perspective	17	17	3	Japan	20	26	5
resources	25	43	5	national systems	10	11	3
scientists	56	113	13	patenting	22	28	3
start-ups	42	78	10	policy	42	75	13
strategies	36	63	8	productivity	23	33	5
technology-transfer	116	446	69	social network analysis	15	17	3
technology-transfer offices	27	33	4	technologies	13	16	3
university	53	120	18	university inventions	18	23	3
university spin-offs	21	34	5				

Cluster 3				Cluster 4			
12 items				12 items			
	links	link strength	occurrences		links	link strength	occurrences
advantage	26	35	4	absorptive-capacity	31	57	9
alliances	37	66	11	academic research	42	69	9
biotechnology industry	47	76	11	biotechnology firms	19	20	3
complementary assets	24	27	3	business	24	28	4
cooperation	20	27	4	capabilities	35	44	7
empirical-analysis	39	50	7	complementarity	22	24	3
geography	19	26	3	impact	52	81	11
industry	79	225	35	innovations	11	15	3
interorganizational collaboration	36	60	8	knowledge spillovers	24	33	4
ivory tower	25	30	3	model	32	50	8
joint ventures	23	29	4	open innovation	18	24	4
knowledge transfer	41	69	11	product development	27	30	5
networks	50	113	16	r-and-d	14	16	3
social networks	20	27	3	research-and-development	88	214	33
social-structure	17	21	3	search	21	25	3
spillovers	29	42	5	smes	16	18	3
strategic alliances	17	17	3	university research	19	22	3
subsidiaries	21	23	3	university-industry collaboration	24	31	4
technology	53	106	17				
united-states	36	49	9				
us	30	51	7				

(continued)

TABLE 12.4 (Continued)
Network Detail – Technology Transfer and Biotechnology – WoS

Cluster 5				Cluster 6			
12 items				11 items			
	links	link strength	occurrences		links	link strength	occurrences
agriculture	8	14	4	Africa	13	13	3
Bayh-Dole act	43	77	10	agricultural biotechnology	29	47	10
biotechnology	135	725	140	agrobacterium-mediated transformation	5	5	3
economic development	11	12	3	boundary	16	20	3
economics	15	20	3	education	7	8	3
faculty	23	37	5	expression	13	16	6
globalization	14	19	3	gene	8	9	3
health	10	13	4	indicators	17	19	3
innovation	132	552	87	intellectual property	45	103	22
institutions	17	23	4	intellectual property rights	16	17	5
licensing	15	18	4	intellectual property-rights	31	40	8
life sciences	43	73	9	international technology transfer	15	16	4
patent	15	20	5	patents	58	112	20
public research	42	77	10	research	15	18	4
quality	8	10	4	transformation	24	36	7
universities	44	78	11				
us universities	14	23	3				

Cluster 9			
items			
	links	link strength	occurrences
efficacy	1	1	3
technology transfer	117	417	97

Source: Elaborated by the authors based on Web of Science [74] and VOSviewer [75].

Cluster 7				Cluster 8			
10 items				9 items			
	links	link strength	occurrences		links	link strength	occurrences
basic research	25	33	4	entrepreneurs	34	46	5
collaboration	57	108	14	incentives	27	34	5
dynamics	31	42	7	integration	18	26	3
entrepreneurial university	33	52	6	investment	31	47	6
experience	32	39	5	organization	22	24	3
framework	26	32	4	spin-offs	28	34	5
government	36	58	6	star scientists	18	22	3
institutional theory	23	30	3	technology-based firms	21	26	3
organizations	32	42	6	UK	25	31	4
science	90	264	39	university-industry relations	26	40	5
spin-off	21	24	4	venture capital	24	29	4
systems	27	38	7				
triple-helix	39	68	8				
university-industry interaction	21	24	3				

TABLE 12.5
Main Applicants and Jurisdictions (Biotechnology)

No.	Applicant	Patents	Jurisdiction	Patents
1	Federalnoe Gosudarstvennoe Byudzhetnoe Obrazovatelnoe Uchrezhdenie Vysshego Obrazovaniya Kubanskij G	46	Russia	3904
2	Cj Cheiljedang Corp	43	United States	419
3	Federalnoe Gosudarstvennoe Byudzhetnoe Uchrezhdenie Natsionalnyj Meditsinskij Issledovatelskij Tsent	39	China	234
4	Ambrx INC	37	European Patents	167
5	Hoffmann La Roche	36	Eurasian Patents	97
6	Fed Byudzhetnoe Uchrezhdenie Nauki Gosudarstvennyj Nauchnyj Tsentr Virusologii I Biotekhnologii Vekt	34	Moldova	92
7	Genentech INC	34	Canada	75
8	Gnosis Spa	33	Australia	73
9	Inst De Microbiologie Si Biotehnologie Al Academiei De Stiinte a Moldovei	33	Korea, Republic of	40
10	Federalnoe Gosudarstvennoe Byudzhetnoe Obrazovatelnoe Uchrezhdenie Vysshego Obrazovaniya Moskovskij	32	Mexico	28

Source: Elaborated by the authors based on Lens [76].

treatment (158), disease (115), and virus (76). Agricultural issues appear in clusters 4 (lemon yellow) and 5 (purple), as well as the portion of the food industry that has traditionally made use of biotechnology (cluster 3, blue) (Figure 12.3 and Table 12.7).

12.6 CONCLUSION

Technology transfer has a wide range of options to promote the use of technological solutions emanating from the field of biotechnology. Patents continue to play a fundamental role in the protection of intellectual property rights. There are numerous alternatives to materialize and commercialize the different cutting-edge inventions produced by biotechnology; among these alternatives are spin-offs, start-ups, and spillovers. There are traditional TT mechanisms that can not only be used in the context of biotechnological development but in multiple other industries such as licensing, technological services, or consultancies. It is essential to understand how TT is being carried out within this sector to optimize its mechanisms and create alternative solutions to benefit the industry and society and, consequently, generate opportunities for the produced knowledge to reach the market and the final stakeholders.

The results of the bibliometric analyses show that developed economies and their institutions dominate the production of documents related to TT and biotechnology. This circumstance becomes apparent when analyzing the top ten positions, where only

TABLE 12.6
Main Technological Sectors According to the IPC Classification (Biotechnology)

No.	Code	Description	Patents
1	C12N1/20	"C12N 1/00 Microorganisms, e.g. protozoa; Compositions thereof (medicinal preparations containing material from protozoa, bacteria or viruses A61K 35/66, from algae A61K 36/02, from fungi A61K 36/06; preparing medicinal bacterial antigen or antibody compositions, e.g. bacterial vaccines, A61K 39/00); Processes of propagating, maintaining or preserving microorganisms or compositions thereof; Processes of preparing or isolating a composition containing a microorganism; Culture media therefor … C12N 1/20 Bacteria; Culture media therefor…".	494
2	C12Q1/68	"C12Q 1/00 Measuring or testing processes involving enzymes, nucleic acids or microorganisms (measuring or testing apparatus with condition measuring or sensing means, e.g. colony counters, C12M 1/34); Compositions therefor; Processes of preparing such compositions … C12Q 1/68 involving nucleic acids…".	483
3	A61P35/00	"Antineoplastic agents".	298
4	C12N15/63	"C12N 15/00 Mutation or genetic engineering; DNA or RNA concerning genetic engineering, vectors, e.g. plasmids, or their isolation, preparation or purification; Use of hosts therefor (mutants or genetically engineered microorganisms C12N 1/00, C12N 5/00, C12N 7/00; new plants A01H; plant reproduction by tissue culture techniques A01H 4/00; new animals A01K 67/00; use of medicinal preparations containing genetic material which is inserted into cells of the living body to treat genetic diseases, gene therapy A61K 48/00; peptides in general C07K) … C12N 15/63 Introduction of foreign genetic material using vectors; Vectors; Use of hosts therefor; Regulation of expression…".	286
5	A61K39/ 395	"A61K 39/00 Medicinal preparations containing antigens or antibodies (materials for immunoassay G01N 33/53) … A61K 39/395 Antibodies (agglutinins A61K 38/36); Immunoglobulins; Immune serum, e.g. antilymphocytic serum".	232
6	C12N5/10	C12N 5/00 Undifferentiated human, animal or plant cells, e.g. cell lines; Tissues; Cultivation or maintenance thereof; Culture media therefor (plant reproduction by tissue culture techniques A01H 4/00) … C12N 5/10 Cells modified by introduction of foreign genetic material, e.g. virus-transformed cells…".	191
7	C12N1/21	"Modified by introduction of foreign genetic material…".	188
8	C07K19/00	"Hybrid peptides (hybrid immunoglobulins composed solely of immunoglobulins C07K 16/46)…".	176
9	C12N15/00	"C12N 15/00 Mutation or genetic engineering; DNA or RNA concerning genetic engineering, vectors, e.g. plasmids, or their isolation, preparation or purification; Use of hosts therefor (mutants or genetically engineered microorganisms C12N 1/00, C12N 5/00, C12N 7/00; new plants A01H; plant reproduction by tissue culture techniques A01H 4/00; new animals A01K 67/00; use of medicinal preparations containing genetic material which is inserted into cells of the living body to treat genetic diseases, gene therapy A61K 48/00; peptides in general C07K) …".	171
10	C12N7/00	"Viruses, e.g. bacteriophages; Compositions thereof; Preparation or purification thereof (medicinal preparations containing viruses A61K 35/76; preparing medicinal viral antigen or antibody compositions, e.g. virus vaccines, A61K 39/00)…".	164

Source: Elaborated by the authors based on Lens [76] and WIPO [77].

TABLE 12.7
Network Detail – Network Map (Co-occurrence) – Biotechnology

Cluster 1				Cluster 2			
Items				Items			
	Links	Total link strength	Occurrences		Links	Total link strength	Occurrences
active component	14	40	11	amdb	4	93	12
addition	65	289	36	antibody	63	942	193
biomass	40	271	43	antigen binding fragment	19	93	13
bioreactor	16	100	17	bind	22	74	10
biotechnological method	22	66	10	biotechnological derivative	9	75	11
biotechnological process	31	195	36	cancer	37	188	37
biotechnological production	44	203	58	cell line	15	125	23
chitosan	18	74	13	conjugate	30	191	28
concentration	68	446	59	diagnostic	38	173	26
condition	56	416	46	disease	70	776	115
container	23	292	33	disorder	29	254	39
controller	11	102	15	dpr	5	120	15
cultivation	58	321	55	fragment	47	218	44
culture	71	595	68	immune response	31	116	17
day	35	166	19	immunology	36	219	31
diaminobutane	8	280	10	kit	49	225	28
distilled water	25	95	13	medicine effect	28	60	18
epidaunorubicin	6	53	11	molecular biology	26	88	17
equipment	19	53	10	monoclonal antibody	29	247	32
fermentation	44	230	23	patient	38	145	22
hour	39	103	11	pharmaceutical composition	44	447	60
lactic acid	21	179	14	pharmaceutical preparation	16	80	15
liquid	50	308	43	plasmid	22	73	14
medium	78	700	106	prevention	44	436	54

Cluster 3				Cluster 4			
Items				Items			
	Links	Total link strength	Occurrences		Links	Total link strength	Occurrences
ability	43	161	29	agriculture substance	45	339	22
asm	13	91	10	escherichia coli	30	792	46
enzyme preparation	15	120	10	expression	62	759	104
food	47	161	16	gene	72	1072	134
food industry	37	149	21	gene therapeutic dna vector	17	925	36
glucide	11	85	10	gene therapeutic DNA vector vtvaf17	14	320	10
industrial microorganisms	24	103	19	genetic engineering	34	290	28
institute	28	265	30	human	42	413	29
lactic acid bacterium	11	56	13	industrial production	29	218	11
lipid	25	94	16	l lysine	3	10	18
local yeast strain	15	93	13	level	42	390	26
microbiology	70	526	74	seq id no	19	697	25
milk	26	132	21	target gene	17	625	21
national collection	27	282	37	therapy	43	327	40
nonpathogenic microorganism	17	112	16	vector	58	938	69
nonpathogenic microorganisms	15	147	18				
number cnmn y	18	117	15				
oil	42	439	64				
oil product	16	159	24				
pathogenic microorganism	23	146	12				
pharmaceutical industry	37	97	12				
producer	57	389	64				
respect	34	114	14				
Russian national collection	24	106	21				

(continued)

TABLE 12.7 (Continued)
Network Detail – Network Map (Co-occurrence) – Biotechnology

Cluster 1				Cluster 2			
Items				Items			
	Links	Total link strength	Occurrences		Links	Total link strength	Occurrences
nutrient medium	56	341	46	recombinant strain	23	111	12
phase	28	116	20	rheumatoid arthritis	9	89	10
process	81	1047	136	Russian collection	20	72	11
purification	48	290	27	Sars Cov	23	346	30
ratio	56	264	39	subject	31	143	20
removal	27	185	14	therapeutic agent	19	57	10
separation	46	402	32	treatment	85	991	158
set	17	48	11	tumor	28	106	13
solution	57	479	53	virology	31	217	21
stage	37	104	13	virus	42	502	76
step	40	374	35				
temperature	60	396	53				
unit	38	215	27				
water	46	253	40				

Cluster 5				Cluster 6			
Items				Items			
	Links	Total link strength	Occurrences		Links	Total link strength	Occurrences
amount	70	475	49	area	50	370	30
biotechnological plant	16	338	12	array	9	172	14
blast furnace	16	276	10	detection	47	231	35
chemical	25	109	14	device	70	1011	94
crude steel	16	350	14	formation	41	361	24
gas	42	950	50	identification	30	150	19
gase	24	222	10	interest	22	64	13
pig iron	14	393	15	mean	56	388	34
plant	71	1350	116	peptidoglycan	4	32	11
plant complex	14	387	16	reduction	43	236	23
power generating plant	14	553	20	sample	51	326	62
sonication plate	5	90	10				
system	81	1567	179				
tube	8	83	15				

Source: Elaborated by the authors based on Lens [76] and VOSviewer [76].

Cluster 3				Cluster 4		
Items				Items		
	Links	Total link strength	Occurrences	Links	Total link strength	Occurrences
saccharomyces cerevisiae yeast strain	11	94	16			
soil	28	178	29			
source	38	339	49			
state collection	25	226	15			
strain	104	2251	401			
surface	43	155	31			
vkpm	21	102	14			
xylanase	19	146	17			
yeast	44	202	31			

Cluster 7			
Items			
	Links	Total link strength	Occurrences
chondroitin	14	207	42
osteoarthritis	10	163	17
position	14	178	23
sulphate	23	223	26

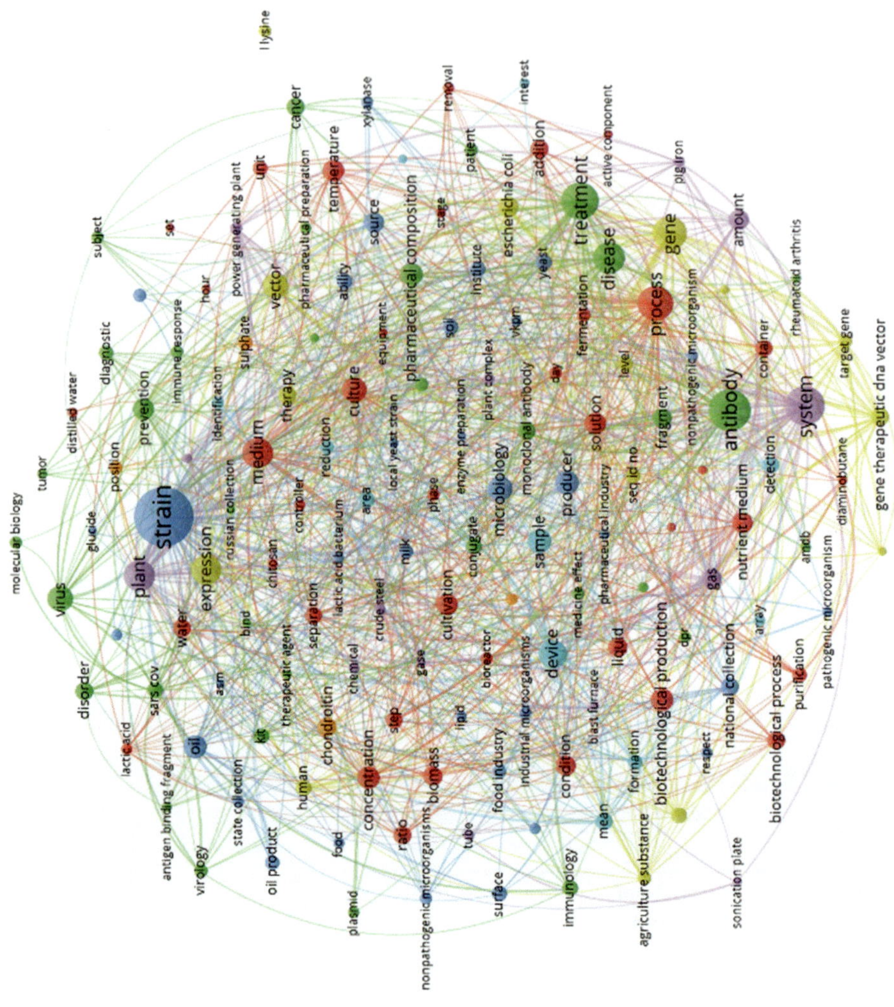

FIGURE 12.4 Network map (co-occurrence) – biotechnology. (Elaborated by the authors based on Lens [76] and VOSviewer [75].)

India appears (fifth position in Scopus) among developing economies, highlighting the need for these to participate more actively, so their context and areas of opportunity are revealed. On the other hand, this topic is increasingly attracting attention in strictly technical areas and journals. For example, in the scenario analyzed in Scopus, *Nature Biotechnology* has published the most significant number of documents related to TT and biotechnology, and *Research Policy*, a journal focused on technology and innovation issues from an economic and social perspective, appears until the third position. In contrast, the results from WoS place this same journal in the first position, and technical journals begin to show until the fourth position.

Concerning the network analysis, even though the bibliographic information obtained from Scopus produced a greater number of nodes and occurrences compared to those obtained using WoS, both analyses reveal important topics associated with TT and biotechnology, such as the actors, tools, and mechanisms used during the transfer process, and there are biotechnological R&D areas that have the potential to strengthen the theoretical and conceptual framework of this field of knowledge. Additionally, although there are multiple sources for a TT exercise, the patent and network analyses revealed the great potential of biotechnology.

These analyses showed the different geographical areas where the multiple biotechnological inventions are being protected, their main applicants, and most importantly, all the technology sectors and relevant issues that are making a difference in the way in which this technological field is evolving and in addition, they contributed to explain the accelerated evolution that the field has experienced, placing it in the frontier of science. Although biotechnology has crucial scientific and technological challenges to meet, which must be addressed hand in hand with multiple disciplines to attain technical and economic feasibility, especially in terms of biosafety, the field must include a social and ethical approach that prioritizes people over aspects strictly concerning income and the market, where the academy must play a fundamental role. The present challenge is to find new, more dynamic, and specific forms of TT catered to each case and each context. Future studies should be focused on increasing the body of case studies to consolidate the empirical knowledge about TT and biotechnology.

ACKNOWLEDGMENTS

We wish to acknowledge the support provided by the National Polytechnic Institute (Instituto Politécnico Nacional) and the Secretariat for Research and Postgraduate Studies, grants number 20220729, 20230213 and 20240867.

REFERENCES

[1] Singh, H. B., Jha, A., and Keswani, C. 2016. Intellectual Property Issues in Biotechnology. CABI, Wallingford, U.K. p. 304.

[2] Tihany, L., and A. Roath. 2002. Technology transfer and institutional development in Central and Eastern Europe. *Journal of World Business* 37, no. 3:188–198. doi: 10.1016/S1090-9516(02)00077-9.

[3] Johnston, K, Kennedy, C., Murdoch, I., Taylor, P., and C. Cook. 2004. The cost-effectiveness of technology transfer using telemedicine. *Health Policy and Planning* 19, no. 5:302–339. doi:10.1093/heapol/czh035.

[4] Póvoa, L., and M. Rapini. 2010. Technology transfer from universities and public research institutes to firms in Brazil: what is transferred and how the transfer is carried out. *Science and Public Policy* 37, no. 2:147–159. doi:10.3152/030234210X496619.

[5] Meseri, O., and S. Maital. 2001. A survey analysis of university-technology transfer in Israel: evaluation of projects and determinants of success. *Journal of Technology Transfer* 26:115–125. doi: 10.1023/A:1007844530539.

[6] Lee, S., Kim, W., Kim, Y. M., and K. J. Oh. 2012. Using AHP to determine intangible priority factors for technology transfer adoption. *Expert Systems with Applications* 39, no. 7:6388–6395. doi: 10.1016/j.eswa.2011.12.030.

[7] Audretsch, D. B., Lehmann, E. E., and M. Wright. 2014. Technology transfer in a global economy. *Journal of Technology Transfer* 39, no. 3:301–312. doi: 10.1007/s10961-012-9283-6.

[8] Al-Tabbaa, O, and S. Ankrah. 2016. Social capital to facilitate 'engineered' university–industry collaboration for technology transfer: a dynamic perspective. *Technological Forecasting and Social Change* 104:1–15. doi: 10.1016/j.techfore.2015.11.027.

[9] Del Giudice, M. E., Carayannis, G., and V. Maggioni. 2017. Global knowledge intensive enterprises and international technology transfer: emerging perspectives from a quadruple helix environment. *Journal of Technology Transfer* 42, no. 2:229–235. doi: 10.1007/s10961-016-9496-1.

[10] Bozeman, B., Rimes, H., and J. Youtie. 2015. The evolving state-of-the-art in technology transfer research: revisiting the contingent effectiveness model. *Research Policy* 44, no. 1:34–49. doi: 10.1016/j.respol.2014.06.008.

[11] Cunningham, J. A., and P. O'Reilly. 2018. Macro, meso and micro perspectives of technology transfer. *Journal of Technology Transfer* 43, no. 3:545–557. doi:10.1007/s10961-018-9658-4.

[12] Daghfous, A. 2004a. Organizational learning, knowledge and technology transfer: a case study. *The Learning Organization* 11, no. 1:67–83. doi: 10.1108/09696470410515733.

[13] Daghfous, A. 2004b. An empirical investigation of the roles of prior knowledge and learning activities in technology transfer. *Technovation* 24, 12:939–953. doi: 10.1016/S0166-4972(03)00059-2.

[14] Cunningham, J. A., Menter, M., andC. Young. 2017. A review of qualitative case methods trends and themes used in technology transfer research. *Journal of Technology Transfer* 42, no. 4:923–956. doi:10.1007/s10961-016-9491-6.

[15] Roxas, S., Piroli, G., andM. Sorrentino. 2011. Efficiency and evaluation analysis of a network of technology transfer brokers. *Technology Analysis & Strategic Management* 23, no. 1: 7–24. doi:10.1080/09537325.2011.537085.

[16] Buratti, N., and L. Penco. 2001. Assisted technology transfer to SMEs: lessons from an exemplary case. *Technovation* 21, no. 1:35–43. doi:10.1016/S0166-4972(00)00015-8.

[17] Landry, R., Amara, N., Cloutier, J. S., and N. Halilem. 2013. Technology transfer organizations: services and business models. *Technovation* 33, no. 12:431–449. doi:10.1016/j.technovation.2013.09.008.

[18] Cooke, P. 2002. Regional innovation systems: general findings and some new evidence from biotechnology clusters. *Journal of Technology Transfer* 27, no. 1:133–145. doi: 10.1023/A:1013160923450.

[19] Meoli, M., Paleari, S., and S. Vismara. 2013. Completing the technology transfer process: M&As of science-based IPOs. *Small Business Economics* 40, no. 2:227–248. doi: 10.1007/s11187-012-9416-1.

[20] Subramanian, A. M., and P. H. Soh. 2010. An empirical examination of the science-technology relationship in the biotechnology industry. *Journal of Engineering and Technology Management* 27, no. 3–4:160–171. doi.org/10.1016/j.jengtecman.2010.06.003 4.

[21] Rampersad, G., Quester, P., and Troshani I. 2010. Managing innovation networks: exploratory evidence from ICT, biotechnology and nanotechnology networks. *Industrial Marketing Management* 39, no. 5:793–805. doi.org/10.1016/j.indmarman.2009.07.002.

[22] Fetterhoff, T. J., and D. Voelkel. 2006. Managing open innovation in biotechnology. *Research-Technology Management* 49, no. 3:14–18. doi: 10.1080/08956308.2006.11657373.

[23] Mendes, L., Amorim-Borher B., and C. Lage. 2013. Patent applications on representative sectors of biotechnology in Brazil: an analysis of the last decade. *Journal of Technology Management & Innovation* 8 no. 4:91–102. doi.org/10.4067/S0718-27242013000500009.
[24] Bozeman, B. 2000. Technology transfer and public policy: a review of research and theory. *Research Policy* 29, no. 4–5:627–655. doi: 10.1016/S0048-7333(99)00093-1.
[25] Junior Vieira, M., Lucato, W. C. Vanalle, R. M., and K. Jagoda. 2014. Effective management of international technology transfer projects: insights from the Brazilian textile industry. *Journal of Manufacturing Technology Management* 25, no. 1:69–99. doi: 10.1108/JMTM-08-2011-0079.
[26] Costantini, V., and P. Liberati. 2014. Technology transfer, institutions and development. *Technological Forecasting & Social Change* 88:26–48. doi: 10.1016/j.techfore.2014.06.014.
[27] De Conink, H., Haake, F., and N. Van Der Linden. 2007. Technology transfer in the clean development mechanism. *Climate Policy* 7, no. 5:444–456. doi: 10.1080/14693062.2007.968567.
[28] Lema, A., and R. Lema. 2013. Technology transfer in the clean development mechanism: insights from wind power. *Global Environmental Change* 23, no. 1:301–313 doi: 10.1016/j.gloenvcha.2012.10.010.
[29] Saggi, K. 2002. Trade, foreign direct investment, and international technology transfer: a survey. *The World Bank Research Observer* 17, no. 2:191–235. doi: 10.1093/wbro/17.2.191.
[30] Coccia, M. 2010. Spatial patterns of technology transfer and measurement of its friction in the geo-economic space. *International Journal of Technology Transfer and Commercialization* 9, no. 3:255–267. doi: 10.1504/IJTTC.2010.030214.
[31] Brewer, T.L. 2008. Climate change technology transfer: a new paradigm and policy agenda. *Climate Policy* 8, no. 5:516–526. doi:10.3763/cpol.2007.0451.
[32] Baek, D.H., Sul, W., Hong, K.P., and H. Kim. 2007. A technology valuation model to support technology transfer negotiations. *R and D Management* 37, no. 2:123–138. doi:10.1111/j.1467-9310.2007.00462.x.
[33] Coccia, M., and S. Rolfo. 2002. Technology transfer analysis in the Italian National Research Council. *Technovation* 2, no. 5:291–299. doi:10.1016/S0166-4972(01)00018-9.
[34] Blalock, G., and P. J. Gertler. 2008. Welfare gains from Foreign Direct Investment through technology transfer to local suppliers. *Journal of International Economics* 74, no. 2:402–421. doi: 10.1016/j.jinteco.2007.05.011.
[35] Lee, H.H., and H.B. Tan. 2006. Technology transfer, FDI and economic growth in the ASEAN region. *Journal of the Asia Pacific Economy* 11, no. 4:394–410. doi:10.1080/13547860600923593.
[36] Franza, R.M., and K.P. Grant. 2006. Improving federal to private sector technology transfer. *Research-Technology Management* 49, no. 3:36–40. doi: 10.1080/08956308.2006.11657376.
[37] Bercovitz, J., and M. Feldman. 2006. Entrepreneurial universities and technology transfer: a conceptual framework for understanding knowledge. based economic development. *Journal of Technology Transfer* 31, no. 1:175–188 doi: 10.1007/s10961-005-5029-z.
[38] Lai, W.H. 2011. Willingness-to-engage in technology transfer in industry–university collaborations. *Journal of Business Research* 64, no. 11:1218–1223. doi:10.1016/j.jbusres.2011.06.026.

[39] Curi, C., Daraio, C., and P. Llerena. 2012. University technology transfer: how (in)efficient are French universities?. *Cambridge Journal of Economics* 36, no. 3:629–654. doi.org/10.1093/cje/bes020.

[40] Wonglimpiyarat, J. 2016. The innovation incubator, university business incubator and technology transfer strategy: the case of Thailand. *Technology in Society* 46:18–27. doi: 10.1016/j.techsoc.2016.04.002.

[41] Schoen, A., van Pottelsberghe de la Potterie, B., and J. Henkel. 2014. Governance typology of universities technology transfer processes. *Journal of Technology Transfer* 39, no. 3:435–453. doi:10.1007/s10961-012-9289-0.

[42] Siegel, D., Wright, M., Chapple, W., and A. Lockett. 2008. Assessing the relative performance of university technology transfer in the US and UK: a stochastic distance function approach. *Economics of Innovation and New Technology* 17, no. 7–8:717–729. doi: 10.1080/10438590701785769.

[43] Huyghe, A., Knockaert, M., Wright, M., and E. Piva. 2014. Technology transfer offices as boundary: the case of a hybrid model. *Small Business Economics* 43, no. 2:289–307. doi: 10.1007/s11187-013-9537-1.

[44] Bubela, T. M., and T. Caulfield. 2010. Role and reality: technology transfer at Canadian universities. *Trends in biotechnology* 28, no. 9:447–451. doi: 10.1016/j.tibtech.2010.06.002.

[45] Owen-Smith, J, and W. W. Powell. 2001. To patent or not: faculty decisions and institutional success at technology transfer. *Journal of Technology Transfer* 26, no. 1:99–114. doi: 10.1023/A:1007892413701.

[46] Siegel, D. S., Veugelers R., and M. Wright. 2007. Technology transfer offices and commercialization of university intellectual property: performance and policy implications. *Oxford Review of Economic Policy* 23, no. 4:640–660. doi: 10.1093/oxrep/grm036.

[47] Huyghe, A., Knockaert, M., Piva, E., and M. Wright. 2016. Are researchers deliberately bypassing the technology transfer office? An analysis of TTO awareness. *Small Business Economics* 47, no. 3:589–607. doi: 10.1007/s11187-016-9757-2.

[48] Anderson, T. R., Tugrul, U. D., and F. F. Lavoie. 2007. Measuring the efficiency of university technology transfer. *Technovation* 27, no. 5:306–318. doi: 10.1016/j.technovation.2006.10.003.

[49] McAdam, R., Keogh W., Galbraith B., and L. Don. 2005. Defining and improving technology transfer business and management processes in university innovation centres. *Technovation* 25, no. 12:1418–1429. doi:10.1016/j.technovation.2004.08.002.

[50] Schmiemann, M., and J. N. Durvy. 2003. New approaches to technology transfer from publicly funded research. *The Journal of Technology Transfer* 28, no. 1:9–15. doi:10.1023/A:1021622501820.

[51] Collins, S., and H. Wakoh. 2000. Universities and technology transfer in Japan: recent reforms in historical perspective. *The Journal of Technology Transfer* 25, no. 2:213–222. doi:10.1023/A:1007884925676.

[52] Haščič, I., and N. Johnstone. 2011. CDM and international technology transfer: empirical evidence on wind power. *Climate Policy* 11, no. 6:1303–1314. doi: 10.1080/14693062.2011.579311.

[53] Ho, M.H.C., Liu, J.S., Lu, W.M., and C. C. Huang. 2014. A new perspective to explore the technology transfer efficiencies in US universities. *Journal of Technology Transfer* 39, no. 2:247–275. doi:10.1007/s10961-013-9298-7.

[54] Salicrup, L., and Fedorková L. 2006. Challenges and opportunities for enhancing biotechnology and technology transfer in developing countries. *Biotechnology Advances* 24, no. 1:69–79. doi: 10.1016/j.biotechadv.2005.06.004.

[55] Costa-Font, J., and E. Mossialos. 2006. The public as a limit to technology transfer: the influence of knowledge and beliefs in attitudes towards biotechnology in the UK. *Journal of Technology Transfer* 31, no. 6:629–645. doi: 10.1007/s10961-006-0019-3.

[56] Bergman, K., and G. D. Graff. 2007. The global stem cell patent landscape: implications for efficient technology transfer and commercial development. *Nature Biotechnology* 25, no. 4:419–424. doi: 10.1038/nbt0407-419.

[57] Mola, E.L., Acevedo B., Silva R., Tormo B., Montero R., and L. Herrera. 2003. Development of Cuban biotechnology. *Journal of Commercial Biotechnology* 9, no. 2:147–152. doi.org/10.1057/palgrave.jcb.3040021.

[58] Albert, A., Begoña, G., and L. Plaza. 2007. Scientific and technological performance evaluation of the Spanish Council for Scientific Research (CSIC) in the field of Biotechnology. *Scientometrics* 70, no. 1:41–51. doi:10.1007/s11192-007-0103-y.

[59] Foltz, J., D., Kim, K., and B. Barham. 2003. A dynamic analysis of university agricultural biotechnology patent production. *American Journal of Agricultural Economics* 85, no. 1:187–197. doi: 10.1111/1467-8276.00112.

[60] Pray, C.E. 2001. Public-private sector linkages in research and development: biotechnology and the seed industry in Brazil, China and India. *American Journal of Agricultural Economics* 83, no. 3:742–747. doi: 10.1111/0002-9092.00201.

[61] Meyers, A.D., and S. Pruthi, 2011. Academic entrepreneurship, entrepreneurial universities and biotechnology. *Journal of Commercial Biotechnology* 17, no. 4:349–357. doi: 10.1057/jcb.2011.22.

[62] Yazdani, K., K.Y. Rashvanlouei, K.Y., and K. Ismail. 2011. Ranking of technology transfer barriers in developing countries; case study of Iran's biotechnology industry. *2011 IEEE International Conference on Industrial Engineering and Engineering Management*, Singapore, pp. 1602–1606 doi: 10.1109/IEEM.2011.6118187.

[63] Daar, A.S., Thorsteinsdóttir, H., Martin, D.K., Smith, A.C., Nast, S., and P.A. Singer. 2002. Top ten biotechnologies for improving health in developing countries. *Nature Genetics* 32, no. 2:229–232. doi: 10.1038/ng1002-229.

[64] Adenle, A.A. 2011. Global capture of crop biotechnology in developing world over a decade. *Journal of Genetic Engineering and Biotechnology* 9, no. 2:83–95. doi: 10.1016/j.jgeb.2011.08.003.

[65] Sáenz, T.W., Thorsteinsdóttir H., and de Souza M.C. 2010. Cuba and Brazil: an important example of South-South collaboration in health biotechnology. *Medicc Review* 12, no. 3:32–35. doi:10.37757/mr2010.v12.n3.8.

[66] Albert, A., and L. Plaza. 2004. The transfer of knowledge from the Spanish public R&D system to the productive sectors in the field of biotechnology. *Scientometrics* 59, no. 1:3–14 doi: 10.1023/b:scie.0000013295.88415.16.

[67] Chen, S.H., and W.T. Lin. 2017. The dynamic role of universities in developing an emerging sector: a case study of the biotechnology sector. *Technological forecasting and social change* 123:283–297. doi: 10.1016/j.techfore.2016.06.006.

[68] Pereira, C.G., Da Silva R.R., Lavoie J.R., and G. S. Porto. 2018. Technological cooperation network in biotechnology: analysis of patents with Brazil as the priority country. *Innovation & Management Review* 15, no. 4:416–434. doi.org/10.1108/INMR-07-2018-0050.

[69] Petruzzelli, A.M., Rotolo, D., and V. Albino. 2015. Determinants of patent citations in biotechnology: an analysis of patent influence across the industrial and organizational boundaries. *Technological Forecasting and Social Change* 91:208–221. doi.org/10.1016/j.techfore.2014.02.018.

[70] Bengoa, A., Maseda, A., Iturralde, T., and G. Aparicio. 2021. A bibliometric review of the technology transfer literature. *The Journal of Technology Transfer* 46, no. 5:1514–1550. doi.org/10.1007/s10961-019-09774-5.
[71] Craiut, L., Bunga, C., Negru, P., Bungau, T., and A.F Radu. 2022. Technology transfer in the context of sustainable development-a bibliometric analysis of publications in the field. *Sustainability* 14, no. 19 (11973):1–16. doi.org/10.3390/su141911973.
[72] Pranckutė, R. 2021. Web of Science (WoS) and Scopus: the Titans of Bibliographic Information in Today's Academic World. *Publications* 9 no. 1(12):1–59. doi.org/10.3390/publications9010012.
[73] Scopus. 2022. Abstract and Citation Database. Available in: www.scopus.com (accessed August 15, 2022).
[74] Web of Science. 2022. Search. Available in: www.webofscience.com/wos/woscc/basic-search (accessed October 25, 2022).
[75] VOSviewer. 2022. Vosviewer (Software). Version 1.6.18.
[76] Lens. 2022. Search, Analyze and Manage Patent and Scholarly Data. Available in: www.lens.org/ (accessed September 19, 2022).
[77] WIPO. 2022. International Patent Classification (IPC). Available in: www.wipo.int/classifications/ipc/en/ (accessed September 19, 2022).

Index

For Product Safety Concerns and Information please contact our EU representative GPSR@taylorandfrancis.com Taylor & Francis Verlag GmbH, Kaufingerstraße 24, 80331 München, Germany